Scattering Matrix Approach to Non-Stationary Quantum Transport

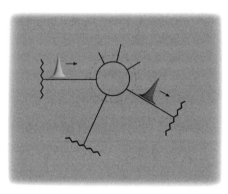

Scattering Matrix Approach to Non-Stationary Quantum Transport

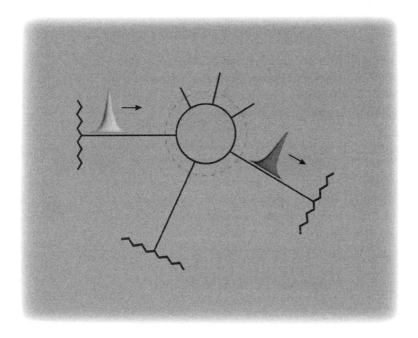

Michael V Moskalets
National Technical University, "Kharkiv Polytechnical Institute", Ukraine

Imperial College Press

Published by

Imperial College Press
57 Shelton Street
Covent Garden
London WC2H 9HE

Distributed by

World Scientific Publishing Co. Pte. Ltd.
5 Toh Tuck Link, Singapore 596224
USA office: 27 Warren Street, Suite 401-402, Hackensack, NJ 07601
UK office: 57 Shelton Street, Covent Garden, London WC2H 9HE

British Library Cataloguing-in-Publication Data
A catalogue record for this book is available from the British Library.

SCATTERING MATRIX APPROACH TO NON-STATIONARY
QUANTUM TRANSPORT

Copyright © 2012 by Imperial College Press

All rights reserved. This book, or parts thereof, may not be reproduced in any form or by any means, electronic or mechanical, including photocopying, recording or any information storage and retrieval system now known or to be invented, without written permission from the Publisher.

For photocopying of material in this volume, please pay a copying fee through the Copyright Clearance Center, Inc., 222 Rosewood Drive, Danvers, MA 01923, USA. In this case permission to photocopy is not required from the publisher.

ISBN-13 978-1-84816-834-3
ISBN-10 1-84816-834-9

Printed by FuIsland Offset Printing (S) Pte Ltd Singapore

I dedicate this book to my parents,
Anna A. Tarasenko and Vasyliy T. Moskalets

Preface

The lectures of Professor Michael Moskalets introduce basic concepts of the scattering approach to transport phenomena in time-dependent, dynamic quantum systems. For stationary problems scattering theory has been widely and successfully used to discuss electronic transport in structures so small that interference effects become important. Such systems are central to nanophysics and mesoscopic physics. The lectures extend this approach to time-dependent scatterers.

Scattering theory derives its success from two sources. The approach captures essential aspects of real experiments and is therefore useful in the laboratory. Scattering theory is often referred to as a formalism but this is almost a misnomer. In the realm of theories, the scattering approach has the advantage that it appeals to our intuition. It is an approach that is clearly not reserved only for physicists with a theoretical inclination.

The lectures provide an introduction to the stationary scattering theory and then bring the reader to the forefront of current research in the transport theory of time-dependent scatterers. Of interest are the charge and heat currents and the noise properties of such systems. Important examples are quantum pumps and mesoscopic capacitors subject to time-dependent potentials.

The lectures are clearly structured and focus on the principal points in the theoretical development. The author is didactical. The lectures develop all the mathematical steps and also provide a physically clear and transparent description of processes in dynamic nanoscopic and mesoscopic systems. The lectures present an excellent record of the current state of the field. This makes these lectures useful not only to students but also to advanced researchers.

Markus Büttiker

Acknowledgments

I am very grateful to Markus Büttiker. Without a collaboration with him that has lasted more than a decade, without numerous discussions with him, without his constant support and encouragement this book would not have been written.

I would like to thank the people with whom I directly collaborate, especially, Liliana Arrachea, Janine Splettstößer and Peter Samuelsson. The many interesting and exciting results that we found together are presented in this book. I also have to thank many people from the department of metal and semiconductor physics of the National Technical University "Kharkiv Polytechnical Institute", Kharkiv, Ukraine, where I work, and from the department of theoretical physics of the University of Geneva, Geneva, Switzerland, which I visited many times while working on this book, for numerous useful and stimulating discussions.

And last but not least, I am very grateful to my wife Natasha for patience, support, and understanding.

Contents

Preface vii

Acknowledgments ix

List of Figures xv

1. Landauer–Büttiker formalism 1
   - 1.1 Scattering matrix 1
     - 1.1.1 Scattering matrix properties 4
   - 1.2 Current operator 9
   - 1.3 Direct current and the distribution function 14
     - 1.3.1 Conservation of a direct current 16
     - 1.3.2 Difference of potentials 18
     - 1.3.3 Difference of temperatures 20
   - 1.4 Examples 21
     - 1.4.1 Scattering matrix 1×1 21
     - 1.4.2 Scattering matrix 2×2 22
     - 1.4.3 Scattering matrix 3×3 23
     - 1.4.4 Scatterer with two leads 25
     - 1.4.5 Scatterer with a potential contact 28
     - 1.4.6 Scatterer embedded in a ring 29

2. Current noise 35
   - 2.1 Nature of a current noise 36
     - 2.1.1 Thermal noise 37
     - 2.1.2 Shot noise 39
     - 2.1.3 Combined noise 40

	2.2	Sample with continuous spectrum	46
	2.2.1	Current correlator	47
	2.2.2	Current correlator in the frequency domain	49
	2.2.3	Spectral noise power for energy-independent scattering	55
	2.2.4	Zero frequency noise power	56
	2.2.5	Fano factor	62
3.	Non-stationary scattering theory	63
	3.1	Schrödinger equation with a potential periodic in time	63
	3.1.1	Perturbation theory	64
	3.1.2	Floquet functions method	67
	3.1.3	Potential oscillating in time and uniform in space	70
	3.2	Floquet scattering matrix	71
	3.2.1	Floquet scattering matrix properties	72
	3.3	Current operator	73
	3.3.1	Alternating current	75
	3.3.2	Direct current	75
	3.4	Adiabatic approximation for the Floquet scattering matrix	77
	3.4.1	Frozen scattering matrix	78
	3.4.2	Zeroth-order approximation	78
	3.4.3	First-order approximation	79
	3.5	Beyond the adiabatic approximation	83
	3.5.1	Scattering matrix in mixed energy-time representation	83
	3.5.2	Dynamic point-like potential	86
	3.5.3	Dynamic double-barrier potential	93
	3.5.4	Unitarity and the sum over trajectories	104
	3.5.5	Current and the sum over trajectories	106
4.	Direct current generated by the dynamic scatterer	111
	4.1	Steady particle flow	112
	4.1.1	Distribution function	112
	4.1.2	Adiabatic regime: Current linear in the pump frequency	114
	4.1.3	Current quadratic in the pump frequency	121

	4.2	Quantum pump effect	122
		4.2.1 Quasi-particle picture of direct current generation	122
		4.2.2 Interference mechanism of direct current generation	123
	4.3	Single-parameter adiabatic direct current generation	126
5.	Alternating current generated by the dynamic scatterer	129
	5.1	Adiabatic alternating current	129
	5.2	External AC bias	133
		5.2.1 Second quantization operators for incident and scattered electrons	136
		5.2.2 Alternating current	140
		5.2.3 Direct current	141
		5.2.4 Adiabatic direct current	142
6.	Noise generated by the dynamic scatterer	147
	6.1	Spectral noise power	147
	6.2	Zero frequency spectral noise power	154
	6.3	Noise in the adiabatic regime	157
		6.3.1 Thermal noise	157
		6.3.2 Low-temperature shot noise	158
		6.3.3 High-temperature shot noise	160
		6.3.4 Shot noise within a wide temperature range	161
		6.3.5 Noise as a function of the pump frequency	162
7.	Energetics of a dynamic scatterer	165
	7.1	DC heat current	165
		7.1.1 Heat generation by the dynamic scatterer	168
		7.1.2 Heat transfer between the reservoirs	168
	7.2	Heat flows in the adiabatic regime	171
		7.2.1 High temperatures	171
		7.2.2 Low temperatures	173
8.	Dynamic mesoscopic capacitor	175
	8.1	General theory for a single-channel scatterer	175
		8.1.1 Scattering amplitudes	176
		8.1.2 Unitarity conditions	176

		8.1.3	Time-dependent current	178
		8.1.4	Dissipation	179
		8.1.5	Dissipation versus squared current	180
	8.2	Chiral single-channel capacitor	181
		8.2.1	Model and scattering amplitude	182
		8.2.2	Unitarity	185
		8.2.3	Gauge invariance	186
		8.2.4	Time-dependent current	189
		8.2.5	High-temperature current	191
		8.2.6	Linear response regime	192
		8.2.7	Non-linear low-frequency regime	196
		8.2.8	Transient current caused by a step potential . . .	198
9.	Quantum circuits with mesoscopic capacitor as a particle emitter			207
	9.1	Quantized emission regime	207
		9.1.1	Simple model for a single-particle state emitted by an adiabatic capacitor	211
	9.2	Shot noise quantization	212
		9.2.1	Probability interpretation of the shot noise	214
	9.3	Two-particle source	216
		9.3.1	Scattering amplitude	216
		9.3.2	Adiabatic approximation	216
		9.3.3	Mean square current	218
		9.3.4	Shot noise of a two-particle source	226
	9.4	Mesoscopic electron collider	229
		9.4.1	Shot noise suppression	231
		9.4.2	Probability analysis	234
	9.5	Noisy mesoscopic electron collider	237
		9.5.1	Current cross-correlator suppression	238
		9.5.2	Probability analysis	240
	9.6	Two-particle interference effect	243
		9.6.1	Model and definitions	243
		9.6.2	Scattering matrix elements	248
		9.6.3	Current cross-correlator	250

Bibliography   261

Index   277

List of Figures

1.1 A mesoscopic sample with scattering matrix $\hat{S}$. The index $\alpha = 1, 2, \ldots, N_r$ numbers electron reservoirs. The arrows directed to (from) the scatterer show a propagation direction for incident (scattered) electrons. The electron flow is calculated at the surface Σ shown as a dashed line. 10

1.2 The distribution function for electrons scattered into the contact $\alpha = 1$. The height of a step at $E = \mu_1$ is $|S_{12}|^2$. The scatterer is connected to two electron reservoirs at zero temperature, $T_1 = T_2 = 0$, and having chemical potentials μ_1 and μ_2. 16

1.3 A single-channel scatterer. a is the amplitude of an incoming wave, b is the amplitude of a reflected wave. A zigzag line denotes an electron reservoir. 21

1.4 A two-channel scatterer. a_α (b_α) are the amplitudes of incoming (scattered) waves, $\alpha = 1, 2$. 22

1.5 A three-channel scatterer. a_α (b_α) are amplitudes of incoming (scattered) waves, $\alpha = 1, 2, 3$. 24

1.6 A mesoscopic scatterer with current carrying $(1, 2)$ and potential (3) leads. 28

1.7 A one-dimensional ring of length L pierced by the magnetic flux Φ with embedded scatterer. A, B are amplitudes of an electron wave function, Eq. (1.78), $\phi = 2\pi\Phi/\Phi_0$. 30

3.1 Scattering of a wave with unit amplitude onto the point-like dynamic potential barrier. The arrows show propagation directions. The labels are amplitudes of the corresponding waves: 1 is an amplitude of an incoming wave, $b_{1,n}^{(-)}$ is an amplitude of a reflected wave, $a_{1,n}^{(+)}$ is an amplitude of a transmitted wave. Only a single (nth) component of the Floquet wave function for scattered electrons is shown. 89

3.2 Two dynamic point-like potentials separated by a ballistic wire of length d. The arrows show propagation directions. The labels are amplitudes of the corresponding waves, see Eqs. (3.99), (3.100), and (3.106). 94

4.1 The non-equilibrium distribution function, $f_\alpha^{(out)}(E)$, for electrons scattered into the lead α at zero temperature. The step width is $\hbar\Omega_0$. The Fermi distribution function at zero temperature is shown by the dashed line. 113

4.2 The parameter space for the frozen scattering matrix in the case of two varying parameters, p_1 and p_2. During one period the point $A(t)$ with coordinates $(p_1(t), p_2(t))$ follows a trajectory $\mathcal{L}$. $\mathcal{F}$ stands for the surface area. The arrow indicates the direction of movement for $\varphi > 0$, see Eq. (4.9). 118

4.3 The creation and scattering of quasi-electron-hole pairs in a dynamic sample with two leads. Under the action of a potential that is periodic in time, $V(t) = V(t+\mathcal{T})$, an electron can absorb one or several energy quanta, $\hbar\Omega_0$. As a result it jumps from the occupied level to the non-occupied one, which can be viewed as the creation of a quasi-electron-hole pair. A quasi-electron (a dark circle) and a hole (a light circle) can leave the scattering region through the same lead (a) or through different leads (b). In the latter case a net charge is transferred from one reservoir to the other. 122

4.4 Photon-assisted scattering amplitudes for an electron propagating through a dynamic double-barrier. An electron can absorb (or emit) an energy quantum $\hbar\Omega_0$ interacting either with a potential $V_1(t)$ or with a potential $V_2(t)$. Therefore, the photon-induced scattering amplitude is a sum of two terms. 124

7.1 The heat flows caused by the dynamic scatterer with two contacts. $I_\alpha^{Q(gen)}$ is the generated heat flowing into the reservoir α and $I^{Q(pump)}$ is the pumped heat. The heat production rate is $I_{tot}^Q = I_1^{Q(gen)} + I_2^{Q(gen)}$. 168

8.1 The model for a single-cavity chiral quantum capacitor driven by a uniform potential $U(t)$. The dotted line denotes a QPC. The arrows indicate the direction of movement of electrons. ... 182

8.2 An equivalent electrical circuit to model the low-frequency response of a quantum capacitor. 192

8.3 The dependence of an emitted charge Q on the height U_0 of a potential step at zero temperature. The Fermi level is centered in the middle between the levels of the cavity. 204

9.1 The mesoscopic electron collider with a single-particle source in one of its branches. The cavity is connected to a linear edge state which in turn is connected to another linear edge state via the central QPC with transmission T_C. The arrows indicate the direction of movement of the electrons. The potential $U(t)$ induced by the back-gate generates an alternating current $I(t)$, which is split at the central QPC into the currents $I_1(t)$ and $I_2(t)$ flowing to the leads. 213

9.2 The model for a double-cavity chiral quantum capacitor. Each cavity is driven by the periodic potential $U_j(t) = U_j(t + \mathcal{T})$, $j = L, R$. Both cavities are connected via the corresponding QPCs with transmission T_L and T_R, respectively, to the same linear edge state. The arrows indicate the direction of movement of electrons. 217

9.3 The mesoscopic electron collider with a two-particle source in one of its branches. The double-cavity quantum capacitor is connected to the linear edge state, which in turn is connected via the central QPC with transmission T_C to another linear edge state. The arrows indicate the direction of movement of the electrons. The potentials $U_L(t)$ and $U_R(t)$ induced by the back-gates generate an alternating current $I(t)$, which is split at the central QPC into the currents $I_1(t)$ and $I_2(t)$ flowing to the leads. 227

9.4 The mesoscopic electron collider with single-particle sources in each of its branches. The two quantum capacitors are connected to the different linear edge states, which in turn are connected together via the central QPC with transmission T_C. The arrows indicate the direction of movement of electrons. The potentials $U_L(t)$ and $U_R(t)$ induced by the back-gates generate alternating currents $I_1(t)$ and $I_2(t)$ at the leads. 230

9.5 The current cross-correlator $\mathcal{P}_{12}$, Eq. (9.73), as a function of the amplitude $U_{L,1}$ of a potential $U_L(t) = U_{L,0} + U_{L,1}\cos(\Omega_0 t + \varphi_L)$ acting upon the left capacitor, see Fig. 9.4. The three lines shown differ in the way that the right capacitor is driven: The right capacitor can be stationary (upper solid black line); driven by the potential $U_R(t) = U_{R,0} + U_{R,1}\cos(\Omega_0 t + \varphi_R)$, which is out of phase, $\varphi_R = \pi$, and has an amplitude $eU_{R,1} = 0.5\Delta_R$ (lower solid green line); or driven by the in-phase potential, $\varphi_R = 0$, with amplitude $eU_{R,1} = 0.5\Delta_R$ (dashed red line). Other parameters are $eU_{L,0} = eU_{R,0} = 0.25\Delta_R$ ($\Delta_L = \Delta_R$), $\varphi_L = 0$, and $T_L = T_R = 0.1$. 233

9.6 The mesoscopic electron collider with noisy single-particle sources. The two noisy flows originating from the quantum point contacts L and R can collide at the quantum point contact C. 238

9.7 The current cross-correlator $\mathcal{P}_{12}$ as a function of the amplitude $U_{L,1}$ of a potential $U_L(t) = U_{L,0} + U_{L,1}\cos\left(\Omega_0 t + \varphi_L\right)$ driving the single-particle source S_L, see Fig. 9.6. The parameters of the single-particle sources S_L and S_R are the same as in Fig. 9.5 but $\varphi_L = \varphi_R$. Other parameters are $T_L = T_R = 0.5$. 239

9.8 A quantum electronic circuit comprising two single-particle sources S_L and S_R and two Mach–Zehnder interferometers with magnetic fluxes Φ_L and Φ_R. 244

Chapter 1

Landauer–Büttiker formalism

According to the Landauer–Büttiker approach [1–6] the transport phenomena in mesoscopic [7, 8] conducting systems can be described with the help of a corresponding quantum-mechanical scattering problem. The mesoscopic system is assumed to be connected to macroscopic contacts acting as reservoirs of electrons, which are scattered by the mesoscopic sample. After scattering the electrons return to the original contact or go to a different one. Thus the problem of calculating such transport characteristics as, for example, electrical conductance or thermal conductance is reduced to solving a quantum-mechanical scattering problem with a potential profile corresponding to the sample under consideration [9] with possibly subsequent statistical averaging [10]. All information concerning transport properties of a sample is encoded in its scattering matrix, $\hat{S}$.

We concentrate on a single-particle scattering matrix. Therefore, we neglect electron-electron interactions and use the Schrödinger equation for spinless electrons as the basic equation. In principle interactions can be easily incorporated on the mean-field level.

1.1 Scattering matrix

In quantum mechanics an electron is characterized by the wave function, $\Psi(t, \mathbf{r})$, dependent on time t and on a spatial coordinate $\mathbf{r}$. If the wave function, $\Psi^{(in)}$, for an electron incident to the scatterer is known then using the Schrödinger equation one can calculate the wave function, $\Psi^{(out)}$, for a scattered electron. One can ask whether one needs to solve the Schrödinger equation for each $\Psi_j^{(in)}$? The answer is no. It is enough to solve the scattering problems for incident states $\psi_\alpha^{(in)}$ constituting the full orthonormal basis. After that, using the superposition principle, one can find the solution for the scattering problem with arbitrary incident state, $\Psi_j^{(in)}$.

To this end we expand an incident electron wave function, $\Psi^{(in)}$, in the basis functions $\psi_\alpha^{(in)}$,

$$\Psi^{(in)} = \sum_\alpha a_\alpha \, \psi_\alpha^{(in)} . \qquad (1.1)$$

Then we expand a wave function for the scattered electron, $\Psi^{(out)}$, in the basis functions $\psi_\alpha^{(out)}$,

$$\Psi^{(out)} = \sum_\beta b_\beta \, \psi_\beta^{(out)} . \qquad (1.2)$$

The set of functions $\psi_\alpha^{(in)}$ and $\psi_\beta^{(out)}$ constitutes the full orthonormal basis.

The problem is to find the coefficients b_β if the set of coefficients a_α is known. First we consider an auxiliary problem, namely the scattering of an electron with wave function $\Psi_1^{(in)} = \psi_1^{(in)}$. In this case the set of coefficients in Eq. (1.1) is the following: $(1, 0, 0, \dots)$. The solution for this scattering problem we write as Eq. (1.2) with coefficients $S_{\beta 1}$,

$$\Psi_1^{(out)} = \sum_\beta S_{\beta 1} \, \psi_\beta^{(out)} . \qquad (1.3)$$

The coefficient $S_{\beta 1}$ is a quantum-mechanical transition amplitude from the state $\psi_1^{(in)}$ to the state $\psi_\beta^{(out)}$. Note if the incident wave function is multiplied by some constant factor A then the wave function for a scattered state is also multiplied by the same factor,

$$\Psi_1^{(in)} = A \, \psi_1^{(in)} \quad \Rightarrow \quad \Psi_1^{(out)} = A \sum_\beta S_{\beta 1} \, \psi_\beta^{(out)} . \qquad (1.4)$$

After solving the scattering problem with incident state $\Psi_\gamma^{(in)} = \psi_\gamma^{(in)}$ we find the coefficients $S_{\beta\gamma}$,

$$\Psi_\gamma^{(out)} = \sum_\beta S_{\beta\gamma} \, \psi_\beta^{(out)} . \qquad (1.5)$$

With coefficients $S_{\alpha\beta}$ we can solve the scattering problem for an arbitrary incident state. Formally the corresponding algorithm is the following:

1. The wave function for an incident state is expanded into the series in basis functions $\psi_\alpha^{(in)}$, Eq. (1.1).

2. The scattered state wave function, $\Psi^{(out)}$, is represented as the sum of partial contributions, $\Psi_\alpha^{(out)}$, due to scattering of partial incident waves, $\Psi_\alpha^{(in)} = a_\alpha \psi_\alpha^{(in)}$,

$$\Psi^{(out)} = \sum_\alpha \Psi_\alpha^{(out)},$$

$$\Psi_\alpha^{(out)} = a_\alpha \sum_\beta S_{\beta\alpha} \psi_\beta^{(out)}.$$

(1.6)

3. The coefficients b_β for the scattered state of interest, $\Psi^{(out)} = \sum_\alpha a_\alpha \sum_\beta S_{\beta\alpha} \psi_\beta^{(out)} \equiv \sum_\beta b_\beta \psi_\beta^{(out)}$, are the following

$$b_\beta = \sum_\alpha S_{\beta\alpha} a_\alpha.$$

(1.7)

Equation (1.7) solves the problem: It expresses the coefficients b_β for the scattered wave function in terms of the coefficients a_α for the incident wave function. It is convenient to treat the quantities, $S_{\beta\alpha}$, of Eq. (1.7) as the elements of some matrix, $\hat{S}$, which is referred to as *the scattering matrix*.[1]

If the coefficients a_α and b_β are collected into vector columns

$$\hat{b} = \begin{pmatrix} b_1 \\ b_2 \\ \vdots \end{pmatrix}, \quad \hat{a} = \begin{pmatrix} a_1 \\ a_2 \\ \vdots \end{pmatrix},$$

(1.8)

then the corresponding relations simplify to

$$\hat{b} = \hat{S}\hat{a}.$$

(1.9)

As we already mentioned, the scattering matrix elements, $S_{\alpha\beta}$, are quantum-mechanical amplitudes for a particle in the state $\psi_\beta^{(in)}$ to be scattered into the state $\psi_\alpha^{(out)}$. The order of indices is important. We use the convention that the first index (for the element $S_{\alpha\beta}$ it is α) corresponds to a scattered state while the second index corresponds to an incident state.

---

[1] The scattering matrix elements are directly related to the corresponding single-particle Green's functions [11, 12]. For the generalization to the periodically driven case see Ref. [13].

1.1.1 Scattering matrix properties

General physical principles put some constraints on the scattering matrix elements.

1.1.1.1 Unitarity

Particle number conservation during scattering requires the scattering matrix to be unitary,

$$\hat{S}^\dagger \hat{S} = \hat{S}\hat{S}^\dagger = \hat{I}. \tag{1.10}$$

Here $\hat{I}$ is a unit matrix of the same dimensions as $\hat{S}$,

$$\hat{I} = \begin{pmatrix} 1 & 0 & 0 & \cdots \\ 0 & 1 & 0 & \cdots \\ 0 & 0 & 1 & \cdots \\ & \cdots & & \end{pmatrix}. \tag{1.11}$$

The elements of the matrix $\hat{S}^\dagger$ are related to the elements of the scattering matrix $\hat{S}$ in the following way: $\left(\hat{S}^\dagger\right)_{\alpha\beta} = \left(\hat{S}\right)^*_{\beta\alpha}$. Therefore, the expanded equation (1.10) reads

$$\sum_{\alpha=1}^{N_r} S^*_{\alpha\beta} S_{\alpha\gamma} = \delta_{\beta\gamma}, \tag{1.12}$$

$$\sum_{\beta=1}^{N_r} S_{\alpha\beta} S^*_{\delta\beta} = \delta_{\alpha\delta}. \tag{1.13}$$

To prove unitarity, for instance, in the case when the wave function is normalized, i.e., it corresponds to scattering of a single particle, we use the integral over space for both the incident wave function and the scattered wave function:

$$\int d^3r \, |\Psi^{(in)}|^2 = \int d^3r \, |\Psi^{(out)}|^2 = 1. \tag{1.14}$$

Then we use Eqs. (1.1) and (1.2). For instance, for $\Psi^{(in)}$ we get,

$$\int d^3r\,|\Psi^{(in)}|^2 = \int d^3r \sum_\alpha a_\alpha \psi_\alpha^{(in)} \left(\sum_\beta a_\beta^* \psi_\beta^{(in)}\right)^*$$

$$= \sum_\alpha \sum_\beta a_\alpha a_\beta^* \int d^3r\, \psi_\alpha^{(in)} \left(\psi_\beta^{(in)}\right)^* = \sum_\alpha \sum_\beta a_\alpha a_\beta^* \delta_{\alpha\beta} \quad (1.15)$$

$$= \sum_\alpha |a_\alpha|^2 = 1\,.$$

Here we took into account that the functions $\psi_\alpha^{(in)}$ are orthonormal,

$$\int d^3r\, \psi_\alpha^{(in)} \left(\psi_\beta^{(in)}\right)^* = \delta_{\alpha\beta}\,, \quad (1.16)$$

where $\delta_{\alpha\beta}$ is the Kronecker symbol,

$$\delta_{\alpha\beta} = \begin{cases} 1, & \alpha = \beta\,, \\ 0, & \alpha \ne \beta\,. \end{cases} \quad (1.17)$$

By analogy we find for $\Psi^{(out)}$:

$$\sum_\alpha |b_\alpha|^2 = 1\,. \quad (1.18)$$

Therefore, from Eqs. (1.15) and (1.18) it follows that

$$\sum_\alpha |a_\alpha|^2 = \sum_\alpha |b_\alpha|^2\,. \quad (1.19)$$

Representing the coefficients a_α and b_α as vector columns, $\hat{a}$ and $\hat{b}$, we write

$$\sum_\alpha |a_\alpha|^2 = \hat{a}^\dagger \hat{a}\,,$$
$$\sum_\alpha |b_\alpha|^2 = \hat{b}^\dagger \hat{b}\,. \quad (1.20)$$

Next we take into account that $\hat{b} = \hat{S}\,\hat{a}$ and, correspondingly, $\hat{b}^\dagger = \hat{a}^\dagger\,\hat{S}^\dagger$ and finally calculate,

$$\hat{b}^\dagger \hat{b} = \hat{a}^\dagger \hat{S}^\dagger \hat{S} \hat{a} = \hat{a}^\dagger \hat{a}. \qquad (1.21)$$

From the last equality the required relation, Eq. (1.10), follows directly.

Note, however, that for the particles with continuous spectrum, which we will consider, the wave function is normalized to the Dirac delta function rather than to unity. In such a case the scattering of particles with fixed incoming flow is a more natural problem. For instance, a plane wave e^{ikx} corresponds to a flow of particles with intensity $v = \hbar k/m$ rather than to a single particle. Charge conservation in this case (under stationary conditions) implies current conservation. Therefore, it is convenient to choose the basis functions normalized to carry a unit flux, see, e.g., Refs [5, 11]. Then we can say more precisely:

Equation (1.9) defines the scattering matrix $\hat{S}$ if the vectors $\hat{b}$ and $\hat{a}$ are calculated using the unit flux basis.

The square of the modulus of a scattering matrix element defines an intensity of a scattered flow if the intensity of an incident flow is unity. Then the unitarity of the scattering matrix reflects particle flow conservation.

1.1.1.2 Micro-reversibility

Micro-reversibility is an invariant of the equations of motion under time reversal. Neither classical physics nor quantum physics makes a distinction between forward time and backward time.

If we change simultaneously, $t \to -t$ and $\mathbf{v} \to -\mathbf{v}$, then the classical equations of motion predict that the particle will move along the same trajectory but in the opposite direction. From the scattering theory point of view movement in the opposite direction means that the scattered particle becomes an incoming one and the incoming particle becomes a scattered one.

Quantum-mechanical formalism deals with states rather than with particles. The additional complication comes from the fact that the wave function is complex. To analyze micro-reversibility in quantum mechanics [14] we consider the Schrödinger equation

$$i\hbar \frac{\partial \Psi}{\partial t} = \mathcal{H}\Psi, \qquad (1.22)$$

where $\mathcal{H}$ is the Hamiltonian dependent on the momentum $\mathbf{p}$ of a particle. Velocity reversal within classical physics is equivalent to a momentum reversal within quantum physics. The Hamiltonian [and, correspondingly, Eq. (1.22)] is invariant under momentum reversal, $\mathcal{H}(\mathbf{p}) = \mathcal{H}(-\mathbf{p})$. While under time reversal the sign on the left hand side (LHS) of Eq. (1.22) is changed. On the other hand if simultaneously we use the complex conjugate equation and take into account that the Hamiltonian is Hermitian, $\mathcal{H}^* = \mathcal{H}$, then we find that the transformed equation for the complex conjugate wave function $\Psi^*(-t)$ is identical to the original equation for $\Psi(t)$,

$$i\hbar \frac{\partial \left(\Psi^* \right)}{\partial(-t)} = \mathcal{H}\left(\Psi^* \right). \tag{1.23}$$

We conclude: If the evolution in forward time is described by the wave function $\Psi(t)$ then the evolution in backward time is described by the complex conjugate function $\Psi^*(-t)$. For scattering theory this means that if initially the incident particle is in the state $\Psi^{(in)}(t)$ and the scattered particle is in the state $\Psi^{(out)}(t)$ then under time reversal the state $\left(\Psi^{(out)}(-t) \right)^*$ is for an incident particle and the state $\left(\Psi^{(in)}(-t) \right)^*$ is for a scattered particle.

Such symmetry results in various properties of the scattering matrix. To clarify these we will consider scattering in both forward and backward times in detail. The initial scattering process: $\Psi^{(in)}(t) = \sum_\alpha a_\alpha \psi_\alpha^{(in)}(t)$ is an incident wave and $\Psi^{(out)}(t) = \sum_\beta b_\beta \psi_\beta^{(out)}(t)$ is a scattered wave. The coefficients a_α and b_β are related through equation (1.9). The scattering process after time reversal: $\left(\Psi^{(out)}(-t) \right)^* = \sum_\beta b_\beta^* \left(\psi_\beta^{(out)}(-t) \right)^*$ is an incident wave and $\left(\Psi^{(in)}(-t) \right)^* = \sum_\alpha a_\alpha^* \left(\psi_\alpha^{(in)}(-t) \right)^*$ is a scattered wave. Under both time reversal and complex conjugation the basis functions for incident and scatterer states replace each other, $\left(\psi_\beta^{(out)}(-t) \right)^* = \psi_\beta^{(in)}(t)$. Therefore, we can write

$$\left(\Psi^{(out)}(-t) \right)^* = \left(\sum_\beta b_\beta \psi_\beta^{(out)}(-t) \right)^* = \sum_\beta b_\beta^* \psi_\beta^{(in)}(t),$$

$$\left(\Psi^{(in)}(-t) \right)^* = \left(\sum_\alpha a_\alpha \psi_\alpha^{(in)}(-t) \right)^* = \sum_\alpha a_\alpha^* \psi_\alpha^{(out)}(t). \tag{1.24}$$

Since the Hamiltonian and the basis functions remain invariant the scattering matrix is invariant as well. Therefore, the coefficients a_α^* and b_β^* in

Eq. (1.24) are related in the same way as the corresponding coefficients (b_β and a_α) in Eqs. (1.1) and (1.2),

$$\hat{a}^* = \hat{S}\hat{b}^*. \tag{1.25}$$

Thus the sets of coefficients $\hat{a}$ and $\hat{b}$ have to fulfill two equations, (1.9) and (1.25). From Eq. (1.9) we find,

$$\hat{a} = \hat{S}^{-1}\hat{b}, \tag{1.26}$$

where $\hat{S}^{-1}$ is an inverse matrix, $\hat{S}\hat{S}^{-1} = \hat{S}^{-1}\hat{S} = \hat{I}$. Comparing Eqs. (1.26) and (1.25) we conclude that $\hat{S}^* = \hat{S}^{-1}$. Further, from the unitarity, Eq. (1.10), it follows that

$$\left.\begin{array}{l}\hat{S}^\dagger\hat{S} = \hat{I}\\ \hat{S}^{-1}\hat{S} = \hat{I}\end{array}\right\} \quad \Rightarrow \quad \hat{S}^\dagger = \hat{S}^{-1}. \tag{1.27}$$

Finally we conclude that micro-reversibility requires the scattering matrix to be invariant under the transposition operation. In other words, the scattering matrix elements are symmetric in their indices,

$$\hat{S} = \hat{S}^T \quad \Rightarrow \quad S_{\alpha\beta} = S_{\beta\alpha}. \tag{1.28}$$

Note the presence of a magnetic field H slightly changes the micro-reversibility condition: In addition to a time and a momentum reversal we need to invert the magnetic field direction, $H \to -H$. It is clear, for instance, from the Hamiltonian of a free particle with mass m and charge e propagating along the axis x in the presence of a magnetic field,

$$\mathcal{H} = \frac{(p_x - eA_x)^2}{2m},$$

where A_x is a vector potential projection onto the axis x. Note that $H = \mathrm{rot}\,\mathbf{A}$. Thus in the presence of a magnetic field Eq. (1.28) is transformed [5]

$$\hat{S}(H) = \hat{S}^T(-H) \quad \Rightarrow \quad S_{\alpha\beta}(H) = S_{\beta\alpha}(-H). \tag{1.29}$$

In particular, the reflection amplitude, $\alpha = \beta$, is an even function for a magnetic field.

1.2 Current operator

Now we consider how the scattering matrix formalism can be applied to transport phenomena in mesoscopic samples. The scattering matrix relies on *the single-electron approximation*. Within this approximation the separate electrons are considered as independent particles whose interaction with other electrons, nuclei, impurities, quasi-particles, etc. can be described via the effective potential energy, $U_{eff}(t, \mathbf{r})$. Such an approach allows a simple and physically transparent description of transport phenomena on a qualitative level and in many cases even on a quantitative level.

Let us consider a mesoscopic sample connected to several, N_r, macroscopic contacts acting as electron reservoirs, Fig. 1.1. Electrons, propagating from some reservoir to the sample, enter it, are scattered inside it, and then leave it to go to the same or a different reservoir. To calculate the current flowing between the sample and the reservoirs we do not need to know what happens with each electron inside the sample. It is enough to look at the incoming and outgoing electron flows. To this end we enclose a sample by a fictitious surface Σ, see Fig. 1.1, and consider electron flows crossing this surface in the directions to the sample or back. In this case we, in fact, deal with the scattering problem: Electrons propagating to the sample are incident, or incoming, particles [we denote them via an upper index (in)], while electrons propagating from the sample are scattered, or outgoing, particles [upper index (out)]. We emphasize that we consider only elastic, i.e., energy-conserving, scattering. To neglect inelastic scattering we assume low temperatures when the phase coherence length, L_φ, is much larger than the size of a sample, $L_\varphi(T) \gg L$.

It is convenient to choose the eigen wave functions for electrons in leads connecting a scatterer to the reservoirs as the basis functions for defining the scattering matrix elements. These wave functions can be represented as the product of longitudinal and transverse terms. For the sake of simplicity we assume that the leads have only one conducting sub-band. Therefore, there is only one type of transverse wave function in each lead. We choose plane waves propagating to the scatterer (wave number $-k$) or from the scatterer (wave number k) as longitudinal wave functions. The former (latter) wave functions comprise the basis for incident, $\psi_\alpha^{(in)}$, (scattered, $\psi_\alpha^{(out)}$) electrons.

To calculate the current flowing between the scatterer and the reservoirs we use the second quantization formalism. This formalism deals with

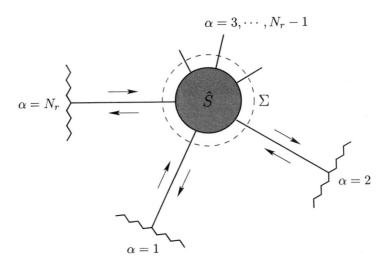

Fig. 1.1 A mesoscopic sample with scattering matrix $\hat{S}$. The index $\alpha = 1, 2, \ldots, N_r$ numbers electron reservoirs. The arrows directed to (from) the scatterer show a propagation direction for incident (scattered) electrons. The electron flow is calculated at the surface Σ shown as a dashed line.

operators creating/annihilating particles in some quantum state. We use different operators corresponding to incident electrons, $\hat{a}_\alpha^\dagger(E)/\hat{a}_\alpha(E)$, and to scattered electrons, $\hat{b}_\alpha^\dagger(E)/\hat{b}_\alpha(E)$. The operator $\hat{a}_\alpha^\dagger(E)$ creates one electron in the state with wave function $\psi_\alpha^{(in)}(E)/\sqrt{\hbar v_\alpha(E)}$, while the operator $\hat{b}_\alpha^\dagger(E)$ creates one electron in the state with wave function $\psi_\alpha^{(out)}(E)/\sqrt{\hbar v_\alpha(E)}$. The factor $1/\sqrt{\hbar v_\alpha(E)}$ takes account of the unit flux normalization. Note the index α can be composite, i.e., it can include, apart from the reservoir's number, the additional sub-indices, for instance, a sub-band number, an electron spin, etc.

Introduced fermionic operators are subject to the following anti-commutation relations:

$$\hat{a}_\alpha^\dagger(E)\,\hat{a}_\beta(E') + \hat{a}_\beta(E')\,\hat{a}_\alpha^\dagger(E) = \delta_{\alpha\beta}\,\delta(E - E'),$$

$$\hat{b}_\alpha^\dagger(E)\,\hat{b}_\beta(E') + \hat{b}_\beta(E')\,\hat{b}_\alpha^\dagger(E) = \delta_{\alpha\beta}\,\delta(E - E').$$

(1.30)

Next we introduce the field operators for electrons in lead α,

$$\hat{\Psi}_\alpha(t, \mathbf{r}) = \frac{1}{\sqrt{2\pi}} \int_0^\infty dE \, e^{-i\frac{E}{\hbar}t} \left\{ \hat{a}_\alpha(E) \frac{\psi_\alpha^{(in)}(E, \mathbf{r})}{\sqrt{\hbar v_\alpha(E)}} + \hat{b}_\alpha(E) \frac{\psi_\alpha^{(out)}(E, \mathbf{r})}{\sqrt{\hbar v_\alpha(E)}} \right\}, \quad (1.31)$$

$$\hat{\Psi}_\alpha^\dagger(t, \mathbf{r}) = \frac{1}{\sqrt{2\pi}} \int_0^\infty dE \, e^{i\frac{E}{\hbar}t} \left\{ \hat{a}_\alpha^\dagger(E) \frac{\psi_\alpha^{(in)*}(E, \mathbf{r})}{\sqrt{\hbar v_\alpha(E)}} + \hat{b}_\alpha^\dagger(E) \frac{\psi_\alpha^{(out)*}(E, \mathbf{r})}{\sqrt{\hbar v_\alpha(E)}} \right\}.$$

Here $v_\alpha(E) = \hbar k_\alpha(E)/m$ is an electron's velocity, $\mathbf{r} = (x, r_\perp)$, with x longitudinal and $r_\perp$ transverse spatial coordinates within the lead α. Note that $1/(\hbar v_\alpha(E))$ is the density of states, $(2\pi)^{-1} dk/dE$, for a one-dimensional conductor.

Using the field operators we write the operator, $\hat{I}_\alpha$, for a current flowing in the lead α

$$\hat{I}_\alpha(t, x) = \frac{i\hbar e}{2m} \int dr_\perp \left\{ \frac{\partial \hat{\Psi}_\alpha^\dagger(t, \mathbf{r})}{\partial x} \hat{\Psi}_\alpha(t, \mathbf{r}) - \hat{\Psi}_\alpha^\dagger(t, \mathbf{r}) \frac{\partial \hat{\Psi}_\alpha(t, \mathbf{r})}{\partial x} \right\}. \quad (1.32)$$

Here the positive direction is from the scatterer to the reservoir.

Next we represent the basis wave functions as the product of transverse and longitudinal parts,

$$\psi^{(in)}(E, \mathbf{r}) = \xi_E(r_\perp) e^{-ik(E)x},$$
$$\psi^{(out)}(E, \mathbf{r}) = \xi_E(r_\perp) e^{ik(E)x}, \quad (1.33)$$

and take into account that the transverse wave functions are normalized,

$$\int dr_\perp |\xi_E(r_\perp)|^2 = 1. \quad (1.34)$$

In what follows we are interested in currents flowing under the bias much smaller than the Fermi energy, μ_0. Therefore, in all equations the main contribution comes from energies within the interval that are much smaller than the energy itself,[2]

---

[2] In the case of a stationary current this restriction can be safely relaxed since the calculation of an expectation value implies $E = E'$. While for calculation of a time-dependent current or a noise and higher current cumulants (even in the stationary case) the restriction (1.35) is important.

$$|E - E'| \ll E \sim \mu_0. \tag{1.35}$$

The last inequality allows us to strongly simplify the equation for a current. We can put, $v(E) \approx v(E')$ and $k(E) \approx k(E')$. Moreover, within the same sub-band the transverse wave functions are the same, $\xi_E = \xi_{E'}$. Note if the functions ξ_E and $\xi_{E'}$ are from different sub-bands then they are orthogonal, $\int dr_\perp \xi_E(r_\perp)\left(\xi_{E'}(r_\perp)\right)^* = 0$. That allows us to split the total current into the sum of contributions from different sub-bands. Therefore, we can assume each lead having only one sub-band.

Substituting Eq. (1.31) into Eq. (1.32) and taking into account Eq. (1.35) we calculate

$$\hat{I}_\alpha(t,x) = \frac{i\hbar e}{2m} \iint dE\, dE' \frac{e^{i\frac{E-E'}{\hbar}t}}{hv_\alpha(E)} \int dr_\perp |\xi_{E,\alpha}(r_\perp)|^2$$

$$\times \left\{ \frac{\partial}{\partial x}\left[\hat{a}_\alpha^\dagger(E)e^{ik_\alpha(E)x} + \hat{b}_\alpha^\dagger(E)e^{-ik_\alpha(E)x}\right]\left(\hat{a}_\alpha(E')e^{-ik_\alpha(E)x} + \hat{b}_\alpha(E')e^{ik_\alpha(E)x}\right) \right.$$

$$\left. - \left(\hat{a}_\alpha^\dagger(E)e^{ik_\alpha(E)x} + \hat{b}_\alpha^\dagger(E)e^{-ik_\alpha(E)x}\right)\frac{\partial}{\partial x}\left[\hat{a}_\alpha(E')e^{-ik_\alpha(E)x} + \hat{b}_\alpha(E')e^{ik_\alpha(E)x}\right] \right\}.$$

Differentiating over x and combining similar terms we finally arrive at the following equation for the current operator [5],

$$\hat{I}_\alpha(t) = \frac{e}{h} \iint dE\, dE'\, e^{i\frac{E-E'}{\hbar}t} \left\{ \hat{b}_\alpha^\dagger(E)\hat{b}_\alpha(E') - \hat{a}_\alpha^\dagger(E)\hat{a}_\alpha(E') \right\}. \tag{1.36}$$

In what follows we use this equation and calculate, in particular, a measurable current, $I_\alpha = \langle \hat{I}_\alpha \rangle$, flowing into the lead α. Here $\langle \ldots \rangle$ stands for quantum-statistical averaging over the state of incoming electrons. To calculate such an average for the products of $\hat{a}^\dagger \hat{a}$ and $\hat{b}^\dagger \hat{b}$ we take into account that the creation and annihilation operators, $\hat{a}_\alpha^\dagger$ and $\hat{a}_\alpha$, correspond to particles propagating from the reservoir. We suppose that the presence of a mesoscopic scatterer does not affect the equilibrium properties of reservoirs. Therefore, the incoming particles are equilibrium particles of macroscopic reservoirs. And for them we can use the standard rules for calculating the quantum-statistical average of the product of creation and annihilation operators. In addition we suppose that electrons at different reservoirs, $\alpha \neq \beta$, are not correlated. Then we can write

$$\langle \hat{a}_\alpha^\dagger(E)\, \hat{a}_\beta(E') \rangle = \delta_{\alpha\beta}\, \delta(E-E')\, f_\alpha(E),$$

$$\langle \hat{a}_\alpha(E)\, \hat{a}_\beta^\dagger(E') \rangle = \delta_{\alpha\beta}\, \delta(E-E')\, \{1 - f_\alpha(E)\},$$
(1.37)

where $f_\alpha(E)$ is the Fermi distribution function [15] for electrons in the reservoir α,

$$f_\alpha(E) = \frac{1}{1 + e^{\frac{E-\mu_\alpha}{k_B T_\alpha}}}.$$
(1.38)

Here k_B is the Boltzmann constant, μ_α is the Fermi energy (the electrochemical potential) and T_α is the temperature of the reservoir α.

In contrast the operators $\hat{b}_\alpha^\dagger$ and $\hat{b}_\alpha$ correspond to scattered particles which, in general, are non-equilibrium particles. To calculate the quantum-statistical average for (the product of) them we need to express them in terms of the operators for incoming particles for which we know how to calculate a corresponding average. To this end we consider both the field operator, $\hat{\Psi}^{(in)}$, corresponding to an incoming wave,

$$\hat{\Psi}^{(in)} = \sum_{\alpha=1}^{N_r} \hat{a}_\alpha \frac{\psi_\alpha^{(in)}}{\sqrt{\hbar v_\alpha}},$$

and the field operator, $\hat{\Psi}^{(out)}$, corresponding to a scattered wave,

$$\hat{\Psi}^{(out)} = \sum_{\beta=1}^{N_r} \hat{b}_\beta \frac{\psi_\beta^{(out)}}{\sqrt{\hbar v_\beta}}.$$

These equations are similar to Eqs. (1.1) and (1.2) excepting the coefficients are now the second quantization operators. Thus each of the operators $\hat{b}_\beta$ is expressed in terms of all the operators $\hat{a}_\alpha$ through the elements of the scattering matrix, which is an $N_r \times N_r$ unitary matrix. By analogy with Eq. (1.7) we write [5]

$$\hat{b}_\alpha = \sum_{\beta=1}^{N_r} S_{\alpha\beta}\, \hat{a}_\beta,$$

$$\hat{b}_\alpha^\dagger = \sum_{\beta=1}^{N_r} S_{\alpha\beta}^*\, \hat{a}_\beta^\dagger.$$
(1.39)

The equations (1.36)–(1.39) constitute the basis of the scattering matrix approach to transport phenomena in mesoscopic physics.

1.3 Direct current and the distribution function

Let us calculate a current, I_α,

$$I_\alpha = \langle \hat{I}_\alpha \rangle, \tag{1.40}$$

flowing into the lead α under the DC bias, $\Delta V_{\alpha\beta} = V_\alpha - V_\beta$. In this case the different reservoirs have different electrochemical potentials,

$$\mu_\alpha = \mu_0 + eV_\alpha. \tag{1.41}$$

Note we include the potential energy eV_α in the μ_α. Then the energy E means the total (kinetic plus potential) energy of an electron. The use of a total energy (instead of a kinetic one) is convenient since it is conserved (in the stationary case) while an electron propagates from one reservoir through the scatterer to another reservoir.

The current operator, $\hat{I}_\alpha(t)$, is given in Eq. (1.36). After averaging Eq. (1.40) reads

$$I_\alpha = \frac{e}{h} \int dE \left\{ f_\alpha^{(out)}(E) - f_\alpha^{(in)}(E) \right\}, \tag{1.42}$$

where we have introduced the distribution functions for incident electrons, $f_\alpha^{(in)}$, and for scattered electrons, $f_\alpha^{(out)}$,

$$\langle \hat{a}_\alpha^\dagger(E) \hat{a}_\alpha(E') \rangle = \delta(E - E') f_\alpha^{(in)}(E),$$
$$\langle \hat{b}_\alpha^\dagger(E) \hat{b}_\alpha(E') \rangle = \delta(E - E') f_\alpha^{(out)}(E). \tag{1.43}$$

The physical meaning for the introduced distribution functions is that the quantity $\frac{dE}{h} f_\alpha^{(in/out)}(E)$ defines the average number of electrons with an energy within the interval dE near E crossing the cross-section of the lead

α in unit time to/from the scatterer. The direct current is obviously the difference of the flows times an electron charge e.

According to Eq. (1.37) the distribution function for incoming electrons is the Fermi function for a corresponding reservoir,

$$f_\alpha^{(in)}(E) = f_\alpha(E). \tag{1.44}$$

To calculate the distribution function for scattered electrons, $f_\alpha^{(out)}(E)$, we use Eqs. (1.39), (1.37) and find,

$$\delta(E - E') f_\alpha^{(out)}(E) \equiv \langle \hat{b}_\alpha^\dagger(E) \hat{b}_\alpha(E') \rangle$$

$$= \sum_{\beta=1}^{N_r} \sum_{\gamma=1}^{N_r} S_{\alpha\beta}^*(E) S_{\alpha\gamma}^*(E') \langle \hat{a}_\beta^\dagger(E) \hat{a}_\gamma(E') \rangle$$

$$= \sum_{\beta=1}^{N_r} \sum_{\gamma=1}^{N_r} S_{\alpha\beta}^*(E) S_{\alpha\gamma}^*(E') \delta(E - E') \delta_{\beta\gamma} f_\beta(E).$$

Therefore, the distribution function, $f_\alpha^{(out)}(E)$, for electrons scattered into the lead α depends on the Fermi functions, $f_\beta(E)$, for all the reservoirs, $\beta = 1, 2, \ldots, N_r$:

$$f_\alpha^{(out)}(E) = \sum_{\beta=1}^{N_r} |S_{\alpha\beta}(E)|^2 f_\beta(E). \tag{1.45}$$

Note if all the reservoirs have the same electrochemical potentials and temperatures (hence the same Fermi functions), $f_\beta = f_0, \forall \beta$, then the distribution function for scattered electrons is the Fermi function as well, i.e., the scattered electrons are in equilibrium. To show this we use the unitarity of the scattering matrix,

$$\hat{S}\hat{S}^\dagger = \hat{I} \quad \Rightarrow \quad \sum_{\beta=1}^{N_r} |S_{\alpha\beta}(E)|^2 = 1, \tag{1.46}$$

and find $f_\alpha^{(out)}(E) = f_0(E) \sum_{\beta=1}^{N_r} |S_{\alpha\beta}(E)|^2 = f_0(E)$. In contrast, if the potentials or temperatures of different reservoirs are different then the scattered electrons are characterized by the non-equilibrium distribution function, Fig. 1.2.

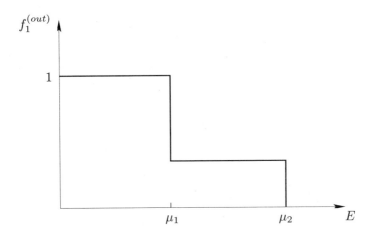

Fig. 1.2 The distribution function for electrons scattered into the contact $\alpha = 1$. The height of a step at $E = \mu_1$ is $|S_{12}|^2$. The scatterer is connected to two electron reservoirs at zero temperature, $T_1 = T_2 = 0$, and having chemical potentials μ_1 and μ_2.

Substituting Eqs. (1.44) and (1.45) into Eq. (1.42) and using Eq. (1.46) we finally calculate a direct current,

$$I_\alpha = \frac{e}{h} \int dE \sum_{\beta=1}^{N_r} |S_{\alpha\beta}(E)|^2 \left\{ f_\beta(E) - f_\alpha(E) \right\}. \quad (1.47)$$

We see that the current flowing into the lead α depends on the difference of the Fermi functions times the corresponding square of the scattering matrix element modulus. If all the reservoirs have the same potentials and temperatures then the current is zero. Otherwise there is a current through the sample.

1.3.1 *Conservation of a direct current*

Let us check whether Eq. (1.47) fulfills a direct current conservation law,

$$\sum_{\alpha=1}^{N_r} I_\alpha = 0, \quad (1.48)$$

which is a direct consequence of no charge accumulation inside the mesoscopic sample. This equation tells us that the sum of the current flowing into all the leads is zero. To avoid misunderstanding we stress that in each lead the positive direction is chosen from the scatterer to the corresponding reservoir. Therefore, the current has a sign "+" or "−" if it is directed from or to the scatterer.

First of all we derive Eq. (1.48). To this end we use the electrical charge continuity equation,

$$\text{div}\,\mathbf{j} + \frac{\partial \rho}{\partial t} = 0\,, \tag{1.49}$$

where $\mathbf{j}$ is a current density vector and ρ is a charge density. We integrate it over the volume enclosed by the surface Σ (see Fig. 1.1). Then transforming the volume integral of a current density divergence into the surface integral of a current density and taking into account that the current flows into the leads only we arrive at the following

$$\sum_{\alpha=1}^{N_r} I_\alpha(t) + \frac{\partial Q}{\partial t} = 0\,. \tag{1.50}$$

Here Q is the charge on the scatterer. In the stationary case under consideration there are only direct currents in the leads and the charge Q is constant. Then Eq. (1.50) results in Eq. (1.48). In the non-stationary case we should average Eq. (1.50) over time. With the following definition of a direct current, $I_\alpha = \lim_{\mathcal{T} \to \infty} \frac{1}{\mathcal{T}} \int_0^{\mathcal{T}} dt\, I_\alpha(t)$, and assuming that the charge $Q(t)$ is bounded we again conclude that Eq. (1.48) is a consequence of Eq. (1.50).

Now we check whether Eq. (1.47) does satisfy Eq. (1.48). We use the unitarity of the scattering matrix in a form slightly different from but still equivalent to Eq. (1.46)

$$\hat{S}^\dagger \hat{S} = \hat{I} \quad \Rightarrow \quad \sum_{\alpha=1}^{N_r} |S_{\alpha\beta}(E)|^2 = 1\,. \tag{1.51}$$

Then from Eq. (1.47) we get

$$\sum_{\alpha=1}^{N_r} I_\alpha = \frac{e}{h} \int dE \sum_{\alpha=1}^{N_r} \sum_{\beta=1}^{N_r} |S_{\alpha\beta}(E)|^2 \left\{ f_\beta(E) - f_\alpha(E) \right\}$$

$$= \frac{e}{h} \int dE \left\{ \sum_{\beta=1}^{N_r} f_\beta(E) \sum_{\alpha=1}^{N_r} |S_{\alpha\beta}(E)|^2 - \sum_{\alpha=1}^{N_r} f_\alpha(E) \sum_{\beta=1}^{N_r} |S_{\alpha\beta}(E)|^2 \right\}$$

$$= \frac{e}{h} \int dE \left\{ \sum_{\beta=1}^{N_r} f_\beta(E) - \sum_{\alpha=1}^{N_r} f_\alpha(E) \right\} = 0\,,$$

as expected. Therefore, we have illustrated the earlier mentioned connection between unitarity and current conservation. Next we will use Eq. (1.47) and calculate a current in two simple but generic cases.

1.3.2 Difference of potentials

Let the reservoirs have different potentials but the same temperature

$$\mu_\alpha = \mu_0 + eV_\alpha, \quad eV_\alpha \ll \mu_0,$$
$$T_\alpha = T_0, \quad \forall \alpha. \tag{1.52}$$

If $|eV_\alpha| \ll k_B T_0$ we can expand

$$f_\alpha = f_0 - eV_\alpha \frac{\partial f_0}{\partial E} + \mathcal{O}(V_\alpha^2),$$

where f_0 is the Fermi function with a chemical potential μ_0 and a temperature T_0. Using this expansion in Eq. (1.47) we calculate a current

$$I_\alpha = \sum_{\beta=1}^{N_r} G_{\alpha\beta} \{V_\beta - V_\alpha\}, \tag{1.53}$$

where we introduce the elements of the conductance matrix

$$G_{\alpha\beta} = G_0 \int dE \left(-\frac{\partial f_0}{\partial E}\right) |S_{\alpha\beta}(E)|^2, \tag{1.54}$$

with $G_0 = e^2/h$ the conductance quantum (for spinless electrons). Taking into account electron spin the conductance quantum should be doubled.

At zero temperature, $T_0 = 0$,

$$-\frac{\partial f_0}{\partial E} = \delta(E - \mu_0),$$

and the integration over energy in Eq. (1.54) becomes trivial. In this case the conductance matrix elements become especially simple [5]

$$G_{\alpha\beta} = G_0 \left|S_{\alpha\beta}(\mu_0)\right|^2. \tag{1.55}$$

It is clear that the linear dependence of a current on the potential difference is kept at a relatively small bias. The corresponding scale is dictated by the energy dependence of the scattering matrix elements, $S_{\alpha\beta}(E)$. To illustrate it we calculate a direct current at zero temperature, $T_0 = 0$, but finite potential, $eV_\alpha \neq 0$. In this case we cannot expand the Fermi function in powers of a potential, therefore, Eq. (1.47) becomes

$$I_\alpha = \frac{G_0}{e} \sum_{\beta=1}^{N_r} \int_{\mu_0 + eV_\alpha}^{\mu_0 + eV_\beta} dE \, |S_{\alpha\beta}(E)|^2. \tag{1.56}$$

If the quantity $G_{\alpha\beta}$ changes only a little within the energy interval $\sim |eV_\beta - eV_\alpha|$ near the Fermi energy μ_0 then we can use $S_{\alpha\beta}(E) \approx S_{\alpha\beta}(\mu_0)$ in Eq. (1.56), which results in linear I–V characteristics, Eq. (1.53).

On the other hand if one cannot ignore the energy dependence of $S_{\alpha\beta}(E)$ then the current becomes a non-linear function of a bias. As a simple example we consider a sample with two leads ($\alpha = 1, 2$) whose scattering properties are governed by the resonance level of a width Γ located at the energy E_1:

$$|S_{12}(E)|^2 = \frac{\Gamma^2}{(E - E_1)^2 + \Gamma^2}. \tag{1.57}$$

For simplicity suppose that $E_1 = \mu_0$. Then substituting the equation above into Eq. (1.56) we find a current

$$I_1 = \frac{e}{h} \Gamma \left\{ \arctan\left(\frac{eV_2}{\Gamma}\right) - \arctan\left(\frac{eV_1}{\Gamma}\right) \right\}. \tag{1.58}$$

If the potentials are small compared to the resonance level width, $|eV_1|, |eV_2| \ll \Gamma$, we recover Ohm's law, $I_{12} = G_0 (V_1 - V_2)$. While in the opposite case, $|eV_1|, |eV_2| \gg \Gamma$, the current is an essentially non-linear function of potentials, $I_1 = (\Gamma^2/h)\left(V_1^{-1} - V_2^{-1}\right)$. Therefore, we see that in this problem the level width Γ is a relevant energy scale.

1.3.3 Difference of temperatures

The temperature difference also can result in a current. This is the so-called *thermoelectric current*. To calculate it we suppose that the reservoirs have the same potentials but their temperatures are different,

$$\mu_\alpha = \mu_0, \quad \forall \alpha,$$
$$T_\alpha = T_0 + \mathcal{T}_\alpha, \quad \mathcal{T}_\alpha \ll T_0. \tag{1.59}$$

Expanding the Fermi functions in Eq. (1.47) in powers of $\mathcal{T}_\alpha$,

$$f_\alpha = f_0 + \mathcal{T}_\alpha \frac{\partial f_0}{\partial T} + \mathcal{O}(\mathcal{T}_\alpha^2),$$

and taking into account that

$$\frac{\partial f_0}{\partial T} = -\frac{E - \mu_0}{T_0} \frac{\partial f_0}{\partial E},$$

we calculate the thermoelectric current flowing into the lead α,

$$I_\alpha = \sum_{\beta=1}^{N_r} G_{\alpha\beta}^{(T)} \{\mathcal{T}_\beta - \mathcal{T}_\alpha\}. \tag{1.60}$$

Here we have introduced the thermoelectric conductance matrix elements,

$$G_{\alpha\beta}^{(T)}(E) = \frac{\pi^2 e}{3h} k_B T_0 \frac{\partial |S_{\alpha\beta}(E)|^2}{\partial E}, \tag{1.61}$$

and used the following integral

$$\int_0^\infty dE \, \frac{e^{\frac{E-\mu_0}{k_B T_0}}}{\left(1 + e^{\frac{E-\mu_0}{k_B T_0}}\right)^2} \left(\frac{E-\mu_0}{k_B T_0}\right)^2 = \frac{\pi^2}{3} k_B T_0.$$

From Eq. (1.61) it follows that if the conductance is energy independent, $G_{\alpha\beta}(E) = \text{const}$, then the thermoelectric conductance (and the thermoelectric current) is zero.

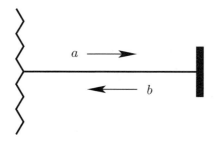

Fig. 1.3 A single-channel scatterer. a is the amplitude of an incoming wave, b is the amplitude of a reflected wave. A zigzag line denotes an electron reservoir.

1.4 Examples

Now we consider several examples to clarify the physical meaning of the scattering matrix elements. The scattering matrix is a square matrix $N_r \times N_r$, where N_r is the number of one-dimensional conducting sub-bands in each lead, connecting a mesoscopic sample to the reservoirs. N_r is the number of *scattering channels*.

1.4.1 *Scattering matrix* 1×1

Such a matrix has only one element, S_{11}, and it describes a sample connected to a single reservoir via a one-dimensional lead, Fig. 1.3. Sometimes such a sample is referred to as *a mesoscopic capacitor*.[3] Unitarity, Eq. (1.10), requires $|S_{11}|^2 = 1$. Therefore, quite generally the scattering matrix 1×1 reads

$$\hat{S} = e^{i\gamma}, \qquad (1.62)$$

where i is an imaginary unity, γ is real. Scattering in this case is reduced to the total reflection of an incident wave. Therefore, the element S_{11} is *the reflection coefficient*. Generally speaking any diagonal element, $S_{\alpha\alpha}$, of the scattering matrix of a higher dimension is a reflection coefficient, since it defines both the amplitude and the phase of a wave returning to the same reservoir where the incident wave originated. In the case under consideration (1×1) the amplitude of the wave remains the same, while

---

[3]More precisely it is one of the capacitor's plates.

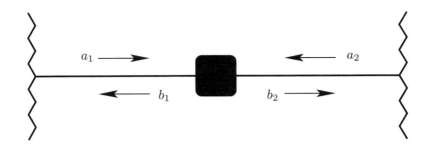

Fig. 1.4 A two-channel scatterer. a_α (b_α) are the amplitudes of incoming (scattered) waves, $\alpha = 1, 2$.

the phase is changed by γ, which is the only quantity encoding information about the properties of the mesoscopic sample. For instance, if the wave is reflected by a hard and infinite potential wall then the phase is changed by $\gamma = \pi$, while if the scatterer is a ring then γ depends on the magnetic flux threading the ring, and so on.

1.4.2 Scattering matrix 2 × 2

This matrix has in general four complex elements, hence there are eight real parameters. However, unitarity, Eq. (1.10), imposes four constraints. As a result there are only four independent parameters. It is convenient to choose the following independent parameters:

1. $R = |S_{11}|^2$ – a reflection probability.
2. γ – a phase relating to an effective charge, Q, of a scatterer via the Friedel sum rule, $Q = e/(2\pi i) \ln(\det \hat{S}) = e\gamma/\pi$ [16, 17].
3. θ – a phase characterizing the reflection asymmetry, $\theta = i \ln(S_{11}/S_{22})/2$.
4. ϕ – a phase characterizing the transmission asymmetry, $\phi = i \ln(S_{12}/S_{21})/2$. This phase depends on an external magnetic field or on an internal magnetic moment of a scatterer.

Therefore, the general expression for the scattering matrix 2×2, describing a sample connected to two electron reservoirs, Fig. 1.4, can be written as follows

$$\hat{S} = e^{i\gamma} \begin{pmatrix} \sqrt{R}\,e^{-i\theta} & i\sqrt{1-R}\,e^{-i\phi} \\ i\sqrt{1-R}\,e^{i\phi} & \sqrt{R}\,e^{i\theta} \end{pmatrix}. \quad (1.63)$$

Note the reflection probability is the same in both scattering channels,

$$|S_{11}|^2 = |S_{22}|^2 = R. \tag{1.64}$$

The same is valid with respect to the transmission probabilities: they are independent of the direction of movement,

$$|S_{12}|^2 = |S_{21}|^2. \tag{1.65}$$

In addition the symmetry condition, Eq. (1.29), restricts the possible dependence of the parameters chosen for the magnetic field. It is easy to see that $\gamma(H)$, $R(H)$, and $\theta(H)$ are even functions, while $\phi(H)$ is an odd function, $\phi(H) = -\phi(-H)$. In particular, if $H = 0$ then $\phi = 0$ and, correspondingly, the transmission amplitude is independent of the movement direction,

$$S_{12}(H = 0) = S_{21}(H = 0). \tag{1.66}$$

We stress that Eq. (1.65) holds also in the presence of a magnetic field.

Turning to the transport properties, we see that the conductance, $G \equiv G_{12} = G_{21}$, of a sample with two leads is an even function of a magnetic field [5, 18]

$$G(H) = G(-H). \tag{1.67}$$

As we will show this property holds also for a sample with two quasi-one-dimensional leads. This symmetry is a consequence of the micro-reversibility of the quantum-mechanical equations of motion that are valid in the absence of inelastic interactions breaking the phase coherence.[4]

1.4.3 Scattering matrix 3×3

Such a matrix describes a scatterer connected to three reservoirs, Fig. 1.5. It has many, namely nine, independent real parameters, which makes it difficult to find a general expression. Usually the particular expressions for the

---

[4]In the non-linear regime and in the presence of electron-electron interactions the current through the two terminal sample is not an even function of a magnetic field [19, 20].

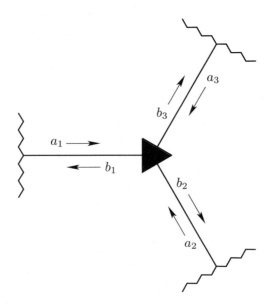

Fig. 1.5 A three-channel scatterer. a_α (b_α) are amplitudes of incoming (scattered) waves, $\alpha = 1, 2, 3$.

scattering matrix elements are used. For instance, following Refs. [21, 22] one can write a one-parametric scattering matrix

$$\hat{S} = \begin{pmatrix} -(a+b) & \sqrt{\epsilon} & \sqrt{\epsilon} \\ \sqrt{\epsilon} & a & b \\ \sqrt{\epsilon} & b & a \end{pmatrix}, \tag{1.68}$$

where $a = (\sqrt{1-2\epsilon} - 1)/2$, $b = (\sqrt{1-2\epsilon} + 1)/2$, and the real parameter ϵ changes within the following interval $0 \leq \epsilon \leq 0.5$. The parameter ϵ characterizes the coupling strength between the lead $\alpha = 1$ and the scatterer. At $\epsilon = 0$ this lead is decoupled completely from the scatterer, $S_{11} = -1$, while electrons freely propagate from the lead $\alpha = 2$ into the lead $\alpha = 3$ and back, $S_{32} = S_{23} = 1$. The limit $\epsilon = 0.5$ corresponds to a reflectionless coupling between the sample and the lead $\alpha = 1$: $S_{11} = 0$.

Sometimes, solving the Schrödinger equation for the junction of three one-dimensional leads, the Griffith boundary conditions are used [23]. These conditions include both the continuity of a wave function and a current conservation at a crossing point. Then a scattering matrix of the type given in Eq. (1.68) with parameter $\epsilon = 4/9$ arises. Other values of the parameter ϵ, for instance, can be understood as related to the presence of some tunnel barrier at the crossing point.

It should be noted that in contrast to the two-lead case, see Eq. (1.64), in the case of three leads, the reflection probabilities $R_{\alpha\alpha} \equiv |S_{\alpha\alpha}|^2$, $\alpha = 1, 2, 3$, for different scattering channels can be different. Moreover, the current flowing between any two leads depends not only on the corresponding transmission probability, $T_{\alpha\beta} \equiv |S_{\alpha\beta}|^2$, $\alpha \neq \beta$, but also on the transmission probabilities to the third lead, $T_{\gamma\alpha}$ and $T_{\gamma\beta}$, $\gamma \neq \alpha, \beta$.

1.4.4 Scatterer with two leads

We will show that the conductance of a mesoscopic sample with two quasi-one-dimensional leads is an even function of a magnetic field. We saw this before, see Eq. (1.67), for the case of two one-dimensional leads when the scattering matrix is a 2×2 unitary matrix. Now we generalize this result onto the case when each lead has several conducting sub-bands [24].

Let one of the leads, say the left, have N_L conducting sub-bands while another one, the right, has N_R conducting sub-bands. The total number of scattering channels is $N_r = N_L + N_R$, therefore, the scattering matrix is an $N_r \times N_r$ unitary matrix. It is convenient to number the scattering channels in such a way that the first N_L scattering channels, $1 \leq \alpha \leq N_L$, correspond to the left lead, while the last N_R scattering channels, $N_L + 1 \leq \alpha \leq N_r$, correspond to the right lead. We assume that the left reservoir has a potential $-V/2$ while the right reservoir has a potential $V/2$. Note for all the sub-bands belonging to the same lead the corresponding potential V_α is the same,

$$V_\alpha = \begin{cases} -\frac{V}{2}, & 1 \leq \alpha \leq N_L, \\ \frac{V}{2}, & N_L \leq \alpha \leq N_r. \end{cases} \tag{1.69}$$

The current, I_α, carried by the electrons of the sub-band α is given by Eq. (1.53). For simplicity we consider a zero temperature case while the conclusion remains valid at finite temperatures also. So we write

$$I_\alpha = G_0 \sum_{\beta=1}^{N_r} |S_{\alpha\beta}|^2 \{V_\beta - V_\alpha\}. \tag{1.70}$$

Here and below the scattering matrix elements are calculated at $E = \mu_0$.

To calculate the current, I_L, flowing within the left lead we need to sum up the contributions from all the sub-bands belonging to the left lead. These are sub-bands with numbers from 1 until N_L. Therefore, the current I_L is

$$I_L = \sum_{\alpha=1}^{N_L} I_\alpha . \tag{1.71}$$

Substituting Eq. (1.70) into Eq. (1.71), we find

$$I_L = V G_0 \sum_{\alpha=1}^{N_L} \sum_{\beta=N_L+1}^{N_r} |S_{\alpha\beta}|^2 . \tag{1.72}$$

Calculating in the same way the current I_R flowing into the right lead it is easy to check that $I_R = -I_L$, as expected. Note the equations for the currents $I_{L/R}$ depend only on the transmission probabilities, $|S_{\alpha\beta}|^2$, between the scattering channels belonging to the different leads. Neither intra-sub-bands reflections nor inter-sub-bands transitions within the same lead affect the current.

The conductance, $G = I_L/V$, is

$$G = G_0 \sum_{\alpha=1}^{N_L} \sum_{\beta=N_L+1}^{N_r} |S_{\alpha\beta}|^2 . \tag{1.73}$$

Our aim is to show that this quantity is an even function of a magnetic field, $G(H) = G(-H)$. To this end we introduce some generalized reflection coefficients for the reservoirs

$$R_{LL} = \sum_{\alpha=1}^{N_L} \sum_{\beta=1}^{N_L} |S_{\alpha\beta}|^2 , \quad R_{RR} = \sum_{\alpha=N_L+1}^{N_r} \sum_{\beta=N_L+1}^{N_r} |S_{\alpha\beta}|^2 , \tag{1.74}$$

and transmission coefficients between the reservoirs

$$T_{LR} = \sum_{\alpha=1}^{N_L} \sum_{\beta=N_L+1}^{N_r} |S_{\alpha\beta}|^2 , \quad T_{RL} = \sum_{\alpha=N_L+1}^{N_r} \sum_{\beta=1}^{N_L} |S_{\alpha\beta}|^2 . \tag{1.75}$$

These coefficients satisfy the following identities,

$$R_{LL} + T_{LR} = \sum_{\alpha=1}^{N_L}\sum_{\beta=1}^{N_L} |S_{\alpha\beta}|^2 + \sum_{\alpha=1}^{N_L}\sum_{\beta=N_L+1}^{N_r} |S_{\alpha\beta}|^2$$

$$= \sum_{\alpha=1}^{N_L}\sum_{\beta=1}^{N_r} |S_{\alpha\beta}|^2 = \sum_{\alpha=1}^{N_L} 1 = N_L,$$

$$R_{LL} + T_{RL} = \sum_{\alpha=1}^{N_L}\sum_{\beta=1}^{N_L} |S_{\alpha\beta}|^2 + \sum_{\alpha=N_L+1}^{N_r}\sum_{\beta=1}^{N_L} |S_{\alpha\beta}|^2$$

$$= \sum_{\beta=1}^{N_L}\sum_{\alpha=1}^{N_r} |S_{\alpha\beta}|^2 = \sum_{\beta=1}^{N_L} 1 = N_L,$$

where we used the unitarity of the scattering matrix, $\sum_{\alpha=1}^{N_r} |S_{\alpha\beta}|^2 = 1$, $\sum_{\beta=1}^{N_r} |S_{\alpha\beta}|^2 = 1$. From the above identities it also follows that

$$T_{LR} = T_{RL}. \tag{1.76}$$

Next we use the symmetry conditions, Eq. (1.29), for the scattering matrix elements in the magnetic field and find

$$T_{LR}(-H) = \sum_{\alpha=1}^{N_L}\sum_{\beta=N_L+1}^{N_r} |S_{\alpha\beta}(-H)|^2 = \sum_{\alpha=1}^{N_L}\sum_{\beta=N_L+1}^{N_r} |S_{\beta\alpha}(H)|^2$$

$$= \sum_{\beta=N_L+1}^{N_r}\sum_{\alpha=1}^{N_L} |S_{\beta\alpha}(H)|^2 = T_{RL}(H).$$

Therefore, we have

$$T_{LR}(-H) = T_{RL}(H). \tag{1.77}$$

Combining together Eqs. (1.76) and (1.77) we finally arrive at the required relation

$$\left.\begin{array}{l} T_{LR} = T_{RL} \\ \\ T_{LR}(-H) = T_{RL}(H) \end{array}\right\} \Rightarrow T_{LR}(H) = T_{LR}(-H),$$

which shows that the conductance, $G = G_0 T_{LR}$, of a sample with two quasi-one-dimensional leads is an even function of a magnetic field.

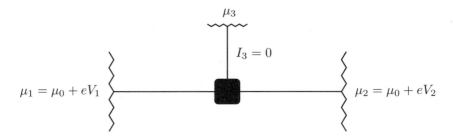

Fig. 1.6 A mesoscopic scatterer with current carrying $(1,2)$ and potential (3) leads.

1.4.5 Scatterer with a potential contact

The phase coherent system represents an entity whose properties are sometimes quite sensitive to the measurement procedure. If one attaches an additional contact, for instance to measure an electric potential inside the mesoscopic sample, then the current flowing through the sample is changed [25, 26].[5]

Let us consider a sample connected to three leads, Fig. 1.6. Two of them, having different electrochemical potentials, $\mu_1 = \mu_0 + eV_1$ and $\mu_2 = \mu_0 + eV_2$, are used to let a current pass through the system. In contrast the third lead acts as a potential contact. As for any potential contacts the current flowing into it is zero, $I_3 = 0$. This condition defines an electrochemical potential, $\mu_3 = \mu_0 + eV_3$, of the third reservoir (which the third lead is connected to) as a function of the bias between the first and the second reservoirs, $V = V_2 - V_1$. One can say that V_3 is a potential of a mesoscopic sample at the point of attachment of the third lead.

Now we calculate the current through the sample. Since $I_3 = 0$ then $I_1 = -I_2$ as for the sample with two leads. Following this analogy we would say that at a given bias V the current depends only on the probability for an electron to go from the first lead to the second lead. However, this is not the case. In the presence of a potential contact (the third lead) the conductance, $G_{12} = I_1/V$, in addition depends on the probability for an electron to be scattered between the current-carrying and the potential leads,

$$I_1 \neq G_0 T_{12} V \quad \Rightarrow \quad G_{12} \neq G_0 T_{12}.$$

---

[5]In Ref. [22] and mentioned above there is an ingenious idea of how to treat inelastic processes within the scattering approach: It consists in attaching to the sample an additional, fictitious lead. This idea, sometimes essentially modified, see, e.g., Ref. [27], is widely used in the literature due to its simplicity and clarity.

Using Eq. (1.53) we write

$$I_1 = G_0\Big(T_{12}(V_2 - V_1) + T_{13}(V_3 - V_1)\Big),$$

$$I_2 = G_0\Big(T_{21}(V_1 - V_2) + T_{23}(V_3 - V_2)\Big),$$

$$I_3 = G_0\Big(T_{31}(V_1 - V_3) + T_{32}(V_2 - V_3)\Big).$$

From the condition $I_3 = 0$ we find

$$V_3 = \frac{T_{31}V_1 + T_{32}V_2}{T_{31} + T_{32}}.$$

Note the potential $V_3 = 0$ in the symmetric case, namely, if $V_1 = -V_2$ and $T_{31} = T_{32}$. Using the equation for V_3, we can find the conductance $G_{12} = I_1/(V_2 - V_1)$:

$$G_{12} = G_0\left\{T_{12} + \frac{T_{13}T_{32}}{T_{31} + T_{32}}\right\}.$$

In the case of a weak coupling between the potential contact and the sample, $T_{31}, T_{32} \ll T_{12}$, we recover the result for the sample with two leads, $G_{12} \approx G_0 T_{12}$.

1.4.6 Scatterer embedded in a ring

We consider two generic cases: (i) a ring with a magnetic flux Φ and (ii) a ring with scatterer having different transmission amplitudes to the left and to the right. For simplicity we suppose the scatterer located at $x = 0$ to be very thin: Its width w is small compared to the length L of the ring. Then we can choose a wave function on the ring threaded by the magnetic flux Φ, Fig. 1.7, as follows

$$\psi(x) = \left(Ae^{ik(x-L)} + Be^{-ikx}\right)e^{i2\pi\frac{x}{L}\frac{\Phi}{\Phi_0}}, \quad 0 \le x < L. \tag{1.78}$$

The scattering matrix is

$$\hat{S} = \begin{pmatrix} S_{11} & S_{12} \\ S_{21} & S_{22} \end{pmatrix}. \tag{1.79}$$

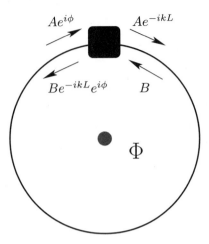

Fig. 1.7 A one-dimensional ring of length L pierced by the magnetic flux Φ with embedded scatterer. A, B are amplitudes of an electron wave function, Eq. (1.78), $\phi = 2\pi\Phi/\Phi_0$.

The scatterer introduces the following boundary conditions ($\alpha = 1$ for $x \to L - 0$ and $\alpha = 2$ for $x \to +0$)

$$Be^{-ikL}e^{i\phi} = Ae^{i\phi}S_{11} + BS_{12},$$

$$Ae^{-ikL} = Ae^{i\phi}S_{21} + BS_{22}. \quad (1.80)$$

Here we have introduced $\phi = 2\pi\Phi/\Phi_0$. We see that the magnetic flux can be fully incorporated into the non-diagonal scattering matrix elements,

$$S'_{12} = S_{12}\, e^{-i\phi}, \quad S'_{21} = S_{21}\, e^{i\phi}. \quad (1.81)$$

Therefore, in what follows we will ignore any magnetic flux and only consider the scattering matrix, Eq. (1.79), with $S_{12} \to S'_{12}$ and $S_{21} \to S'_{21}$.

1.4.6.1 Spectrum

Now we consider the spectrum of free electrons in a ring with an embedded scatterer. The dispersion equation is defined by the consistency condition for Eq. (1.80). We rewrite this equation as follows (note that we incorporated ϕ into $S'_{\alpha\beta}$, $\alpha \neq \beta$)

$$AS_{11} - B(e^{-ikL} - S'_{12}) = 0,$$
$$A(e^{-ikL} - S'_{21}) - BS_{22} = 0.$$
(1.82)

The consistency condition means that the corresponding determinant is zero,

$$\det \equiv (e^{-ikL} - S'_{21})(e^{-ikL} - S'_{12}) - S_{11}S_{22} = 0. \quad (1.83)$$

To solve it we make the following substitution,

$$S'_{12} = te^{-i\phi}, \quad S'_{21} = te^{i\phi}. \quad (1.84)$$

Next we divide Eq. (1.83) by $S'_{12}S'_{21} = t^2$ and use the equality, $S_{11}S'^*_{21} = -S'_{12}S^*_{22}$, following from the unitarity of the scattering matrix. Then we arrive at the following

$$\left(\frac{e^{-ikL}}{t} - e^{i\phi}\right)\left(\frac{e^{-ikL}}{t} - e^{-i\phi}\right) = -\frac{|S_{22}|^2}{|S'_{21}|^2}. \quad (1.85)$$

Note the amplitude t can be complex.

Further, since the right hand side (RHS) of Eq. (1.85) is definitely real the left hand side (LHS) of the same equation has to be real as well. After decoupling the real part from the imaginary part we obtain two equations,

$$\left[\mathrm{Re}\left(\frac{e^{-ikL}}{t}\right) - \cos(\phi)\right]^2 + \sin^2(\phi) - \left[\mathrm{Im}\left(\frac{e^{-ikL}}{t}\right)\right]^2 = -\frac{R}{T}, \quad (1.86a)$$

$$\mathrm{Im}\left(\frac{e^{-ikL}}{t}\right)\left[\mathrm{Re}\left(\frac{e^{-ikL}}{t}\right) - \cos(\phi)\right] = 0. \quad (1.86b)$$

Here we introduced $|S_{22}|^2 = R \geq 0$ and $|S'_{12}|^2 \equiv |t|^2 = T \geq 0$. From Eq. (1.86a) we conclude that $\mathrm{Im}\left(e^{-ikL}/t\right) \neq 0$ otherwise the LHS of

Eq. (1.86a) would be positive whereas the RHS is strictly negative. Therefore, from Eq. (1.86b) we conclude that the dispersion equation is the following

$$\text{Re}\left(\frac{e^{-ikL}}{t}\right) = \cos(\phi), \qquad (1.87)$$

as is well known from the literature [28, 29].

One can check directly that Eq. (1.86a) is consistent with Eq. (1.87).

1.4.6.2 *Circulating current*

The current carried by an electron in the state with a wave function given by Eq. (1.78) is the following,

$$I = \frac{e\hbar k}{m}\left(|A|^2 - |B|^2\right). \qquad (1.88)$$

Note the magnetic flux Φ does not enter this equation. Therefore, this equation can be used regardless of whether there is a magnetic flux through the ring or the scattering matrix is merely asymmetric, $S'_{12} \ne S'_{21}$.

To calculate the current, Eq. (1.88), we use both the normalization condition,

$$\int_0^L dx |\psi|^2 \equiv |A|^2 + |B|^2 = 1, \qquad (1.89)$$

and one of the equations of the system (1.82), say, the second one,

$$B = A\frac{e^{-ikL} - S'_{21}}{S_{22}} \equiv A\frac{e^{-ikL} - te^{i\phi}}{S_{22}}. \qquad (1.90)$$

Substituting Eqs. (1.89) and (1.90) into Eq. (1.88) we find

$$I = \frac{e\hbar k}{mL}\frac{1-|F|^2}{1+|F|^2}, \quad |F|^2 = \frac{T}{R}\left|\frac{e^{-ikL}}{t} - e^{i\phi}\right|^2. \qquad (1.91)$$

Note at $\phi = 0$, i.e., in the symmetric case $S'_{12} = S'_{21}$, the current, Eq. (1.91), is identically zero, because $|F|^2 = 1$. The latter follows from Eqs. (1.86) and (1.87). The dispersion equation, Eq. (1.87), gives $\text{Re}\left(e^{-ikL}/t\right) = 1$. Then at $\phi = 0$ we find from Eq. (1.86a), $\left[\text{Im}\left(e^{-ikL}/t\right)\right]^2 = R/T$. Therefore, $|F|^2 = T\left[\text{Im}\left(e^{-ikL}/t\right)\right]^2/R = TR/(TR) = 1$.

If the scatterer is not symmetric, $S'_{12} \neq S'_{21}$ (i.e., $\phi \neq 0$), then the current is not zero. Using the dispersion equation (1.87), $\text{Re}\left(e^{-ikL}/t\right) = \cos(\phi)$, we calculate $|F|^2$

$$\frac{R}{T}|F|^2 = \left[\text{Im}\left(\frac{e^{-ikL}}{t}\right)\right]^2 + \sin^2(\phi) - 2\text{Im}\left(\frac{e^{-ikL}}{t}\right)\sin(\phi). \tag{1.92}$$

Then from Eqs.(1.86) we find

$$\left[\text{Im}\left(\frac{e^{-ikL}}{t}\right)\right]^2 = \sin^2(\phi) + \frac{R}{T}.$$

Substituting the equation above into Eq. (1.92) and then into Eq. (1.91) we calculate the current

$$I = -\frac{e\hbar k}{mL} \frac{T\sin(\phi)}{T\sin(\phi) + \dfrac{R}{\sin(\phi) - \text{Im}\left(\dfrac{e^{-ikL}}{t}\right)}}. \tag{1.93}$$

If we denote $t = it_0 e^{i\chi}$ then the dispersion equation gives: $\sin(kL + \chi) = -t_0\cos(\phi)$. We write a solution as follows: $k_n L + \chi = \pi n + (-1)^n \arcsin[t_0\cos(\phi)]$. In this case we calculate, $\text{Im}\left(e^{-ik_n L}/t\right) = -\cos(k_n L + \chi)/t_0$. Then the current, Eq. (1.93), reads,

$$I_n = -\frac{e\hbar k_n}{mL} \frac{\sqrt{T}\sin(\phi)}{\sqrt{T}\sin(\phi) + \dfrac{R}{\sqrt{T}\sin(\phi) + \cos(k_n L + \chi)}}, \tag{1.94}$$

where we use $t_0 = \sqrt{T}$.

Note in the equation above ϕ is either an enclosed magnetic flux or an asymmetry in transmission to the left and to the right, Eq. (1.84), caused, for instance, by the internal magnetic moment. In general R and $T = 1 - R$ can depend on k_n.

Chapter 2

Current noise

One of the manifestations of a charge quantization is a fluctuating current: The instantaneous value of a current, I, deviates from its average value, $\langle I \rangle$. The magnitude of the fluctuations, or the noise value, is characterized by the mean square fluctuations

$$\langle \delta I^2 \rangle = \left\langle \left(I - \langle I \rangle\right)^2 \right\rangle. \tag{2.1}$$

On the other hand this quantity can be represented as the difference between the average square current, $\langle I^2 \rangle$, and the square of an average current

$$\langle \delta I^2 \rangle = \langle I^2 \rangle - \langle I \rangle^2. \tag{2.2}$$

Below we concentrate on two primary sources of noise in mesoscopic physics. First is the thermal noise, or the Nyquist–Johnson noise, due to the finite temperature, $T_0 > 0$, of the reservoirs, see, e.g., Refs. [15, 30]. This noise exists even in equilibrium. If the sample is connected to reservoirs with equal potentials then the average current through such a sample is zero, $\langle I \rangle = 0$. Nevertheless, there is a fluctuating current with non-zero mean square fluctuations

$$\frac{\langle \delta I^2 \rangle^{(th)}}{\Delta \nu} = 2k_B T_0 G, \tag{2.3}$$

where G is the conductance of the sample, $\Delta \nu$ is a frequency bandwidth within which the current fluctuations are measured. This noise is due to fluctuations of the occupation numbers of quantum states in the macroscopic reservoirs, see, e.g., Ref. [15], that results in fluctuating electron flows incident to the scatterer. At zero temperature the quantum state occupation numbers do not fluctuate and, therefore, the thermal noise is absent.

Second is the shot noise, see Ref. [31]. As was first shown by Schottky [32], who investigated current flows in electronic lamps, the probabilistic character of the propagation of electrons through the system results in current fluctuations. In mesoscopic samples the shot noise arises due to the quantum-mechanical probabilistic nature of scattering. The shot noise arises only in a non-equilibrium case when the current flows through the sample. If the bias V is applied to the sample then the average current is $\langle I \rangle = GV$. This current fluctuates even at zero temperature

$$\frac{\langle \delta I^2 \rangle^{(sh)}}{\Delta \nu} = |e\langle I \rangle| \left(1 - T_{12}\right). \tag{2.4}$$

The term T_{12}, the probability for an electron traveling from one reservoir and being scattered into another one, reflects the probabilistic nature of the shot noise. Moreover, taking into account that $\langle I \rangle = VG$ and $G \sim T_{12}$, one can easily show that the shot noise is at a maximum if the reflection and the transmission probabilities are equal, $R_{11} = T_{12} = 1/2$. Then we conclude: The larger the uncertainty in the scattering outcome the larger the shot noise. If the outcome of scattering is definite, i.e., an electron is always either transmitted through the sample, $T_{12} = 1$, or reflected from the sample, $R_{11} = 1$, the shot noise is zero [33, 34].

We stress the two mentioned sources of noise are not independent. The presence of a current changes the thermal noise and the shot noise is modified at finite temperatures. This fact points out that the physics underlying the thermal noise and the shot noise is of the same nature. Before we present a formal theory of current fluctuations we give simple physical arguments illustrating the appearance of a current noise in mesoscopic systems.

2.1 Nature of a current noise

We consider an extremely simplified model, a single-channel sample connecting two reservoirs and transmitting only electrons with energy E. To clarify the physics we first consider separately cases with either thermal noise or shot noise.

2.1.1 Thermal noise

Let electrons with energy E propagate ballistically, $T_{12}(E) = T_{21}(E) = 1$, while electrons with any other energy do not propagate at all, $T_{12}(E') = T_{21}(E') = 0$, $\forall E' \neq E$. Then the electrons propagating in a channel, say, from the first reservoir to the second, carry a current

$$\langle I_\rightarrow \rangle = I_0 \, P_\rightarrow \,, \tag{2.5}$$

where $I_0 = ev/\mathcal{L}$ is a current supported by the state $\Psi_\rightarrow(E)$ of an electron with energy E, e is an electron charge, v is an electron velocity, $\mathcal{L}^{-1}$ is an electron density for a unit length, and $P_\rightarrow$ is the probability that the state $\Psi_\rightarrow(E)$ is occupied. Since in the ballistic case any electron propagating to the second reservoir originated in the first reservoir, the probability $P_\rightarrow$ is equal to the occupation probability for electrons with energy E within the first reservoir. The latter is given by the Fermi distribution function, $f_1(E)$, see Eq. (1.38),

$$P_\rightarrow = f_1(E) \,. \tag{2.6}$$

The occupation probability can be defined as the ratio of the time, $\Delta t_\rightarrow$, when the state $\Psi_\rightarrow(E)$ is occupied and the total time (the observation time), $\mathcal{T} \to \infty$,

$$P_\rightarrow = \lim_{\mathcal{T} \to \infty} \frac{\Delta t_\rightarrow}{\mathcal{T}} \,. \tag{2.7}$$

Using this definition we can say that during a time $\Delta t_\rightarrow$ there is a current $I_\rightarrow(t) = I_0$ in the channel, while during the rest of the time, $\mathcal{T} - \delta t_\rightarrow$, there is no current, $I_\rightarrow(t) = 0$. Therefore, the current varies in time. Using Eq. (2.7) we calculate the mean current,

$$\langle I_\rightarrow \rangle = \lim_{\mathcal{T} \to \infty} \frac{1}{\mathcal{T}} \int_0^\mathcal{T} dt \, I_\rightarrow(t) = \lim_{\mathcal{T} \to \infty} \frac{I_0 \Delta t_\rightarrow}{\mathcal{T}} = I_0 P_\rightarrow \,, \tag{2.8}$$

which agrees with Eq. (2.5). The mean square current is

$$\langle I_\rightarrow^2 \rangle = \lim_{\mathcal{T} \to \infty} \frac{1}{\mathcal{T}} \int_0^\mathcal{T} dt \, I_\rightarrow^2(t) = \lim_{\mathcal{T} \to \infty} \frac{I_0^2 \Delta t_\rightarrow}{\mathcal{T}} = I_0^2 P_\rightarrow \,. \tag{2.9}$$

And finally, using Eq. (2.2), we calculate the mean square current fluctuations,

$$\langle \delta I_\rightarrow^2 \rangle = I_0^2 P_\rightarrow \left(1 - P_\rightarrow\right). \tag{2.10}$$

We see the current fluctuations are absent, $\langle \delta I_\rightarrow^2 \rangle = 0$, in those cases when the state of interest, $\Psi_\rightarrow(E)$, is either always occupied, $P_\rightarrow = 1$, or always empty, $P_\rightarrow = 0$. In contrast if the presence of an electron in a current-carrying state has a probabilistic character, $0 < P_\rightarrow < 1$, the current fluctuates.

Let us express $\langle \delta I_\rightarrow^2 \rangle$, Eq. (2.10), in terms of the temperature T_1 of the reservoir where electrons originate. To this end we use Eq. (2.6) and take into account the following identity for the Fermi function,

$$f_1(E)\bigl(1 - f_1(E)\bigr) = \left(-\frac{\partial f_1(E)}{\partial E}\right) k_B T_1. \tag{2.11}$$

As a result we get

$$\langle \delta I_\rightarrow^2 \rangle = I_0^2 \left(-\frac{\partial f_1(E)}{\partial E}\right) k_B T_1. \tag{2.12}$$

Thus the fluctuations under consideration vanish at zero temperature, $T_1 = 0$, as expected for thermal noise, see Eq. (2.3).

Next we take into account electrons propagating in the opposite direction, i.e., from the second reservoir to the first one. Then we calculate the mean total current, $\langle I \rangle$, and current fluctuations, $\langle \delta I^2 \rangle$, of the total current, $I(t) = I_\rightarrow(t) - I_\leftarrow(t)$,

$$\langle I \rangle = \langle I_\rightarrow \rangle - \langle I_\leftarrow \rangle = I_0 \left\{ f_1(E) - f_2(E) \right\},$$

$$\langle \delta I^2 \rangle = \langle \delta I_\rightarrow^2 \rangle + \langle \delta I_\leftarrow^2 \rangle \tag{2.13}$$

$$= I_0^2 \Bigl[f_1(E)\{1 - f_1(E)\} + f_2(E)\{1 - f_2(E)\} \Bigr],$$

where $f_2(E)$ is the Fermi distribution function for electrons in the second reservoir. Calculating $\langle \delta I^2 \rangle$ we took into account that in the ballistic case electrons propagating from the first reservoir to the second one and back originate from different reservoirs, which are assumed to be uncorrelated. Therefore, the corresponding fluctuating currents, $I_\to(t)$ and $I_\leftarrow(t)$, are statistically independent and have to be averaged independently, $\langle I_\to(t) I_\leftarrow(t) \rangle = \langle I_\to(t) \rangle \langle I_\leftarrow(t) \rangle$.

If both reservoirs have the same temperature, $T_1 = T_2 \equiv T_0$, and potential, then the corresponding distribution functions are the same as well, $f_1(E) = f_2(E) \equiv f_0(E)$. In this case Eq. (2.13) gives

$$\langle I \rangle = 0,$$

$$\langle \delta I_\to^2 \rangle = 2 I_0^2 \left(-\frac{\partial f_0(E)}{\partial E} \right) k_B T_0 .$$

(2.14)

We see that the current is zero, as expected without bias. However, the mean square of current fluctuations is not zero due to fluctuations of the occupation of quantum states in the macroscopic reservoirs with finite temperature, $T_0 > 0$.

2.1.2 Shot noise

Now we analyze a zero temperature case where the thermal noise vanishes. However, additionally we assume that there is a scatterer in the otherwise ballistic channel, see Fig. 1.4. This scatterer is characterized by having the same probability of transmitting electrons with energy E from one side to the other and back, $T_{12}(E) = T_{21}(E)$. Let us assume also that the reservoirs have different potentials. More precisely, we assume that electrons with energy E are present in the first reservoir only: $\mu_2 + eV_2 < E < \mu_1 + eV_1 \Rightarrow f_1(E) = 1$, $f_2(E) = 0$.

From the first reservoir electrons with velocity v and linear density $1/\mathcal{L}$ fall onto the scatterer. They hit the scatterer with frequency $v/\mathcal{L}$. Each electron can be either transmitted or reflected. In the former case an electron reaches the second reservoir and causes a current pulse, $I_\to(t) = I_0$. There is no current in the latter case, $I_\to(t) = 0$, since the electron returns to its original reservoir. The quantity, $T_{21}(E)$, being the probability for an electron to tunnel through the scatterer, defines a relative time period, $\Delta t_\to$, when the current flows between the reservoirs,

$$T_{21} = \lim_{\mathcal{T} \to \infty} \frac{\Delta t_\to}{\mathcal{T}} .$$

(2.15)

Repeating the reasoning of Section 2.1.1 we can calculate the mean current and the mean square current fluctuations, see Eqs. (2.7)–(2.10):

$$\langle I \rangle = I_0 \, T_{21}(E),$$
$$\langle \delta I^2 \rangle = I_0 \, \langle I \rangle \left\{ 1 - T_{21}(E) \right\}. \tag{2.16}$$

Comparing Eq. (2.10) with Eq. (2.16) we conclude that the structure of the expressions for the thermal noise and for the shot noise is the same. The difference is only in the source of stochasticity: In the former case it comes from the distribution function of electrons in macroscopic reservoirs, while in the latter case it comes from the quantum-mechanical scattering processes.

2.1.3 Combined noise

Finally we consider a case where both thermal noise and shot noise are present. We assume that the channel with a scatterer is connected to the reservoirs having non-zero temperatures and different potentials. In this case the probability, $P_\rightarrow$, of an electron, moving from the first reservoir to the second one, contributes to the current, is a product of two factors, namely, a probability, $f_1(E)$, that the state with energy E is occupied in the first reservoir and the probability, $T_{21}(E)$, that an electron tunnels through the scatterer

$$P_\rightarrow = T_{21}(E) \, f_1(E). \tag{2.17}$$

In the same way,

$$P_\leftarrow = T_{12}(E) \, f_2(E). \tag{2.18}$$

Thus the total current, $\langle I \rangle = \langle I_\rightarrow \rangle - \langle I_\leftarrow \rangle$, flowing through the channel is equal to

$$\langle I \rangle = I_0 \, T_{12}(E) \left\{ f_1(E) - f_2(E) \right\}, \tag{2.19}$$

where we used $T_{12}(E) = T_{21}(E)$.

Next we consider current fluctuations. If the currents $I_\rightarrow(t)$ and $I_\leftarrow(t)$ are statistically independent then we can say, see Eq. (2.13), that $\langle \delta I^2 \rangle$ is equal to the sum of $\langle \delta I_\rightarrow^2 \rangle$ and $\langle \delta I_\leftarrow^2 \rangle$, where

$$\langle \delta I_\rightarrow^2 \rangle = I_0^2 \, P_\rightarrow \left(1 - P_\rightarrow\right) = I_0^2 \, T_{12}(E) \, f_1(E) \{1 - T_{12}(E) \, f_1(E)\},$$
$$\langle \delta I_\leftarrow^2 \rangle = I_0^2 \, P_\leftarrow \left(1 - P_\leftarrow\right) = I_0^2 \, T_{12}(E) \, f_2(E) \{1 - T_{12}(E) \, f_2(E)\}. \tag{2.20}$$

However, as we show, this is not the case,

$$\langle \delta I^2 \rangle \neq \langle \delta I_\rightarrow^2 \rangle + \langle \delta I_\leftarrow^2 \rangle. \tag{2.21}$$

This is because the currents $I_\rightarrow(t)$ and $I_\leftarrow(t)$ are correlated. These correlations arising between the scattered electrons are a manifestation of the Pauli exclusion principle. Due to this principle two electrons cannot be in the same state. Let us consider the state corresponding to an electron propagating from the scatterer to the first reservoir. There are two ways to arrive at this state: Either an electron incident from the first reservoir is reflected, or an electron incident from the second reservoir is transmitted. Since this state cannot be occupied by two electrons we conclude that the result of scattering of an electron originating in the first reservoir depends on the result of scattering of an electron originating in the second reservoir. Therefore, the initially uncorrelated electrons in the two reservoirs after scattering at the same obstacle become correlated. Hence the currents carrying scattered electrons are correlated. In particular, these correlations result in the vanishing of the shot noise if there is an equal electron flow falling upon the scatterer from both sides.

To take into account the above mentioned correlations due to the Pauli principle we should describe the electrons quantum mechanically. We use the second-quantization formalism and introduce creation/annihilation operators, $\hat{a}_\alpha^\dagger / \hat{a}_\alpha$, for electrons with energy E incident from the reservoir $\alpha = 1, 2$, and operators $\hat{b}_\alpha^\dagger / \hat{b}_1$ for electrons scattered into the reservoir α. The reflection and transmission at the obstacle are described with the help of the unitary 2×2 scattering matrix $\hat{S}$. As we showed before, the operators for scattered and for incident electrons are related as follows,

$$\hat{b}_\alpha = \sum_{\beta=1}^{2} S_{\alpha\beta} \hat{a}_\beta, \quad \hat{b}_\alpha^\dagger = \sum_{\beta=1}^{2} S_{\alpha\beta}^* \hat{a}_\beta^\dagger. \tag{2.22}$$

For definiteness we calculate the current and its fluctuations on the left of the scatterer. We choose the positive direction to be from the scatterer to the first reservoir. Then the current operator, $\hat{I}_1$, reads

$$\hat{I}_1 = I_0(\hat{b}_1^\dagger \hat{b}_1 - \hat{a}_1^\dagger \hat{a}_1). \tag{2.23}$$

The measured current, I_1, and its mean square fluctuations, $\langle \delta I_1^2 \rangle$, are the following

$$I_1 = \langle \hat{I}_1 \rangle, \quad \langle \delta I_1^2 \rangle = \langle \hat{I}_1^2 \rangle - \langle \hat{I}_1 \rangle^2. \tag{2.24}$$

where $\langle \ldots \rangle$ stands for a quantum-statistical average over the incoming state with energy E. To calculate it we take into account that the product $\hat{n}_\alpha = \hat{a}_\alpha^\dagger \hat{a}_\alpha$ is a particle number density operator. Averaging $\hat{n}_\alpha$ quantum-mechanically over the state with energy E we get a particle number density n_α in this state in the reservoir α. After statistical averaging of the particle number density we arrive at the Fermi distribution function, f_α, of the reservoir $\alpha = 1, 2$, where an incident electron (described by the operator a_α) originated. Taking into account that electrons at different reservoirs are statistically independent, i.e., $\langle a_\alpha^\dagger a_\beta \rangle = 0$, $\alpha \neq \beta$, we have

$$\langle \hat{a}_\alpha^\dagger \hat{a}_\beta \rangle = \delta_{\alpha\beta} f_\alpha, \quad f_\alpha = \frac{1}{1 + e^{\frac{E-\mu_\alpha}{k_B T_\alpha}}}, \quad \alpha = 1, 2. \tag{2.25}$$

Also we take into account the anti-commutation relation for the Fermi particle operators,

$$\hat{a}_\alpha^\dagger \hat{a}_\beta + \hat{a}_\beta \hat{a}_\alpha^\dagger = \delta_{\alpha\beta}. \tag{2.26}$$

First, we calculate a current,

$$\langle \hat{I}_1 \rangle = I_0 \langle \hat{b}_1^\dagger \hat{b}_1 - \hat{a}_1^\dagger \hat{a}_1 \rangle = I_0 \left\langle \sum_{\beta=1}^{2} S_{1\beta}^* \hat{a}_\beta^\dagger \sum_{\gamma=1}^{2} S_{1\gamma} \hat{a}_\gamma - \hat{a}_1^\dagger \hat{a}_1 \right\rangle$$

$$= I_0 \left\{ \sum_{\beta=1}^{2} \sum_{\gamma=1}^{2} S_{1\beta}^* S_{1\gamma} \langle \hat{a}_\beta^\dagger \hat{a}_\gamma \rangle - \langle \hat{a}_1^\dagger \hat{a}_1 \rangle \right\} = I_0 \left\{ \sum_{\beta=1}^{2} |S_{1\beta}|^2 f_\beta - f_1 \right\}.$$

Using the unitarity of the scattering matrix, $|S_{11}|^2 + |S_{12}|^2 = 1$, and introducing the transmission probability, $T_{12} = |S_{12}|^2$, we finally find

$$\langle \hat{I}_1 \rangle = I_0 T_{12}(f_2 - f_1). \tag{2.27}$$

This equation is different from the current in the ballistic case, Eq. (2.13), by the factor $T_{12} < 1$, which reduces the current due to a partial reflection of the electron flow from the scatterer.

Next we calculate the mean square current fluctuations, $\langle \delta I_1^2 \rangle$. To simplify calculations we write the current operator $\hat{I}_1$ in terms of operators for incident electrons,

$$\hat{b}_1 = S_{11}\hat{a}_1 + S_{12}\hat{a}_2, \quad \hat{b}_1^\dagger = S_{11}^*\hat{a}_1^\dagger + S_{12}^*\hat{a}_2^\dagger,$$

$$\hat{I}_1/I_0 = \hat{b}_1^\dagger \hat{b}_1 - \hat{a}_1^\dagger \hat{a}_1 = \left(S_{11}^*\hat{a}_1^\dagger + S_{12}^*\hat{a}_2^\dagger\right)(S_{11}\hat{a}_1 + S_{12}\hat{a}_2) - \hat{a}_1^\dagger \hat{a}_1$$

$$= T_{12}(\hat{a}_2^\dagger \hat{a}_2 - \hat{a}_1^\dagger \hat{a}_1) + S_{11}^* S_{12}\hat{a}_1^\dagger \hat{a}_2 + S_{12}^* S_{11}\hat{a}_2^\dagger \hat{a}_1.$$

Note the last two terms do not contribute to the measured current, $I_1 = \langle \hat{I}_1 \rangle$, since after averaging they give zero, see Eq. (2.25). However, these terms are responsible for current fluctuations.

The square of the current operator, $\hat{I}_1^2 = \left(\hat{I}_1\right)^2$:

$$\hat{I}_1^2/I_0^2 = \left(T_{12}(\hat{a}_2^\dagger \hat{a}_2 - \hat{a}_1^\dagger \hat{a}_1) + S_{11}^* S_{12}\hat{a}_1^\dagger \hat{a}_2 + S_{12}^* S_{11}\hat{a}_2^\dagger \hat{a}_1\right)^2$$

$$= T_{12}^2 \left(\hat{a}_2^\dagger \hat{a}_2 \hat{a}_2^\dagger \hat{a}_2 + \hat{a}_1^\dagger \hat{a}_1 \hat{a}_1^\dagger \hat{a}_1 - \hat{a}_2^\dagger \hat{a}_2 \hat{a}_1^\dagger \hat{a}_1 - \hat{a}_1^\dagger \hat{a}_1 \hat{a}_2^\dagger \hat{a}_2\right)$$

$$+ R_{11} T_{12} \left(\hat{a}_1^\dagger \hat{a}_2 \hat{a}_2^\dagger \hat{a}_1 + \hat{a}_2^\dagger \hat{a}_1 \hat{a}_1^\dagger \hat{a}_2\right)$$

$$+ T_{12} S_{11}^* S_{12} \left(\hat{a}_2^\dagger \hat{a}_2 \hat{a}_1^\dagger \hat{a}_2 + \hat{a}_1^\dagger \hat{a}_2 \hat{a}_2^\dagger \hat{a}_2 - \hat{a}_1^\dagger \hat{a}_1 \hat{a}_1^\dagger \hat{a}_2 - \hat{a}_1^\dagger \hat{a}_2 \hat{a}_1^\dagger \hat{a}_1\right)$$

$$+ T_{12} S_{12}^* S_{11} \left(\hat{a}_2^\dagger \hat{a}_2 \hat{a}_2^\dagger \hat{a}_1 + \hat{a}_1^\dagger \hat{a}_1 \hat{a}_2^\dagger \hat{a}_2 - \hat{a}_1^\dagger \hat{a}_1 \hat{a}_2^\dagger \hat{a}_1 - \hat{a}_2^\dagger \hat{a}_1 \hat{a}_1^\dagger \hat{a}_1\right)$$

$$+ \left(S_{11}^* S_{12}\right)^2 \hat{a}_1^\dagger \hat{a}_2 \hat{a}_1^\dagger \hat{a}_2 + \left(S_{12}^* S_{11}\right)^2 \hat{a}_2^\dagger \hat{a}_1 \hat{a}_2^\dagger \hat{a}_1.$$

Here we introduced the reflection coefficient, $R_{11} = |S_{11}|^2$. Note the terms in the last three lines give zero after averaging since they include a different number of creation and annihilation operators with the same indices. To average remaining terms we use Eq. (2.26),

$$\langle \hat{a}_\alpha^\dagger \hat{a}_\alpha \hat{a}_\alpha^\dagger \hat{a}_\alpha \rangle = \langle \hat{a}_\alpha^\dagger (1 - \hat{a}_\alpha^\dagger \hat{a}_\alpha) \hat{a}_\alpha \rangle = \langle \hat{a}_\alpha^\dagger \hat{a}_\alpha \rangle - \langle \hat{a}_\alpha^\dagger \hat{a}_\alpha^\dagger \hat{a}_\alpha \hat{a}_\alpha \rangle = f_\alpha - 0 = f_\alpha,$$

$$\langle \hat{a}_\alpha^\dagger \hat{a}_\alpha \hat{a}_\beta^\dagger \hat{a}_\beta \rangle = \langle \hat{a}_\alpha^\dagger \hat{a}_\alpha \rangle \langle \hat{a}_\beta^\dagger \hat{a}_\beta \rangle = f_\alpha f_\beta, \quad \alpha \neq \beta,$$

$$\langle \hat{a}_\alpha^\dagger \hat{a}_\beta \hat{a}_\beta^\dagger \hat{a}_\alpha \rangle = \langle \hat{a}_\alpha^\dagger (1 - \hat{a}_\beta^\dagger \hat{a}_\beta) \hat{a}_\alpha \rangle = \langle \hat{a}_\alpha^\dagger \hat{a}_\alpha \rangle - \langle \hat{a}_\alpha^\dagger \hat{a}_\beta^\dagger \hat{a}_\beta \hat{a}_\alpha \rangle$$

$$= f_\alpha - \langle \hat{a}_\alpha^\dagger \hat{a}_\alpha \hat{a}_\beta^\dagger \hat{a}_\beta \rangle = f_\alpha - \langle \hat{a}_\alpha^\dagger \hat{a}_\alpha \rangle \langle \hat{a}_\beta^\dagger \hat{a}_\beta \rangle = f_\alpha(1 - f_\beta), \quad \alpha \neq \beta.$$

With these equations we calculate,

$$\langle \hat{I}_1^2 \rangle / I_0^2 = T_{12}^2(f_2 + f_1 - 2f_1 f_2) + R_{11} T_{12} \{f_1(1 - f_2) + f_2(1 - f_1)\}.$$

And finally we find the mean square current fluctuations,

$$\langle \delta I_1^2 \rangle / I_0^2 = \langle I_1^2 \rangle / I_0^2 - \langle I_1 \rangle^2 / I_0^2$$

$$= T_{12}^2(f_2 + f_1 - 2f_1 f_2) + R_{11} T_{12} \{f_1(1 - f_2) + f_2(1 - f_1)\} - T_{12}^2(f_2 - f_1)^2$$

$$= T_{12}^2 \{f_1(1 - f_1) + f_2(1 - f_2)\} + R_{11} T_{12} \{f_1(1 - f_2) + f_2(1 - f_1)\}. \tag{2.28}$$

Let us analyze where the different terms in this equation originate.

First we consider the term with squared transmission probability, $T_{12}^2 \{f_1(1 - f_1) + f_2(1 - f_2)\}$. This term originates from averaging those pairs of creation and annihilation operators which contribute to the current. Since the current is due to electrons transmitted from one reservoir to another, we can attribute this part of the noise to fluctuations in incident electron flows. The effect of scattering in this case is rather trivial: It reduces the electron flow by the factor T_{12} and, correspondingly, it reduces the noise (a squared current) by the factor T_{12}^2. This is evident for electrons flowing from the second reservoir and transmitted through the scatterer; we have already calculated their contribution to the current I_1. However, the same is also true for electrons flowing from the first reservoir, since their current is reduced by the factor $T_{21} = T_{12} = 1 - R_{11}$ due to reflection at the scatterer. As a result the (part of the) mean square of current fluctuations due to fluctuations of the occupation numbers of states in the reservoirs is proportional to the transmission probability square. Since these fluctuations are present at non-zero temperatures only, this part could be considered as the thermal noise in the system under consideration (the scatterer connected to reservoirs). Comparing it to Eq. (2.13) we see that these two results are consistent at $T_{12} = 1$. However, at $T_{12} < 1$ this part of the noise is different from what we called the thermal noise, Eq. (2.3), since the conductance G is proportional to the transmission probability, $G = G_0 T_{12}$, not its square.

To resolve an apparent contradiction and to find the correct expression for the thermal noise, i.e., for the part of the noise vanishing at zero temperature, we should consider the remaining part in Eq. (2.28) due to reflection at the scatterer, $R_{11}T_{12}\{f_1(1-f_2) + f_2(1-f_1)\}$. This part originates from averaging those operators that do not contribute to the current and, therefore, which do not correspond to any real single-particle processes. However, they correspond to some two-particle processes. To clarify we introduce the notion of a hole whose distribution function is $1 - f_\alpha$. Then one can say that there are two kind of particles incident to the scatterer: There is either an electron (with probability f_α) or a hole (with probability $1 - f_\alpha$). Then the corresponding part of the noise is due to two-particle processes: An electron/hole incoming from the first reservoir is reflected (with probability R_{11}) while a hole/electron incoming from the second reservoir is transmitted (with probability T_{12}). Apparently these processes do not contribute to the current. Notice the fluctuations in the reservoirs and fluctuations due to scattering are statistically independent, therefore, they contribute additively into the mean square current fluctuations. This fact justifies the splitting seen in Eq. (2.28). On the other hand one can rearrange these terms in another way,

$$\langle \delta I_1^2 \rangle / I_0^2 = T_{12}^2 \{f_1(1-f_1) + f_2(1-f_2)\}$$

$$+ R_{11}T_{12}\{f_1(1 - f_1 + f_1 - f_2) + f_2(1 - f_2 + f_2 - f_1)\}$$

$$= (T_{12}^2 + R_{11}T_{12})\{f_1(1-f_1) + f_2(1-f_2)\}$$

$$+ R_{11}T_{12}\{f_1(f_1 - f_2) + f_2(f_2 - f_1)\}$$

$$= T_{12}\{f_1(1 - f_1) + f_2(1 - f_2)\} + R_{11}T_{12}(f_2 - f_1)^2 \,.$$

One can see that the first term vanishes at zero temperature, therefore, we call it the thermal noise. The second term vanishes with the vanishing of the current, Eq. (2.27), flowing through the scatterer. Therefore, following Schottky one can attribute it to the stochasticity in scattering of indivisible particles at the obstacle. We call such noise the shot noise. Thus we write

$$\langle \delta I_1^2 \rangle / I_0^2 = \langle \delta I_1^2 \rangle^{(th)} / I_0^2 + \langle \delta I_1^2 \rangle^{(sh)} / I_0^2 \,, \qquad (2.29)$$

where

$$\langle \delta I_1^2 \rangle^{(th)} / I_0^2 = T_{12}\{f_2(1 - f_2) + f_1(1 - f_1)\} \,,$$

$$\langle \delta I_1^2 \rangle^{(sh)} / I_0^2 = R_{11}T_{12}(f_2 - f_1)^2 \,.$$

Notice in this equation the thermal noise is proportional to the first power of the transmission probability, T_{12}, in agreement with Eq. (2.3). The shot noise is proportional to the product of the transmission and reflection probabilities, and by virtue of Eq. (2.27) is consistent with Eq. (2.4). Moreover, equation (2.29) reproduces correctly the equations for the thermal noise and for the shot noise in the particular cases we considered earlier.

2.2 Sample with continuous spectrum

Now using the scattering matrix approach we present a formal theory for current fluctuations in a mesoscopic sample connected via one-dimensional leads to N_r reservoirs. The essential difference from the simple model considered above is that the incident electrons are particles with a continuous spectrum. This fact complicates calculations but qualitatively the answer remains the same.

2.2.1 Current correlator

The mathematical quantity which is usually considered in connection with noise is a correlation function of currents,

$$P_{\alpha\beta}(t_1, t_2) = \frac{1}{2} \left\langle \Delta \hat{I}_\alpha(t_1) \Delta \hat{I}_\beta(t_2) + \Delta \hat{I}_\beta(t_2) \Delta \hat{I}_\alpha(t_1) \right\rangle. \qquad (2.30)$$

The operator $\Delta \hat{I}_\alpha = \hat{I}_\alpha - \left\langle \hat{I}_\alpha \right\rangle$ describes the deviation of an instantaneous current, $\hat{I}_\alpha$, from its mean value, $\left\langle \hat{I}_\alpha \right\rangle$. The quantity $P_{\alpha\alpha}$ is referred to as *the current auto-correlator*, while the quantity $P_{\alpha\beta}$, $\alpha \neq \beta$, is referred to as *the current cross-correlator*.

At $t_1 = t_2$ and $\alpha = \beta$ equation (2.30) defines the mean square fluctuations of a current within a lead α, $P_{\alpha\alpha}(t_1, t_1) = \left\langle \Delta \hat{I}_\alpha^2 \right\rangle$, which strictly speaking diverges due to quantum fluctuations in the system with continuous unbounded spectrum. To overcome this difficulty usually in experiments the fluctuations are measured within frequency window $\Delta\omega$.

To calculate the spectral contents of the fluctuations we go from the real-time to the frequency representation,

$$P_{\alpha\beta}(\omega_1, \omega_2) = \int_{-\infty}^{\infty} dt_1 \, e^{i\omega_1 t_1} \int_{-\infty}^{\infty} dt_2 \, e^{i\omega_2 t_2} P_{\alpha\beta}(t_1, t_2), \qquad (2.31)$$

$$P_{\alpha\beta}(t_1, t_2) = \int_{-\infty}^{\infty} \frac{d\omega_1}{2\pi} e^{-i\omega_1 t_1} \int_{-\infty}^{\infty} \frac{d\omega_2}{2\pi} e^{-i\omega_2 t_2} P_{\alpha\beta}(\omega_1, \omega_2). \qquad (2.32)$$

Note in the stationary case the correlation function depends on the difference of times only, $P_{\alpha\beta}(t_1, t_2) = P_{\alpha\beta}(t_1 - t_2)$, and in the frequency representation is

$$P_{\alpha\beta}(\omega_1, \omega_2) = 2\pi \, \delta(\omega_1 + \omega_2) \, \mathcal{P}_{\alpha\beta}(\omega_1), \qquad (2.33)$$

where $\delta(X)$ is the Dirac delta function. The spectral noise power, $\mathcal{P}_{\alpha\beta}(\omega_1)$, is related to the correlator $P_{\alpha\beta}(t_1 - t_2) = P_{\alpha\beta}(t)$ in the following way

$$\mathcal{P}_{\alpha\beta}(\omega) = \int_{-\infty}^{\infty} dt\, e^{i\omega t}\, P_{\alpha\beta}(t), \qquad (2.34)$$

$$P_{\alpha\beta}(t) = \int_{-\infty}^{\infty} \frac{d\omega}{2\pi} e^{-i\omega t}\, \mathcal{P}_{\alpha\beta}(\omega). \qquad (2.35)$$

As we have already mentioned the quantity $P_{\alpha\alpha}(t=0)$, defining the mean square current fluctuations, diverges. However, if we restrict the frequency interval, $\pm\Delta\omega/2$, where the current fluctuations are measured, then we obtain a finite quantity,

$$\langle \delta I_\alpha^2 \rangle = \int_{-\Delta\omega/2}^{\Delta\omega/2} \frac{d\omega}{2\pi} \mathcal{P}_{\alpha\alpha}(\omega). \qquad (2.36)$$

Further simplification arises if the scattering properties of a sample depend on energy only a little. In this case the spectral noise power, $\mathcal{P}_{\alpha\beta}(\omega)$, depends weakly on frequency, $\mathcal{P}_{\alpha\beta}(\omega) \approx \mathcal{P}_{\alpha\beta}(0)$, and we can evaluate Eq. (2.36),

$$\frac{\langle \delta I_\alpha^2 \rangle}{\Delta\nu} = \mathcal{P}_{\alpha\alpha}(0), \qquad (2.37)$$

where $\Delta\nu = \Delta\omega/(2\pi)$. In the same way the cross-correlator of currents flowing into the leads α and β measured within the frequency window $\Delta\omega$ becomes

$$\frac{\langle \delta I_\alpha \delta I_\beta \rangle}{\Delta\nu} = \mathcal{P}_{\alpha\beta}(0). \qquad (2.38)$$

We see that the mean square of current fluctuations is defined by the zero frequency noise power. Below we calculate $\mathcal{P}_{\alpha\beta}(0)$ and confirm Eqs. (2.3) and (2.4).

2.2.2 Current correlator in the frequency domain

Let us calculate the quantity $P_{\alpha\beta}(\omega_1, \omega_2)$ and show that, indeed, it can be represented as Eq. (2.33).

Substituting Eq. (2.30) into Eq. (2.31) we get

$$P_{\alpha\beta}(\omega_1, \omega_2) = \frac{1}{2} \left\langle \Delta \hat{I}_\alpha(\omega_1) \Delta \hat{I}_\beta(\omega_2) + \Delta \hat{I}_\beta(\omega_2) \Delta \hat{I}_\alpha(\omega_1) \right\rangle, \quad (2.39)$$

where $\Delta \hat{I}_\alpha(\omega) = \hat{I}_\alpha(\omega) - \left\langle \hat{I}_\alpha(\omega) \right\rangle$, and $\hat{I}_\alpha(\omega)$ is a current operator in the frequency representation. To calculate it we apply the Fourier transformation to Eq. (1.36) and find the following

$$\hat{I}_\alpha(\omega) = e \int_0^\infty dE \left\{ \hat{b}_\alpha^\dagger(E) \hat{b}_\alpha(E + \hbar\omega) - \hat{a}_\alpha^\dagger(E) \hat{a}_\alpha(E + \hbar\omega) \right\}. \quad (2.40)$$

For convenience we represent a current as the sum of two contributions. The first one is due to scattered electrons, while the second one is due to incident electrons. To distinguish these contributions we use the upper indices (out) and (in) for the former and latter contributions, respectively. So, the total current is $\hat{I}_\alpha(\omega) = \hat{I}_\alpha^{(out)}(\omega) + \hat{I}_\alpha^{(in)}(\omega)$, where

$$\hat{I}_\alpha^{(out)}(\omega) = e \int_0^\infty dE \, \hat{b}_\alpha^\dagger(E) \hat{b}_\alpha(E + \hbar\omega), \quad (2.41)$$

$$\hat{I}_\alpha^{(in)}(\omega) = -e \int_0^\infty dE \, \hat{a}_\alpha^\dagger(E) \hat{a}_\alpha(E + \hbar\omega). \quad (2.42)$$

Then the current correlator, $P_{\alpha\beta}(\omega_1, \omega_2)$, is the sum of four terms,

$$P_{\alpha\beta}(\omega_1, \omega_2) = \sum_{i,j=in,out} P_{\alpha\beta}^{(i,j)}(\omega_1, \omega_2),$$

$$P_{\alpha\beta}^{(i,j)}(\omega_1, \omega_2) = \frac{1}{2} \left\langle \Delta \hat{I}_\alpha^{(i)}(\omega_1) \Delta \hat{I}_\beta^{(j)}(\omega_2) + \Delta \hat{I}_\beta^{(j)}(\omega_2) \Delta \hat{I}_\alpha^{(i)}(\omega_1) \right\rangle. \quad (2.43)$$

We calculate each of these terms separately.

2.2.2.1 Correlator for incoming currents

The part of the current correlation function dependent only on incoming currents is

$$P_{\alpha\beta}^{(in,in)}(\omega_1,\omega_2) = e^2 \int\!\!\!\int_0^\infty dE_1\, dE_2\, \frac{J_{\alpha\beta}^{(in,in)}(E_{1,2},\omega_{1,2}) + J_{\beta\alpha}^{(in,in)}(E_{2,1},\omega_{2,1})}{2}, \tag{2.44}$$

where

$$J_{\alpha\beta}^{(in,in)}(E_{1,2},\omega_{1,2}) = \left\langle \left\{ \hat{a}_\alpha^\dagger(E_1)\hat{a}_\alpha(E_1+\hbar\omega_1) - \left\langle \hat{a}_\alpha^\dagger(E_1)\hat{a}_\alpha(E_1+\hbar\omega_1) \right\rangle \right\} \right.$$
$$\left. \times \left\{ \hat{a}_\beta^\dagger(E_2)\hat{a}_\beta(E_2+\hbar\omega_2) - \left\langle \hat{a}_\beta^\dagger(E_2)\hat{a}_\beta(E_2+\hbar\omega_2) \right\rangle \right\} \right\rangle.$$

Taking into account that the average of the product of four operators is the sum of products of pair correlators we finally find

$$J_{\alpha\beta}^{(in,in)}(E_{1,2},\omega_{1,2}) = \left\langle \hat{a}_\alpha^\dagger(E_1)\hat{a}_\beta(E_2+\hbar\omega_2) \right\rangle \left\langle \hat{a}_\alpha(E_1+\hbar\omega_1)\hat{a}_\beta^\dagger(E_2) \right\rangle.$$

Using Eq. (1.37), we calculate pair correlators,

$$\left\langle \hat{a}_\alpha^\dagger(E_1)\hat{a}_\beta(E_2+\hbar\omega_2) \right\rangle = \delta_{\alpha\beta}\,\delta(E_1 - E_2 - \hbar\omega_2)\, f_\alpha(E_1),$$

$$\left\langle \hat{a}_\alpha(E_1+\hbar\omega_1)\hat{a}_\beta^\dagger(E_2) \right\rangle = \delta_{\alpha\beta}\,\delta(E_1 + \hbar\omega_1 - E_2)\,\{1 - f_\alpha(E_1+\hbar\omega_1)\},$$

and correspondingly,

$$J_{\alpha\beta}^{(in,in)}(E_{1,2},\omega_{1,2}) = \delta_{\alpha\beta}\,\delta(E_1 - E_2 - \hbar\omega_2)\,\delta(E_1 + \hbar\omega_1 - E_2)$$
$$\times f_\alpha(E_1)\,\{1 - f_\alpha(E_1+\hbar\omega_1)\}.$$

In the same way we get

$$J^{(in,in)}_{\beta\alpha}(E_{2,1},\omega_{2,1}) = \delta_{\alpha\beta}\,\delta(E_1+\hbar\omega_1-E_2)\,\delta(E_1-E_2-\hbar\omega_2)$$
$$\times f_\alpha(E_1+\hbar\omega_1)\,\{1-f_\alpha(E_1)\}.$$

Substituting these equations into Eq. (2.44) and integrating over the energy E_2, we represent this part of the current correlation function in the following way

$$P^{(in,in)}_{\alpha\beta}(\omega_1,\omega_2) = 2\pi\,\delta(\omega_1+\omega_2)\,\mathcal{P}^{(in,in)}_{\alpha\beta}(\omega_1),$$

(2.45)

$$\mathcal{P}^{(in,in)}_{\alpha\beta}(\omega_1) = \delta_{\alpha\beta}\frac{e^2}{h}\int_0^\infty dE_1\,F_{\alpha\alpha}(E_1,E_1+\hbar\omega_1).$$

Here we have introduced the following short notation,

$$F_{\alpha\beta}(E,E') = \frac{1}{2}\Big\{f_\alpha(E)\big[1-f_\beta(E')\big]+f_\beta(E')\big[1-f_\alpha(E)\big]\Big\}. \quad (2.46)$$

It follows from Eq. (2.45) that the currents flowing in the different leads, $\alpha\neq\beta$, to the scatterer are uncorrelated, $P^{(in,in)}_{\alpha\neq\beta}=0$. This is a consequence of our assumption that electrons at different reservoirs are uncorrelated.

2.2.2.2 Correlator for incoming and outgoing currents

The part of the correlator dependent on an incoming current in the lead α and an outgoing current in the lead β

$$P^{(in,out)}_{\alpha\beta}(\omega_1,\omega_2) = -e^2\iint_0^\infty dE_1\,dE_2\,\frac{J^{(in,out)}_{\alpha\beta}(E_{1,2},\omega_{1,2})+J^{(out,in)}_{\beta\alpha}(E_{2,1},\omega_{2,1})}{2}.$$

(2.47)

To calculate, for instance,

$$J^{(in,out)}_{\alpha\beta}(E_{1,2},\omega_{1,2}) = \Big\langle\big\{\hat{a}^\dagger_\alpha(E_1)\,\hat{a}_\alpha(E_1+\hbar\omega_1)-\big\langle\hat{a}^\dagger_\alpha(E_1)\,\hat{a}_\alpha(E_1+\hbar\omega_1)\big\rangle\big\}$$
$$\times\big\{\hat{b}^\dagger_\beta(E_2)\,\hat{b}_\beta(E_2+\hbar\omega_2)-\big\langle\hat{b}^\dagger_\beta(E_2)\,\hat{b}_\beta(E_2+\hbar\omega_2)\big\rangle\big\}\Big\rangle$$
$$=\big\langle\hat{a}^\dagger_\alpha(E_1)\,\hat{b}_\beta(E_2+\hbar\omega_2)\big\rangle\big\langle\hat{a}_\alpha(E_1+\hbar\omega_1)\,\hat{b}^\dagger_\beta(E_2)\big\rangle,$$

we express b-operators in terms of a-operators, see Eq. (1.39),

$$\hat{b}^\dagger_\beta(E) = \sum_{\gamma=1}^{N_r} S^*_{\beta\gamma}(E)\hat{a}^\dagger_\gamma(E), \quad \hat{b}_\beta(E) = \sum_{\gamma=1}^{N_r} S_{\beta\gamma}(E)\hat{a}_\gamma(E),$$

and calculate pair correlators,

$$\left\langle \hat{a}^\dagger_\alpha(E_1)\hat{b}_\beta(E_2+\hbar\omega_2) \right\rangle = \delta(E_1 - E_2 - \hbar\omega_2)\, S_{\beta\alpha}(E_2+\hbar\omega_2)\, f_\alpha(E_1),$$

$$\left\langle \hat{a}_\alpha(E_1+\hbar\omega_1)\hat{b}^\dagger_\beta(E_2) \right\rangle = \delta(E_1 + \hbar\omega_1 - E_2)\, S^*_{\beta\alpha}(E_2)\, \{1 - f_\alpha(E_1+\hbar\omega_1)\}.$$

After that we find,

$$J^{(in,out)}_{\alpha\beta}(E_{1,2},\omega_{1,2}) = \delta(E_1 - E_2 - \hbar\omega_2)\,\delta(E_1 + \hbar\omega_1 - E_2)$$

$$\times\, S_{\beta\alpha}(E_2+\hbar\omega_2)\, S^*_{\beta\alpha}(E_2)\, f_\alpha(E_1)\, \{1 - f_\alpha(E_1+\hbar\omega_1)\}.$$

Similar calculations give us the second term in Eq. (2.47):

$$J^{(out,in)}_{\beta\alpha}(E_{2,1},\omega_{2,1}) = \delta(E_1 + \hbar\omega_1 - E_2)\,\delta(E_1 - E_2 - \hbar\omega_2)$$

$$\times\, S^*_{\beta\alpha}(E_2)\, S_{\beta\alpha}(E_2+\hbar\omega_2)\, f_\alpha(E_1+\hbar\omega_1)\, \{1 - f_\alpha(E_1)\}.$$

Using these equations in Eq. (2.47) and integrating over E_2, we calculate,

$$P^{(in,out)}_{\alpha\beta}(\omega_1,\omega_2) = 2\pi\,\delta(\omega_1+\omega_2)\,\mathcal{P}^{(in,out)}_{\alpha\beta}(\omega_1),$$

(2.48)

$$\mathcal{P}^{(in,out)}_{\alpha\beta}(\omega_1) = -\frac{e^2}{h}\int_0^\infty dE_1\, F_{\alpha\alpha}(E_1, E_1+\hbar\omega_1)\, S^*_{\beta\alpha}(E_1+\hbar\omega_1)\, S_{\beta\alpha}(E_1).$$

This equation shows us that the current carried by the electrons scattered into the lead β is correlated with a current carried by the electrons incoming from the reservoir α. In fact these correlations are due to electrons scattered from the lead α into the lead β. This is indicated by the corresponding scattering matrix elements, $S_{\beta\alpha}$.

In the same way we calculate the third element of Eq. (2.43):

$$P_{\alpha\beta}^{(out,in)}(\omega_1,\omega_2) = 2\pi\,\delta(\omega_1+\omega_2)\,\mathcal{P}_{\alpha\beta}^{(out,in)}(\omega_1),$$
(2.49)

$$\mathcal{P}_{\alpha\beta}^{(out,in)}(\omega_1) = -\frac{e^2}{h}\int_0^\infty dE_1\,F_{\beta\beta}(E_1, E_1+\hbar\omega_1)\,S_{\alpha\beta}^*(E_1)\,S_{\alpha\beta}(E_1+\hbar\omega_1).$$

This term is due to correlations between electrons coming from the reservoir β and electrons scattered in the reservoir α.

2.2.2.3 Correlator for outgoing currents

Finally we calculate the last element of Eq. (2.43):

$$P_{\alpha\beta}^{(out,out)}(\omega_1,\omega_2) = \frac{e^2}{2}\iint_0^\infty dE_1\,dE_2 \qquad (2.50)$$

$$\times \left\{ \left\langle \hat{b}_\alpha^\dagger(E_1)\,\hat{b}_\beta(E_2+\hbar\omega_2) \right\rangle \left\langle \hat{b}_\alpha(E_1+\hbar\omega_1)\,\hat{b}_\beta^\dagger(E_2) \right\rangle \right.$$

$$\left. + \left\langle \hat{b}_\beta^\dagger(E_2)\,\hat{b}_\alpha(E_1+\hbar\omega_1) \right\rangle \left\langle \hat{b}_\beta(E_2+\hbar\omega_2)\,\hat{b}_\alpha^\dagger(E_1) \right\rangle \right\}.$$

To calculate a pair correlator with b-operators we use Eqs. (1.39), (1.37) and obtain, for example,

$$\left\langle \hat{b}_\alpha^\dagger(E_1)\,\hat{b}_\beta(E_2+\hbar\omega_2) \right\rangle = \delta(E_1 - E_2 - \hbar\omega_2)$$

$$\times \sum_{\gamma=1}^{N_r} S_{\alpha\gamma}^*(E_1)\,S_{\beta\gamma}(E_2+\hbar\omega_2)\,f_\gamma(E_1),$$

$$\left\langle \hat{b}_\alpha(E_1+\hbar\omega_1)\,\hat{b}_\beta^\dagger(E_2) \right\rangle = \delta(E_1 + \hbar\omega_1 - E_2)$$

$$\times \sum_{\delta=1}^{N_r} S_{\alpha\delta}(E_1+\hbar\omega_1)\,S_{\beta\delta}^*(E_2)\,\{1 - f_\delta(E_2)\}.$$

Other pair correlators are calculated in a similar way. Then Eq. (2.50) results in the following:

$$P_{\alpha\beta}^{(out,out)}(\omega_1,\omega_2) = 2\pi\,\delta(\omega_1+\omega_2)\,\mathcal{P}_{\alpha\beta}^{(out,out)}(\omega_1)\,, \tag{2.51}$$

$$\mathcal{P}_{\alpha\beta}^{(out,out)}(\omega_1) = \frac{e^2}{h}\int_0^\infty dE_1 \sum_{\gamma=1}^{N_r}\sum_{\delta=1}^{N_r} F_{\gamma\delta}(E_1, E_1+\hbar\omega_1)$$
$$\times S_{\alpha\gamma}^*(E_1)\,S_{\beta\gamma}(E_1)\,S_{\alpha\delta}(E_1+\hbar\omega_1)\,S_{\beta\delta}^*(E_1+\hbar\omega_1)\,.$$

Note the correlator of scattered currents depends on the Fermi functions for all the reservoirs. In addition it depends not only on the amplitudes of scattering between the leads α and β, where the currents are measured, but rather on all the possible scattering amplitudes. It emphasizes a non-locality inherent to phase-coherent systems.

Summing up Eqs. (2.45), (2.48), (2.49), and (2.51), we arrive at Eq. (2.33), where

$$\mathcal{P}_{\alpha\beta}(\omega) = \frac{e^2}{h}\int_0^\infty dE\bigg\{ F_{\alpha\alpha}(E, E+\hbar\omega)\left[\delta_{\alpha\beta} - S_{\beta\alpha}^*(E+\hbar\omega)\,S_{\beta\alpha}(E)\right]$$

$$- F_{\beta\beta}(E, E+\hbar\omega)\,S_{\alpha\beta}^*(E)\,S_{\alpha\beta}(E+\hbar\omega) \tag{2.52}$$

$$+ \sum_{\gamma=1}^{N_r}\sum_{\delta=1}^{N_r} F_{\gamma\delta}(E, E+\hbar\omega)\,S_{\alpha\gamma}^*(E)\,S_{\beta\gamma}(E)\,S_{\alpha\delta}(E+\hbar\omega)\,S_{\beta\delta}^*(E+\hbar\omega)\bigg\}.$$

The frequency dependence of noise is due to internal and external factors. The internal factor is the energy dependence of the scattering amplitudes. The external factors, represented by the combination of the Fermi functions, $F_{\gamma\delta}(E, E+\hbar\omega)$, are chemical potentials and temperatures of the reservoirs. The joint effect of internal and external factors is rather sample specific. However, in some simple cases the effect of bias and temperature can be analyzed.

2.2.3 Spectral noise power for energy-independent scattering

Let the reservoirs have different potentials but the same temperature,

$$eV_{\alpha\beta} = \mu_\alpha - \mu_\beta; \quad T_\alpha = T_0, \quad \forall \alpha. \tag{2.53}$$

We assume the bias and temperature are small compared to the Fermi energy,

$$|eV_{\alpha\beta}|, \, k_B T_0 \ll \mu_0. \tag{2.54}$$

Suppose also that the scattering matrix varies with energy only a little within the energy window of order $k_B T_0$, $|eV_{\alpha\beta}|$ near the Fermi energy μ_0. Then the scattering matrix elements in Eq. (2.52) can be calculated at the Fermi energy, $E \approx E + \hbar\omega = \mu_0$. The integration over energy becomes trivial,

$$\int_0^\infty dE \, F_{\alpha\beta}(E, E + \hbar\omega) = \frac{eV_{\alpha\beta} + \hbar\omega}{2} \coth\left(\frac{eV_{\alpha\beta} + \hbar\omega}{2k_B T_0}\right), \tag{2.55}$$

and we calculate the spectral noise power,

$$\mathcal{P}_{\alpha\beta}(\omega) = \frac{e^2}{h} \left\{ \frac{\hbar\omega}{2} \coth\left(\frac{\hbar\omega}{2k_B T_0}\right) \left[\delta_{\alpha\beta} - |S_{\beta\alpha}(\mu_0)|^2 - |S_{\alpha\beta}(\mu_0)|^2\right] \right.$$

$$\left. + \sum_{\gamma=1}^{N_r} \sum_{\delta=1}^{N_r} \frac{eV_{\gamma\delta} + \hbar\omega}{2} \coth\left(\frac{eV_{\gamma\delta} + \hbar\omega}{2k_B T_0}\right) S^*_{\alpha\gamma}(\mu_0) S_{\beta\gamma}(\mu_0) S_{\alpha\delta}(\mu_0) S^*_{\beta\delta}(\mu_0) \right\}. \tag{2.56}$$

Let us consider the particular case of $N_r = 2$ [35]. Then we find

$$\mathcal{P}_{11}(\omega) = \frac{e^2}{h} \left\{ \hbar\omega \coth\left(\frac{\hbar\omega}{2k_B T_0}\right) T_{12}^2 \right.$$

$$\left. + R_{11} T_{12} \left[\frac{eV + \hbar\omega}{2} \coth\left(\frac{eV + \hbar\omega}{2k_B T_0}\right) + \frac{eV - \hbar\omega}{2} \coth\left(\frac{eV - \hbar\omega}{2k_B T_0}\right) \right] \right\}, \tag{2.57}$$

where $V = V_{12} = -V_{21}$, $T_{12} = |S_{12}(\mu_0)|^2$, and $R_{11} = |S_{11}(\mu_0)|^2 = 1 - T_{12}$. Note in the two-terminal case the calculated quantity defines all other correlation functions: $\mathcal{P}_{12} = \mathcal{P}_{21} = -\mathcal{P}_{22} = -\mathcal{P}_{11}$.

The noise depends on the frequency ω at which the current is measured, on the bias V, and on the temperature T_0. If one of these factors exceeds the others then we get

$$\mathcal{P}_{11}(\omega) = \begin{cases} 2k_B T_0 G, & k_B T_0 \gg |eV|,\ \hbar\omega, \\ |eI|R_{11}, & |eV| \gg \hbar\omega,\ k_B T_0, \\ \frac{e^2}{2\pi}|\omega|T_{12}, & \hbar\omega \gg k_B T_0,\ |eV|, \end{cases} \quad (2.58)$$

where $G = (e^2/h)T_{12}$ is the conductance and $I = VG$ is the current through the sample. The first line represents thermal noise, which is linear in temperature. The coefficient 2 arises because the two reservoirs have the same temperature. If the temperatures were different then we would make a replacement, $2T_0 \to T_1 + T_2$. Taking into account Eq. (2.37) we see that this equation is exactly Eq. (2.3). The second line in Eq. (2.58) corresponds to a regime where shot noise dominates. It reproduces Eq. (2.4). And, finally, the third line represents the so-called *quantum noise* dependent on the measurement frequency ω [36]. This last contribution is responsible for the divergence of the mean square current fluctuations $\langle I_1^2 \rangle = \mathcal{P}_{11}(t=0)$, see Eq. (2.35).

It follows from Eq. (2.58) that the frequency dependence of noise can be ignored if

$$\hbar\omega \ll \max\{k_B T_0, |eV_{\alpha\beta}|\}, \quad \forall \alpha, \beta. \quad (2.59)$$

In this case the quantum noise becomes negligible and the main sources of current fluctuations are thermal and shot noise. At $T_0 \sim 10^{-2}\,\text{K}$ or $V \sim 10^{-6}\,\text{V}$ the quantum noise can be ignored up to the frequencies $\omega \sim 10^9\,\text{Hz}$.

2.2.4 Zero frequency noise power

If the measurement is carried out at a sufficient number of low frequencies, Eq. (2.59), then the value of current fluctuations is defined by the noise power at zero frequency, $\omega = 0$, see Eq. (2.37). The quantity $\mathcal{P}_{\alpha\alpha}(0)$ is usually referred to as *the noise power*.

Let us represent the quantity $\mathcal{P}_{\alpha\beta}(0)$, Eq. (2.52), as the sum of two terms such that one of them vanishes at zero temperature, while the other one vanishes in the absence of a current through the sample. To this end we write

$$F_{\gamma\delta}(E,E) = \frac{1}{2}\Big\{F_{\gamma\gamma}(E,E) + F_{\delta\delta}(E,E) + \big[f_\gamma(E) - f_\delta(E)\big]^2\Big\},$$

and substitute it into Eq. (2.52) with $\omega = 0$. Then in the term with factor $F_{\gamma\gamma}(E,E)$ we sum over δ and, taking into account Eq. (1.13), find

$$\sum_{\gamma=1}^{N_r} F_{\gamma\gamma} S^*_{\alpha\gamma} S_{\beta\gamma} \sum_{\delta=1}^{N_r} S_{\alpha\delta} S^*_{\beta\delta} = \delta_{\alpha\beta} \sum_{\gamma=1}^{N_r} F_{\gamma\gamma} |S_{\alpha\gamma}|^2 .$$

The term with factor $F_{\delta\delta}(E,E)$ reads exactly the same. So then we write [4]

$$\mathcal{P}_{\alpha\beta}(0) = \mathcal{P}^{(th)}_{\alpha\beta} + \mathcal{P}^{(sh)}_{\alpha\beta}, \tag{2.60}$$

where

$$\mathcal{P}^{(th)}_{\alpha\beta} = \frac{e^2}{h} \int_0^\infty dE \Big\{ \delta_{\alpha\beta} \Big[F_{\alpha\alpha}(E,E) + \sum_{\gamma=1}^{N_r} F_{\gamma\gamma}(E,E) |S_{\alpha\gamma}(E)|^2 \Big]$$
$$- F_{\alpha\alpha}(E,E) |S_{\beta\alpha}(E)|^2 - F_{\beta\beta}(E,E) |S_{\alpha\beta}(E)|^2 \Big\}, \tag{2.61}$$

$$\mathcal{P}^{(sh)}_{\alpha\beta} = \frac{e^2}{h} \int_0^\infty dE \sum_{\gamma=1}^{N_r} \sum_{\delta=1}^{N_r} \frac{[f_\gamma(E) - f_\delta(E)]^2}{2} S^*_{\alpha\gamma}(E) S_{\beta\gamma}(E) S_{\alpha\delta}(E) S^*_{\beta\delta}(E). \tag{2.62}$$

The quantity $\mathcal{P}^{(th)}_{\alpha\alpha}$ can be referred to as *the thermal noise power*. This quantity vanishes at zero temperature, since at $T_\alpha = 0$ it is $F_{\alpha\alpha}(E,E) = 0$, $\forall \alpha$. The quantity $\mathcal{P}^{(sh)}_{\alpha\alpha}$ is called *the shot noise power* since it vanishes in the absence of a current through the system. Recall that the current is driven by the Fermi function difference.

It should be noted that as $\mathcal{P}_{\alpha\beta}^{(th)}$ and $\mathcal{P}_{\alpha\beta}^{(sh)}$ depend on both the temperature and the bias voltage, this emphasizes the universal probabilistic nature of noise. However, there is an essential difference between equilibrium (thermal) noise and non-equilibrium (shot) noise. Thermal noise depends on the probabilities $|S_{\alpha\beta}|^2$ as the conductance $G_{\alpha\beta}$, Eq. (1.55). This is a consequence of the fluctuation-dissipation theorem, see, e.g., Ref. [15]. However, shot noise depends on different combinations of the scattering matrix elements. In general this allows us to extract additional information concerning the properties of a sample from shot noise measurements. Below we consider general properties of the noise power.

2.2.4.1 Noise power conservation law

The sum of the zero-frequency current correlation function power over either incoming or outgoing indices is zero [5]

$$\sum_{\alpha=1}^{N_r} \mathcal{P}_{\alpha\beta}(0) = \sum_{\beta=1}^{N_r} \mathcal{P}_{\alpha\beta}(0) = 0. \quad (2.63)$$

These conservation laws are quite analogous to the direct current conservation law, (1.48). They are due to particle number conservation at scattering (due to unitarity of the scattering matrix).

Remarkably thermal noise and shot noise are subject to these conservation laws separately. So using Eq. (1.51) we find for thermal noise, Eq. (2.61),[1]

$$\sum_{\alpha=1}^{N_r} \mathcal{P}_{\alpha\beta}^{(th)} \sim \sum_{\alpha=1}^{N_r} \delta_{\alpha\beta} \left[F_{\alpha\alpha}(E,E) + \sum_{\gamma=1}^{N_r} F_{\gamma\gamma}(E,E) \left|S_{\alpha\gamma}(E)\right|^2 \right]$$

$$- \sum_{\alpha=1}^{N_r} F_{\alpha\alpha}(E,E) \left|S_{\beta\alpha}(E)\right|^2 - F_{\beta\beta}(E,E) \sum_{\alpha=1}^{N_r} \left|S_{\alpha\beta}(E)\right|^2$$

$$= F_{\beta\beta}(E,E) + \sum_{\gamma=1}^{N_r} F_{\gamma\gamma}(E,E) \left|S_{\beta\gamma}(E)\right|^2$$

$$- \sum_{\alpha=1}^{N_r} F_{\alpha\alpha}(E,E) \left|S_{\beta\alpha}(E)\right|^2 - F_{\beta\beta}(E,E) = 0.$$

---

[1] We drop an integration over energy since the conservation laws hold not only integrally but also separately for each energy.

Current noise

In the same way using Eq. (1.46) we show that $\sum_{\beta=1}^{N_r} \mathcal{P}_{\alpha\beta}^{(th)}(0) = 0$.
In the case of shot noise, Eq. (2.62), we use Eq. (1.12) and get

$$\sum_{\alpha=1}^{N_r} \mathcal{P}_{\alpha\beta}^{(sh)} \sim \sum_{\gamma=1}^{N_r}\sum_{\delta=1}^{N_r} \frac{[f_\gamma(E) - f_\delta(E)]^2}{2} S_{\beta\gamma}(E) S_{\beta\delta}^*(E) \sum_{\alpha=1}^{N_r} S_{\alpha\gamma}^*(E) S_{\alpha\delta}(E)$$

$$= \sum_{\gamma=1}^{N_r}\sum_{\delta=1}^{N_r} \frac{[f_\gamma(E) - f_\delta(E)]^2}{2} S_{\beta\gamma}(E) S_{\beta\delta}^*(E) \delta_{\gamma\delta} = 0\,.$$

Then with Eq. (1.13) we also prove, $\sum_{\beta=1}^{N_r} \mathcal{P}_{\alpha\beta}^{(sh)}(0) = 0$.

The conservation laws, Eq. (2.63), show that the auto-correlator and cross-correlators at zero frequency are not independent of each other. Some of them can be calculated if the others were measured.

2.2.4.2 *Sign rule for the noise power*

The auto-correlator is positive (or zero) while the cross-correlator is negative (or zero) [5],

$$\mathcal{P}_{\alpha\alpha}(0) \geq 0\,, \tag{2.64a}$$

$$\mathcal{P}_{\alpha\beta}(0) \leq 0\,, \ \alpha \neq \beta\,. \tag{2.64b}$$

The positiveness of $\mathcal{P}_{\alpha\alpha}(0)$ is clear, since this quantity is a mean square of a real quantity, Eq. (2.37). The negative sign of a cross-correlator is a consequence, first, of an indivisibility of electrons and, second, of the Pauli exclusion principle requiring (spinless) electrons with some energy, which pass one by one through the one-dimensional lead. Therefore, we can consider the scattering of a single electron with a given energy and forget about the other electrons. Let us consider scattering of an electron flow moving toward the sample in the lead γ. Electrons from this flow can be scattered to any lead δ with probability $|S_{\delta\gamma}(E)|^2$. In particular some electrons will be scattered into the leads α and β. These electrons define the mean currents, $\langle I_\alpha^{(\gamma)} \rangle$ and $\langle I_\beta^{(\gamma)} \rangle$. On the other hand each particular electron can be scattered to only one lead. This can be either lead α, or β, or any other lead δ. In any case the current pulse due to scattering of this particular electron arises only in one lead. Therefore, the product of instantaneous currents in any two leads, for example in α and β, is zero, $I_\alpha^{(\gamma)} I_\beta^{(\gamma)} = 0$. Then we immediately conclude that the cross-correlator of currents in leads α and β due to single electrons originating with energy E from the reservoir γ is negative, $\mathcal{P}_{\alpha\beta}^{(\gamma)}(E) \sim \langle I_\alpha^{(\gamma)} I_\beta^{(\gamma)} \rangle - \langle I_\alpha^{(\gamma)} \rangle \langle I_\beta^{(\gamma)} \rangle \sim 0 - |S_{\alpha\gamma}(E)|^2 |S_{\beta\gamma}(E)|^2 \leq 0$. In different reservoirs and at different energies electrons are statistically independent. Therefore, we can sum up correlation functions $\mathcal{P}_{\alpha\beta}^{(\gamma)}(E)$ over γ and integrate over E. Then we arrive at Eq. (2.64b).

Let us show that thermal noise, Eq. (2.61), and shot noise, Eq. (2.62), do satisfy the sign rule, Eqs. (2.64). We will omit an integration over energy because it does not affect the sign of the current correlation function. First we consider thermal noise. The auto-correlator gives

$$\mathcal{P}_{\alpha\alpha}^{(th)} \sim F_{\alpha\alpha}(E,E) + \sum_{\gamma=1}^{N_r} F_{\gamma\gamma}(E,E) |S_{\alpha\gamma}(E)|^2 - 2F_{\alpha\alpha}(E,E) |S_{\alpha\alpha}(E)|^2$$

$$= F_{\alpha\alpha}(E,E) \left[1 - |S_{\alpha\alpha}(E)|^2\right] + \sum_{\gamma \neq \alpha=1}^{N_r} F_{\gamma\gamma}(E,E) |S_{\alpha\gamma}(E)|^2 \geq 0.$$

Here we took into account $0 \leq F_{\alpha\alpha}(E,E) \leq 1$ and $|S_{\alpha\alpha}(E)|^2 \leq 1$. For the cross-correlator, $\alpha \neq \beta$, we find an expression, which is definitely negative, $\mathcal{P}_{\alpha \neq \beta}^{(th)} \sim -|S_{\beta\alpha}|^2 f_\alpha [1 - f_\alpha] - |S_{\alpha\beta}|^2 f_\beta [1 - f_\beta] \leq 0$.

Next we consider shot noise. The auto-correlator is definitely positive, $\mathcal{P}_{\alpha\alpha}^{(sh)} \sim \sum_{\gamma=1}^{N_r} \sum_{\delta=1}^{N_r} \frac{1}{2} (f_\gamma - f_\delta)^2 |S_{\alpha\gamma}|^2 |S_{\alpha\delta}|^2 \geq 0$. To calculate the cross-correlator we use, $(f_\gamma - f_\delta)^2 = f_\gamma^2 + f_\delta^2 - 2f_\gamma f_\delta$, use Eq. (1.13), and get

$$\mathcal{P}_{\alpha \neq \beta}^{(sh)} \sim \frac{1}{2} \sum_{\gamma=1}^{N_r} \sum_{\delta=1}^{N_r} \left(f_\gamma^2 + f_\delta^2 - 2f_\gamma f_\delta\right) S_{\alpha\gamma}^* S_{\beta\gamma} S_{\alpha\delta} S_{\beta\delta}^*$$

$$= \frac{1}{2} \sum_{\gamma=1}^{N_r} f_\gamma^2 S_{\alpha\gamma}^* S_{\beta\gamma} \sum_{\delta=1}^{N_r} S_{\alpha\delta} S_{\beta\delta}^* + \frac{1}{2} \sum_{\gamma=1}^{N_r} S_{\alpha\gamma}^* S_{\beta\gamma} \sum_{\delta=1}^{N_r} f_\delta^2 S_{\alpha\delta} S_{\beta\delta}^*$$

$$- \sum_{\gamma=1}^{N_r} f_\gamma S_{\alpha\gamma}^* S_{\beta\gamma} \sum_{\delta=1}^{N_r} f_\delta S_{\alpha\delta} S_{\beta\delta}^* = - \left| \sum_{\gamma=1}^{N_r} f_\gamma S_{\alpha\gamma}^* S_{\beta\gamma} \right|^2 \leq 0.$$

In the second line we used $\sum_\delta S_{\alpha\delta} S_{\beta\delta}^* = \delta_{\alpha\beta} = 0$. Thus the sign rule for the current correlator power at zero frequency has been proven.

To illustrate the above general properties we consider a simple example.

2.2.4.3 Scatterer with two leads

From Eq. (2.63) it follows that for $N_2 = 2$ the whole noise power matrix, $\hat{\mathcal{P}}(0)$, is defined by only a single element. This is true for thermal noise and for shot noise separately,

$$\mathcal{P}_{11}^{(th)} = \mathcal{P}_{22}^{(th)} = -\mathcal{P}_{12}^{(th)} = -\mathcal{P}_{21}^{(th)} \equiv \mathcal{P}^{(th)},$$

$$\mathcal{P}_{11}^{(sh)} = \mathcal{P}_{22}^{(sh)} = -\mathcal{P}_{12}^{(sh)} = -\mathcal{P}_{21}^{(sh)} \equiv \mathcal{P}^{(sh)},$$

(2.65a)

where

$$\mathcal{P}^{(th)} = \frac{e^2 k_B}{h} \int_0^\infty dE \left(-T_1 \frac{\partial f_1(E)}{\partial E} - T_2 \frac{\partial f_2(E)}{\partial E} \right) T_{12}(E), \quad (2.65b)$$

$$\mathcal{P}^{(sh)} = \frac{e^2}{h} \int_0^\infty dE \left[f_1(E) - f_2(E) \right]^2 T_{12}(E) R_{11}(E). \quad (2.65c)$$

Here $T_{12}(E) = |S_{12}(E)|^2$, $R_{11}(E) = 1 - T_{12}(E)$ are, respectively, the transmission and reflection probabilities for electrons with energy E. While transforming the expression for $\mathcal{P}^{(th)}$ we used the following identity for the Fermi distribution function,

$$F_{\alpha\alpha}(E,E) \equiv f_\alpha(E)[1 - f_\alpha(E)] = -k_B T_\alpha \frac{\partial f_\alpha(E)}{\partial E}. \quad (2.66)$$

From Eqs. (2.65) it follows that the character of the dependence of the transmission coefficient on energy is crucial for the dependence of noise on both the temperature and the bias voltage. For instance, if the transmission coefficient $T_{12}(E)$ changes only a little within a relevant energy window (maximum of two, the reservoir temperature and the bias) then the thermal noise is linear in the reservoir temperatures T_1, T_2 and it is independent of the bias: $\mathcal{P}^{(th)} = k_B (T_1 + T_2) G$, where $G = (e^2/h) T_{12}(\mu_0)$. In contrast, the shot noise, $\mathcal{P}^{(sh)}$, is a non-linear function of both the temperature and the bias. And only in the limit of a large bias, $|eV| \gg k_B T_1, k_B T_2$, does the shot noise become merely proportional to the current, $I = VG$, $\mathcal{P}^{(sh)} = |eI| R_{11}(\mu_0)$.

2.2.5 Fano factor

The Fano factor, F, is the ratio of shot noise to the direct current times the charge of the carriers, see, e.g., Ref. [31]:

$$F = \frac{P^{(sh)}}{|qI|}. \tag{2.67}$$

As was shown by Schottky [32], for statistically independent carriers the Fano factor is unity. In the presence of correlations or interactions between carriers the Fano factor is generally different from unity.

In mesoscopic physics also one can introduce the Fano factor. However, it follows from Eqs. (1.47) (for $N_r = 2$) and (2.65c), that in general $F \neq 1$. Even in the simplest case, $T(E) = $ const and $eV \gg k_B T$, the Fano factor $F = 1 - T_{12}$. As $T_{12} \to 0$ the quantity $F \approx 1$, therefore, one can say that in the case of a small conductance, $G/G_0 = T_{12} \ll 1$, the current is carried by statistically independent particles. With increasing conductance the factor Fano decreases, due to correlations between carriers. These correlations are a consequence of the Pauli exclusion principle forcing electrons to pass through a lead one by one.

Chapter 3

Non-stationary scattering theory

Applying a time-dependent bias or by varying in time the properties of a sample we can create conditions where time-dependent currents flow through the system. Our aim is to consider how the non-stationary transport can be described within the scattering matrix formalism.

To calculate the scattering matrix elements, which are quantum-mechanical amplitudes, we need to solve the Schrödinger equation. Therefore, we first consider methods of solution of the non-stationary Schrödinger equation and then we analyze the properties of the scattering matrix for a non-stationary sample. We are interested in a particular case where the dependence on time is periodic.

3.1 Schrödinger equation with a potential periodic in time

Let us consider the Schrödinger equation for the wave function Ψ of a particle with mass m in the case of a time-dependent Hamiltonian, $H(t, \vec{r})$,

$$i\hbar \frac{\partial \Psi(t, \vec{r})}{\partial t} = H(t, \vec{r}) \Psi(t, \vec{r}),$$

$$H(t, \vec{r}) = H_0(\vec{r}) + V(t, \vec{r}).$$

(3.1)

Here we have split the Hamiltonian into two parts, time-independent, $H_0(\vec{r})$, and dependent on time, $V(t, \vec{r})$. The corresponding boundary conditions are assumed to be stationary. We suppose that the solution to the stationary problem with Hamiltonian $H_0(\vec{r})$,

$$\Psi(t,\vec{r}) = e^{-\frac{iEt}{\hbar}}\psi(\vec{r})\,,$$

$$H_0(\vec{r})\psi(\vec{r}) = E\psi(\vec{r})\,,$$

(3.2)

and with the same boundary conditions is known. That is, we have already found all the eigenfunctions, $\psi_n(\vec{r})$, and eigen energies, E_n,

$$H_0(\vec{r})\,\psi_n(\vec{r}) = E_n\psi_n(\vec{r})\,. \qquad (3.3)$$

Note that $\Psi_n(t,\vec{r}) = e^{-\frac{iE_n t}{\hbar}}\psi_n(\vec{r})$. The index n (not necessarily an integer) numbers the states belonging to both discrete and continuous parts of a spectrum.

We compare two methods for solving the non-stationary problem. The first method is the perturbation theory of P. A. M. Dirac [37], see, e.g., Ref. [14], which is applicable for a weak time-dependent potential with arbitrary dependence on time. The second one, based on the Floquet theorem, see, e.g., Refs. [38, 39], is applied to potentials that are periodic in time with arbitrary strength.

3.1.1 Perturbation theory

Let the time-dependent potential be small,

$$V(t,\vec{r}) \to 0\,, \qquad (3.4)$$

and, therefore, it can be considered as a perturbation that changes only slightly the state of a quantum system with Hamiltonian $H_0(\vec{r})$.

We are looking for a solution to Eq. (3.1) as a series of stationary eigen wave functions,

$$\Psi(t,\vec{r}) = \sum_n a_n(t)\,\Psi_n(t,\vec{r})\,. \qquad (3.5)$$

Substituting Eq. (3.5) into Eq. (3.1) and using Eq. (3.3) we find,

$$i\hbar \sum_n \Psi_n(t,\vec{r}) \frac{da_n(t)}{dt} = \sum_n a_n(t)\, V(t,\vec{r})\, \Psi_n(t,\vec{r}). \tag{3.6}$$

Further we multiply both parts of this equation by $\Psi_k^*(t,\vec{r})$ and integrate over space. Because the eigenfunctions of the Hamiltonian are orthogonal,

$$\int d^3r\, \psi_k^*(\vec{r})\, \psi_n(\vec{r}) = \delta_{n,k},$$

we arrive at the following equation for the coefficients a_k:

$$i\hbar \frac{da_k(t)}{dt} = \sum_n V_{kn}(t)\, a_n(t), \tag{3.7}$$

where the perturbation matrix elements are

$$V_{kn}(t) = \int d^3r\, \psi_k^*(\vec{r})\, V(t,\vec{r})\, \psi_n(\vec{r})\, e^{i\frac{E_k - E_n}{\hbar}t}. \tag{3.8}$$

To find the coefficients $a_n(t)$ we need to solve the system of an infinite number of differential equations of the first order, Eq. (3.7).

Up to now we have not used the fact that the perturbation is weak. Now we use it to solve the system of equations to the linear order in $V(t,\vec{r})$. To be more precise we consider the following problem: The perturbation $V(t,\vec{r})$ is switched on at $t = 0$. We consider a particle which was in the state $\Psi_m(t,\vec{r})$ with energy E_m at $t \leq 0$. We need to calculate its wave function $\Psi^{(m)}(t,\vec{r})$ at $t > 0$.

We will use an upper index (m) to show an initial state. So, we have a problem with the following initial conditions,

$$\Psi^{(m)}(t=0,\vec{r}) = \Psi_m(t=0,\vec{r}) \Rightarrow \begin{cases} a_m^{(m)}(0) = 1, \\ a_n^{(m)}(0) = 0,\, n \neq m, \end{cases}$$

where $a_n^{(m)}(t)$ are the coefficients in Eq. (3.5) for the wave function of interest, $\Psi^{(m)}(t,\vec{r})$. After the perturbation is switched on the coefficients become functions of time, $a_n^{(m)}(t)$, which we will consider as a series in powers of a small parameter $V(t,\vec{r})$. In the linear order we have

$$a_m^{(m)}(t) = 1 + a_m^{(m,1)}(t),$$

$$a_n^{(m)}(t) = 0 + a_n^{(m,1)}(t), \, n \neq m.$$
(3.9)

Substituting these equations into Eq. (3.7) and keeping only linear terms in V we find

$$i\hbar \frac{da_k^{(m,1)}(t)}{dt} = V_{km}(t).$$
(3.10)

This linear first-order equation can be easily integrated,

$$a_k^{(m,1)}(t) = -\frac{i}{\hbar} \int_0^t dt' \, V_{km}(t').$$
(3.11)

According to the basic principles of quantum mechanics the absolute value square, $|a_k^{(m)}(t)|^2$, defines the probability of observing a particle in the state $\Psi_k(t,\vec{r})$ with energy E_k at time t. Note at initial time $t=0$ the particle was in the state with energy E_m. The change of the particle's energy is due to the interaction with a time-dependent potential $V(t,\vec{r})$. The particle can either gain energy, $E_k > E_m$, or lose it, $E_k < E_m$.

Now we clarify when the potential can be treated as small, Eq. (3.4). Let us consider a potential that is uniform in space and periodic in time, $V(t,\vec{r}) = U(t) R(\vec{r})$, where

$$U(t) = 2U \cos(\Omega_0 t).$$
(3.12)

Then we can solve Eq. (3.11),

$$a_k^{(m,1)}(t) = -U R_{km} \left(\frac{e^{i(\omega_{km} - \Omega_0)t} - 1}{\hbar(\omega_{km} - \Omega_0)} + \frac{e^{i(\omega_{km} + \Omega_0)t} - 1}{\hbar(\omega_{km} + \Omega_0)} \right),$$
(3.13)

where $R_{km} = \int d^3r \, \psi_k^*(\vec{r}) R(\vec{r}) \psi_n(\vec{r})$ and $\hbar \omega_{km} = E_k - E_m$. Perturbation theory is correct if the absolute value of $a_{k \neq m}^{(m)}(t)$ is small compared to unity:

$$\frac{V_{km}}{\hbar(\omega_{km} \pm \Omega_0)} \sim \frac{UR_{km}}{\hbar(\omega_{km} \pm \Omega_0)} \ll 1. \tag{3.14}$$

In this case a particle with a large probability stays in its initial state and the effect of a time-dependent potential is very small as expected.

If the perturbation frequency, Ω_0, is close to some difference, $\pm(E_{k_0} - E_m)/\hbar$, then equation (3.14) can be easily violated and perturbation theory fails. In such a case the time-dependent potential will cause a particle to pass over from the initial state $\Psi_m(t,\vec{r})$ to the state $\Psi_{k_0}(t,\vec{r})$ and back, because the coefficients $a_m^{(m)}(t)$ and $a_{k_0}^{(m)}(t)$ are of the same order.

Substituting Eqs. (3.13) and (3.9) into Eq. (3.5), for the function $\Psi^{(m)}(t,\vec{r})$, we finally calculate

$$\Psi^{(m)}(t,\vec{r}) = e^{-i\frac{E_m}{\hbar}t} \sum_n \psi_n(\vec{r})$$

$$\times \left\{ \delta_{nm} - \frac{UR_{nm}\left(e^{-i\Omega_0 t} - e^{-i\omega_{nm}t}\right)}{\hbar(\omega_{nm} - \Omega_0)} - \frac{UR_{nm}\left(e^{i\Omega_0 t} - e^{-i\omega_{nm}t}\right)}{\hbar(\omega_{nm} + \Omega_0)} \right\}. \tag{3.15}$$

Thus we have found that the periodic perturbation results in additional terms in the expression for the wave function, which correspond to the initial energy shifted by $\pm\hbar\Omega_0$. It is easy to understand that the spectral contents of the perturbation define energies of additional sidebands of a wave function.

3.1.2 Floquet functions method

This method overcomes the restrictions of Eq. (3.14) and allows us to consider an arbitrary potential that is periodic in time. The main idea is to use the Floquet theorem. According to this theorem a solution to the Schrödinger equation with a Hamiltonian that is periodic in time,

$$H(t,\vec{r}) = H(t+\mathcal{T},\vec{r}), \tag{3.16}$$

can be written as follows

$$\Psi(t,\vec{r}) = e^{-i\frac{E}{\hbar}t}\phi(t,\vec{r})\,,$$

$$\phi(t,\vec{r}) = \phi(t+\mathcal{T},\vec{r})\,. \tag{3.17}$$

To outline the proof of this theorem we consider the general solution $\Psi(t,\vec{r})$ to Eq. (3.1) with the Hamiltonian of Eq. (3.16). Let us shift time by one period, $t \to t+\mathcal{T}$. Then the wave function $\Psi(t+\mathcal{T},\vec{r})$ is also a solution to the same equation,

$$i\hbar\frac{\partial \Psi(t+\mathcal{T},\vec{r})}{\partial t} = H(t+\mathcal{T},\vec{r})\,\Psi(t+\mathcal{T},\vec{r})$$

$$= H(t,\vec{r})\,\Psi(t+\mathcal{T},\vec{r})\,.$$

Therefore, these two general solutions have to be proportional to each other,

$$\Psi(t+\mathcal{T},\vec{r}) = C\,\Psi(t,\vec{r})\,. \tag{3.18}$$

Because the wave function is normalized,

$$\int d^3r |\Psi(t,\vec{r})|^2 = 1\,,$$

$$\int d^3r |\Psi(t+\mathcal{T},\vec{r})|^2 = |C|^2 \int d^3r |\Psi(t,\vec{r})|^2 = 1\,,$$

we find for the constant C,

$$|C|^2 = 1 \quad \Rightarrow \quad C = e^{-i\alpha}\,. \tag{3.19}$$

The general expression for the function subject to Eq. (3.18) with a coefficient given by Eq. (3.19) is the following,

$$\Psi(t,\vec{r}) = e^{-i\frac{\alpha}{\mathcal{T}}t}\phi(t,\vec{r})\,,$$

$$\phi(t,\vec{r}) = \phi(t+\mathcal{T},\vec{r})\,. \tag{3.20}$$

Let us check that Eq. (3.18) holds,

$$\Psi(t+\mathcal{T}) = e^{-i\frac{\alpha}{\mathcal{T}}(t+\mathcal{T})}\phi(t+\mathcal{T}) = e^{-i\alpha}\left\{e^{-i\frac{\alpha}{\mathcal{T}}t}\phi(t)\right\} = e^{-i\alpha}\Psi(t).$$

Finally introducing $E = \hbar\alpha/\mathcal{T}$ instead of α we see that Eq. (3.20) reduces to Eq. (3.17). The Floquet theorem has been proven.

Next we expand a function that is periodic in time $\phi(t,\vec{r})$ into the Fourier series,

$$\phi(t,\vec{r}) = \sum_{q=-\infty}^{\infty} e^{-iq\Omega_0 t}\psi_q(\vec{r}), \qquad (3.21a)$$

$$\psi_q(\vec{r}) = \int_0^{\mathcal{T}} \frac{dt}{\mathcal{T}} e^{iq\Omega_0 t}\phi(t,\vec{r}), \qquad (3.21b)$$

where $\Omega_0 = 2\pi/\mathcal{T}$. Then the Floquet wave function, Eq. (3.17), becomes

$$\Psi(t,\vec{r}) = e^{-i\frac{E}{\hbar}t}\sum_{q=-\infty}^{\infty} e^{-iq\Omega_0 t}\psi_q(\vec{r}). \qquad (3.22)$$

In the case of a stationary Hamiltonian the solution corresponding to energy E has a factor $e^{-i\frac{E}{\hbar}t}$. Therefore, in the stationary case for Eq. (3.22) only the term with $q = 0$ survives. In the case of a time-dependent Hamiltonian the energy is a quantity that is not uniquely defined. For instance, if we change E in Eq. (3.22) by any number p of energy quanta $\hbar\Omega_0$, $E \to E + p\hbar\Omega_0$, then we arrive at the same wave function. To show this we need only to redefine function $\psi_q(\vec{r})$ changing its indices, $q \to q+p$. Because the quantity E is defined up to energy quantum $\hbar\Omega_0$, then it is referred to as *the quasi-energy* or *the Floquet energy*. In each particular problem the quantity E is fixed as convenient. For numerical calculations people often use $0 \leq E < \hbar\Omega_0$. On the other hand, when exploring the problem of how some stationary state evolves under the action of a periodic potential, it is convenient to choose E equal to the energy of this initial stationary state. We will follow this latter method when we consider the scattering of electrons with fixed energy E from a dynamic sample.

Comparing Eqs. (3.15) and (3.22) we conclude that the Floquet theorem predicts the existence of multi-photon processes when the energy changes by several quanta $\hbar\Omega_0$ in addition to the single-photon processes for a weak perturbation. The Floquet theorem gives an ansatz for the solution to the Schrödinger equation with a periodic Hamiltonian. The unknown function $\psi_q(\vec{r})$ is a solution to some stationary problem. It should be noted that in the general case the functions $\psi_q(\vec{r})$ with different q are not independent. Therefore, the non-stationary problem is reduced to a multi-channel stationary problem.

3.1.3 Potential oscillating in time and uniform in space

Let us consider a simple exactly solvable example to show that the solution of a periodic problem is really of a Floquet function type and at a weak perturbation only single-photon processes are allowed. So we consider the Schrödinger equation with a uniform potential, Eq. (3.12),

$$i\hbar \frac{\partial \Psi(t,\vec{r})}{\partial t} = \{H_0 + 2U\cos(\Omega_0 t)\}\Psi(t,\vec{r}). \quad (3.23)$$

The solution to this equation reads

$$\Psi(t,\vec{r}) = e^{-i\left\{\frac{E}{\hbar}t + \frac{2U}{\hbar\Omega_0}\sin(\Omega_0 t)\right\}} \psi_E(\vec{r}), \quad (3.24)$$

where $\psi_E(\vec{r})$ is a solution to the following stationary equation,

$$H_0 \psi_E(\vec{r}) = E\,\psi_E(\vec{r}). \quad (3.25)$$

Next we use the following Fourier series,

$$e^{-i\alpha \sin(\Omega_0 t)} = \sum_{q=-\infty}^{\infty} e^{-iq\Omega_0 t} J_q(\alpha), \quad (3.26)$$

where J_q is the Bessel function of the first kind of the qth order, and rewrite Eq. (3.24) as follows,

$$\Psi(t,\vec{r}) = e^{-i\frac{E}{\hbar}t} \sum_{q=-\infty}^{\infty} e^{-iq\Omega_0 t} J_q\left(\frac{2U}{\hbar\Omega_0}\right) \psi_E(\vec{r}). \quad (3.27)$$

Comparing the equation above with Eq. (3.22) we see that the obtained solution is the Floquet function with $\psi_q(\vec{r}) = J_q(2U/\hbar\Omega_0)\psi_E(\vec{r})$.

Let us analyze Eq. (3.27) for small amplitudes, $U/(\hbar\Omega_0) \ll 1$. So we expand the Bessel functions using a Taylor series in powers of a small parameter $\alpha = 2U/(\hbar\Omega_0)$,

$$J_0(\alpha) \approx 1 - \alpha^2/4, \quad J_{\pm 1}(\alpha) \approx \pm\alpha/2, \quad J_{\pm|n|} \sim \pm\alpha^{|n|}, \quad |n| > 1.$$

Then up to terms linear in U the solution to Eq. (3.27) becomes

$$\Psi(t,\vec{r}) \approx e^{-i\frac{E}{\hbar}t}\psi_E(\vec{r})\left\{1 + \frac{Ue^{-i\Omega_0 t}}{\hbar\Omega_0} - \frac{Ue^{i\Omega_0 t}}{\hbar\Omega_0}\right\}.$$

This equation is exactly Eq. (3.15) with $R_{nm} = \delta_{nm}$ and $\psi_m(\vec{r}) = \psi_E(\vec{r})$.

3.2 Floquet scattering matrix

The main difference between a dynamic scatterer and a stationary one is that the former can change the energy of an incident electron. We are interested in the particular case when the parameters of a scatterer vary periodically in time. This variation can be caused by some external (classical) influence affecting the scattering properties of a sample. For instance, it could be an electric potential forming a barrier for propagating electrons.

We assume that the Hamiltonian describing the interaction of the electrons with the scatterer depends periodically on time. Then the wave function of a scattered electron is of the Floquet function type, Eq. (3.22), having components corresponding to different energies. We choose an energy E of an incident electron as the Floquet energy. Then the absolute value square of its qth sideband integrated over space defines the probability of absorbing, $q > 0$, or emitting, $q < 0$, energy $|q|\hbar\Omega_0$ during scattering.

From the scattering theory point of view the fact that the scattering properties are varied periodically in time results in a scattering matrix dependent on two energies, incident and scattered, which differ from each other by the integer number of the energy quantum $\hbar\Omega_0$. Such a scattering matrix is referred to as *the Floquet scattering matrix*, $\hat{S}_F$. The element $S_{F,\alpha\beta}(E_n, E)$ is a photon-assisted propagation amplitude times $\sqrt{k_n/k}$, where $k_n = \sqrt{2mE_n/\hbar^2}$. This amplitude describes the process when an electron with energy E incident from the lead β is scattered into the lead α and its energy is changed to $E_n = E + n\hbar\Omega_0$ [40]. As we did in the stationary case, we define the scattering amplitudes so as to describe transitions between the states with fixed energies carrying a unit flux. These states are eigenstates for the Hamiltonian in leads that are assumed to be stationary.

3.2.1 Floquet scattering matrix properties

3.2.1.1 Unitarity

Because the particle flow is conserved during scattering, the Floquet scattering matrix is unitary [41],

$$\sum_n \sum_{\alpha=1}^{N_r} S_{F,\alpha\beta}^*(E_n, E_m) S_{F,\alpha\gamma}(E_n, E) = \delta_{m0}\delta_{\beta\gamma}, \quad (3.28a)$$

$$\sum_n \sum_{\beta=1}^{N_r} S_{F,\gamma\beta}(E_m, E_n) S_{F,\alpha\beta}^*(E, E_n) = \delta_{m0}\delta_{\alpha\gamma}. \quad (3.28b)$$

In the sum over n we keep only those terms that correspond to current-carrying states (with $E_n > 0$). Therefore, $n > -[E/\hbar\Omega_0]$, where $[X]$ stands for the integer part of X. In the case where

$$\epsilon = \frac{\hbar\Omega_0}{E} \ll 1 \quad (3.29)$$

the sum over n in Eq. (3.28) runs from $-\infty$ to ∞. In the following we assume this to be the case.

Note the negative values, $E_n < 0$, correspond to the states localized on the scatterer. These states do not contribute to the current. Strictly speaking the transitions between these localized states and the current-carrying states, $E > 0$, are also described by Floquet scattering matrix elements. However, in the steady state such transitions do not contribute to the current. Therefore, they do not enter Eqs. (3.28). Below we use only that part of the Floquet scattering matrix corresponding to transitions between delocalized states and for brevity we name it the Floquet scattering matrix.

3.2.1.2 Micro-reversibility

The invariance of the equations of motion under time reversal put constraints on the Floquet scattering matrix elements. As we considered earlier, see Section 1.1.1.2, in the stationary case the Schrödinger equation remains invariant under $t \to -t$ if we simultaneously reverse the magnetic field direction and replace the wave function by its complex conjugate. Note that the incoming and outgoing scattering channels are interchanged.

In the case of dynamic scattering the time reversal can change the time-dependent Hamiltonian. Let us assume that the Hamiltonian depends on N_p parameters $p_i(t)$, $i = 1, \ldots, N_p$, which are all periodic in time,

$$p_i(t) = p_{i,0} + p_{i,1} \cos(\Omega_0 t + \varphi_i). \tag{3.30}$$

Then under the time reversal, $t \to -t$, the Hamiltonian remains invariant if in addition we change the signs of all the phases, $\varphi_i \to -\varphi_i$, $\forall i$. Thus micro-reversibility results in the following symmetry conditions [42]

$$S_{F,\alpha\beta}(E, E_n; H, \{\varphi\}) = S_{F,\beta\alpha}(E_n, E; -H, \{-\varphi\}), \tag{3.31}$$

where $\{\varphi\}$ is a set of phases φ_i.

3.3 Current operator

To calculate the current operator, Eq. (1.36), one needs to express the operators for scattered electrons, $\hat{b}_\alpha(E)$, in terms of operators for incident electrons, $\hat{a}_\alpha(E)$. These operators annihilate an electron in the state with definite energy. Taking into account that during scattering an electron can change its energy by several energy quanta $\hbar\Omega_0$, we arrive at the following generalization of Eq. (1.39) for scattering that is periodic in time [40]

$$\hat{b}_\alpha(E) = \sum_{n=-\infty}^{\infty} \sum_{\beta=1}^{N_r} S_{F,\alpha\beta}(E, E_n) \, \hat{a}_\beta(E_n), \tag{3.32a}$$

$$\hat{b}_\alpha^\dagger(E) = \sum_{n=-\infty}^{\infty} \sum_{\beta=1}^{N_r} S_{F,\alpha\beta}^*(E, E_n) \, \hat{a}_\beta^\dagger(E_n). \tag{3.32b}$$

Note the summation over the energy scattering channels is quite similar to the summation over the orbital scattering channels. The equations above together with the unitarity conditions, Eq. (3.28), guarantee anti-commutation relations for b-operators similar to those for a-operators, (1.30).

It is natural to assume that scattering properties, which vary periodically in time, result in periodic currents flowing in the system [43]. This guess remains true even in the absence of both a bias voltage and a temperature difference. To analyze periodic currents it is convenient to use the frequency representation,

$$\hat{I}_\alpha(t) = \int_{-\infty}^{\infty} \frac{d\omega}{2\pi} e^{-i\omega t} \hat{I}_\alpha(\omega), \qquad (3.33a)$$

$$\hat{I}_\alpha(\omega) = \int_{-\infty}^{\infty} dt\, e^{i\omega t} \hat{I}_\alpha(t). \qquad (3.33b)$$

Using Eq. (1.36) we calculate

$$\hat{I}_\alpha(\omega) = e \int_0^\infty dE \left\{ \hat{b}_\alpha^\dagger(E) \hat{b}_\alpha(E + \hbar\omega) - \hat{a}_\alpha^\dagger(E) \hat{a}_\alpha(E + \hbar\omega) \right\}, \qquad (3.34)$$

where we used

$$\int_{-\infty}^{\infty} dt\, e^{i\frac{E-E'+\hbar\omega}{\hbar}t} = 2\pi\hbar\, \delta(E - E' + \hbar\omega), \qquad (3.35)$$

and

$$\int_0^\infty dE'\, \delta(E - E' + \hbar\omega) X(E') = X(E + \hbar\omega), \qquad (3.36)$$

with $X = \hat{b}_\alpha(E'), \hat{a}_\alpha(E')$.

3.3.1 Alternating current

Substituting Eqs. (3.32) into Eq. (3.34) and averaging over the equilibrium state of the reservoirs, we calculate a current spectrum, $I_\alpha(\omega) = \langle \hat{I}_\alpha(\omega) \rangle$ [44]

$$I_\alpha(\omega) = \sum_{l=-\infty}^{\infty} 2\pi\delta(\omega - l\Omega_0) I_{\alpha,l}, \qquad (3.37a)$$

$$I_{\alpha,l} = \frac{e}{h} \int_0^\infty dE \left\{ \sum_{\beta=1}^{N_r} \sum_{n=-\infty}^{\infty} S_{F,\alpha\beta}^*(E, E_n) S_{F,\alpha\beta}(E_l, E_n) f_\beta(E_n) - \delta_{l0} f_\alpha(E) \right\}. \qquad (3.37b)$$

Taking into account Eq. (3.28b) we rewrite $I_{l,\alpha}$ as follows

$$I_{\alpha,l} = \frac{e}{h} \int_0^\infty dE \sum_{\beta=1}^{N_r} \sum_{n=-\infty}^{\infty} S_{F,\alpha\beta}^*(E_n, E) S_{F,\alpha\beta}(E_{l+n}, E) \left\{ f_\beta(E) - f_\alpha(E_n) \right\}, \qquad (3.38)$$

where we additionally replaced $E_n \to E$ and $n \to -n$. The convenience of the last equation containing the difference of the Fermi functions becomes evident in the case of a slow variation of the scatterer parameters, $\Omega_0 \to 0$, when a current can be expanded in powers of Ω_0.

Substituting Eq. (3.37a) into Eq. (3.33a) we finally arrive at a time-dependent current,

$$I_\alpha(t) = \sum_{l=-\infty}^{\infty} e^{-il\Omega_0 t} I_{\alpha,l}, \qquad (3.39)$$

which is certainly periodic in time, $I_\alpha(t) = I_\alpha(t + 2\pi/\Omega_0)$.

3.3.2 Direct current

Of special interest is the case when the current $I_\alpha(t)$ has a time-independent part. We emphasize that while an alternating current is always generated by the dynamic scatterer, the direct current exists only under specific conditions, which we will discuss later. Now we just give general expressions for the direct current, the term with $l = 0$ in Eq. (3.39).

Using $l = 0$ in Eq. (3.37b) we find,

$$I_{\alpha,0} = \frac{e}{h} \int_0^\infty dE \left\{ \sum_{n=-\infty}^{\infty} \sum_{\beta=1}^{N_r} |S_{F,\alpha\beta}(E, E_n)|^2 f_\beta(E_n) - f_\alpha(E) \right\}. \quad (3.40)$$

The direct current is subject to the conservation law, Eq. (1.48). To show this we transform the expression above as follows. In the part with a factor $f_\beta(E_n)$ we shift $E \to E - n\hbar\Omega_0$ and $n \to -n$ [40][1]

$$I_{\alpha,0} = \frac{e}{h} \int_0^\infty dE \sum_{n=-\infty}^{\infty} \sum_{\beta=1}^{N_r} \left\{ |S_{F,\alpha\beta}(E_n, E)|^2 f_\beta(E) - f_\alpha(E) \right\}. \quad (3.41)$$

Then using Eq. (3.28a) one can easily check that $\sum_{\alpha=0}^{N_r} I_{\alpha,0} = 0$.

Another expression for a direct current can be found if we substitute Eq. (3.28b) with $m = 0$ and $\alpha = \gamma$ into Eq. (3.41) as a unity in front of $f_\alpha(E)$ and make a shift $E \to E - n\hbar\Omega_0$ and a substitution $n \to -n$ [40]:

$$I_{\alpha,0} = \frac{e}{h} \int_0^\infty dE \sum_{n=-\infty}^{\infty} \sum_{\beta=1}^{N_r} |S_{F,\alpha\beta}(E_n, E)|^2 \left\{ f_\beta(E) - f_\alpha(E_n) \right\}. \quad (3.42)$$

From this equation it follows that (for $\hbar\Omega_0 \ll \mu$) only electrons with an energy close to the Fermi energy contribute to the current. This is because only for such electrons is the difference of the Fermi functions noticeable, $f_\beta(E) - f_\alpha(E + n\hbar\Omega_0) \neq 0$. Note the energy window, where the current flows, is defined by the maximum of the following quantities: the energy quantum $\hbar\Omega_0$ dictated by the driving frequency, a possible bias $|eV_{\alpha\beta}|$, and the temperature, $k_B T_\alpha$.

And finally the intuitively clear expression for a current can be derived in the same way as Eq. (3.42) was derived from Eq. (3.41). We use Eq. (3.28a) instead of Eq. (3.28b) and replace $\alpha \to \beta$ and $\beta = \gamma \to \alpha$ [40]:

$$I_{\alpha,0} = \frac{e}{h} \int_0^\infty dE \sum_{n=-\infty}^{\infty} \sum_{\beta=1}^{N_r}$$
$$\left\{ |S_{F,\alpha\beta}(E_n, E)|^2 f_\beta(E) - |S_{F,\beta\alpha}(E_n, E)|^2 f_\alpha(E) \right\}. \quad (3.43)$$

---

[1] The limits of integration over energy are not changed, because, as we already mentioned, only those elements of the Floquet scattering matrix for which both $E > 0$ and $E_n > 0$ contribute to the current.

This equation represents a direct current in the lead α as a difference of two electron flows. The first one is composed of the flows incident from the various leads β and scattered with probability $\left|S_{F,\alpha\beta}(E_n, E)\right|^2$ into the lead α. And the second one is incident from the lead α and with probability $\left|S_{F,\beta\alpha}(E_n, E)\right|^2$ scattered into the various leads β.

We emphasize all the equations (3.40)–(3.43) are equivalent. Which of them is used is dictated by convenience in each particular case.

3.4 Adiabatic approximation for the Floquet scattering matrix

To calculate the Floquet scattering matrix elements one needs to solve the non-stationary Schrödinger equation, which is, in the general case, more complicated than solving the stationary problem. The reason for this is, in particular, that the stationary scattering matrix $\hat{S}$ has only $N_r \times N_r$ elements while the Floquet scattering matrix $\hat{S}_F$ has much more elements, $N_r \times N_r \times (2n_{\max} + 1)^2$, where $n_{\max}$ is the maximum number of energy quanta $\hbar\Omega_0$ that an electron can absorb/emit while interacting with a dynamic scatterer. Formally an electron can change its energy by $n \to \infty$ energy quanta $\hbar\Omega_0$. However, in practice there is some number $n_{\max}$ such that the probability of absorbing/emitting $n_{\max} + 1$ or more energy quanta is negligible within a given accuracy. For instance, if the amplitude δU of an oscillating potential is small compared with $\hbar\Omega_0$ then $n_{\max} = 1$, that is, only single-photon processes are relevant. In contrast if $\delta U \gg \hbar\Omega_0$ then $n_{\max} \gg 1$.

In general the multi-photon processes become important if the parameters of a scatterer vary slowly. Therefore, at $\Omega_0 \to 0$ we should calculate a huge number of scattering amplitudes that can be impossible in practice. On the other hand it is natural to expect that the scattering properties of a sample with parameters varying slowly enough should be close to the scattering properties of a strictly stationary sample. Because at $\mathcal{T} = 2\pi/\Omega_0 \to \infty$ any finite time spent by an electron within the scattering region is always small compared to $\mathcal{T}$ and an electron will not feel that the scatterer is dynamic. However, as we show below, there is a principal difference between the properties of a dynamic scatterer and the properties of a stationary scatterer [41, 42]. For instance, a dynamic scatterer can generate a direct current in the absence of a bias applied to the reservoirs.

3.4.1 Frozen scattering matrix

Let us consider the stationary scattering matrix $\hat{S}$ dependent on several parameters, $p_i \in \{p\}, i = 1, 2, \ldots, N_p$, which are varied periodically in time, Eq. (3.30). Then the matrix $\hat{S}$ becomes a periodic function of time, $\hat{S}(t, E) = \hat{S}(\{p(t)\}; E)$, $\hat{S}(t, E) = \hat{S}(t + \mathcal{T}, E)$. We stress the matrix $\hat{S}(t)$ does not describe scattering onto a dynamic scatterer. Its physical meaning is the following: Let us fix all the parameters at a time $t = t_0$ and not change them any more. Then the matrix $\hat{S}(t_0, E)$ describes scattering from such a frozen scatterer. Therefore, we can say that the matrix $\hat{S}(t, E)$ is a *frozen scattering matrix*. We stress the variable t here is a parameter, relating to a given variation of the properties of a scatterer, rather than a true dynamic time entering the equation of motion.

As we pointed out the frozen scattering matrix $\hat{S}(t, E)$ does not have a direct relation to scattering from a dynamic sample, because it depends on a single energy only. However, at $\Omega_0 \to 0$ there exists some relation between the frozen and the Floquet scattering matrices. It becomes more clear if we expand $\hat{S}_F$ in powers of Ω_0,

$$\hat{S}_F = \sum_{q=0}^{\infty} (\hbar\Omega_0)^q \hat{S}_F^{(q)}. \tag{3.44}$$

Below we relate the first and the second terms in this *adiabatic expansion* to the frozen scattering matrix.

3.4.2 Zeroth-order approximation

To zeroth order, $q = 0$ in Eq. (3.44), all the terms proportional to Ω_0 (or its higher powers) should be dropped. Within this accuracy the initial energy, E, and the final energy, $E_n = E + n\hbar\Omega_0$, are the same. Therefore, the term $\hat{S}_F^{(0)}$ depends, in fact, only on a single energy similar to the frozen scattering matrix. To establish a connection between these two matrices we take into account the following. The element $S_{F,\alpha\beta}(E_n, E)$ describes the scattering process when an electron energy is changed: $\Psi_{E_n,\alpha}^{(out)} \sim S_{F,\alpha\beta}(E_n, E) \Psi_{E,\beta}^{(in)}$, with $\Psi_{E,\beta}^{(in)} \sim e^{-iEt/\hbar}$ and $\Psi_{E_n,\alpha}^{(out)} \sim e^{-iE_n t/\hbar} = e^{-iEt/\hbar} e^{-in\Omega_0 t}$. On the other hand if we consider scattering from the frozen scatterer, $\Psi_{E,\alpha}^{(out)} \sim S_{\alpha\beta}(t, E) \Psi_{E,\beta}^{(in)}$, and use the following Fourier expansion

$$\hat{S}(t, E) = \sum_{n=-\infty}^{\infty} e^{-in\Omega_0 t} \hat{S}_n(E), \qquad (3.45)$$

then one can see that the part of the wave function of a scattered electron proportional to $S_{\alpha\beta,n}$ has a time-dependent phase factor $e^{-iE_n t/\hbar}$, the same as that of $S_{F,\alpha\beta}(E_n, E)$. These simple arguments allow us to conclude that to zeroth order in Ω_0 the Floquet scattering matrix elements are equal to the Fourier coefficients of the frozen scattering matrix,

$$\hat{S}_F^{(0)}(E_n, E) = \hat{S}_n(E), \qquad (3.46a)$$

$$\hat{S}_F^{(0)}(E, E_n) = \hat{S}_{-n}(E). \qquad (3.46b)$$

To prove that this approximation does not violate unitarity we substitute the equations above into Eq. (3.28). Then after the inverse Fourier transformation we find

$$\hat{S}(t, E)\hat{S}^\dagger(t, E) = \hat{S}^\dagger(t, E)\hat{S}(t, E) = \hat{I}, \qquad (3.47)$$

which is completely consistent with the unitarity condition, Eq. (1.10), for the stationary scattering matrix.

3.4.3 First-order approximation

Up to terms of the first order in Ω_0 the initial energy, E, is different from the final energy, E_n. The simplest generalization of Eq. (3.46) would be the same relation but with the frozen scattering matrix calculated at the middle energy, $(E + E_n)/2$. However, it is easy to check that such a matrix is not unitary. To recover unitarity we need to introduce an additional term, $\hbar\Omega_0 \hat{A}_n(E)$, where $\hat{A}_n(E)$ is a Fourier transform of some matrix $\hat{A}(t, E)$. Therefore, we arrive at the following ansatz for the first order in Ω_0 corrections to the frozen scattering matrix, the term with $q = 1$ in Eq. (3.44),

$$\hbar\Omega_0 \hat{S}_F^{(1)}(E_n, E) = \frac{n\hbar\Omega_0}{2} \frac{\partial \hat{S}_n(E)}{\partial E} + \hbar\Omega_0 \hat{A}_n(E), \qquad (3.48a)$$

$$\hbar\Omega_0 \hat{S}_F^{(1)}(E, E_n) = \frac{n\hbar\Omega_0}{2} \frac{\partial \hat{S}_{-n}(E)}{\partial E} + \hbar\Omega_0 \hat{A}_{-n}(E). \qquad (3.48b)$$

Notice the right hand side (RHS) of Eq. (3.48a) is calculated at the energy of an incident electron, while the RHS of Eq. (3.48b) is calculated at the energy of a scattered electron.

The equations (3.48) identify the actual expansion parameter in Eq. (3.44). This, so-called *adiabaticity parameter*, is

$$\varpi = \frac{\hbar\Omega_0}{\delta E} \ll 1, \qquad (3.49)$$

where δE is a characteristic energy scale over which the stationary scattering matrix changes significantly. For instance, if the energy, E, of an incident electron is close to the transmission resonance energy then δE is a width of a resonance. While if E is far from resonance then δE is of the order of the distance between resonances. In the case when the scatterer does not show a resonance transmission then, as a rule, δE is of order E. We emphasize that such a definition of adiabaticity is in general different from the one usually used in quantum mechanics. The latter definition requires a small energy quantum $\hbar\Omega_0$ compared to the difference between the energy levels.

The matrix $\hat{A}$ in Eqs. (3.48) cannot be expressed in terms of the frozen scattering matrix $\hat{S}$. However, the unitarity of the Floquet scattering matrix leads to a relation between these two matrices [41]. To find it we use

$$S_{F,\alpha\beta}(E_n, E) = S_{\alpha\beta,n}(E) + \frac{n\hbar\Omega_0}{2}\frac{\partial S_{\alpha\beta,n}}{\partial E} + \hbar\Omega_0 A_{\alpha\beta,n} + \mathcal{O}\left(\varpi^2\right), \quad (3.50)$$

in Eq. (3.28a):

$$\sum_{n=-\infty}^{\infty}\sum_{\alpha=1}^{N_r}\left\{S^*_{\alpha\gamma,n-m}(E) + \frac{(n+m)\hbar\Omega_0}{2}\frac{\partial S^*_{\alpha\gamma,n-m}(E)}{\partial E} + \hbar\Omega_0 A^*_{\alpha\gamma,n-m}(E)\right\}$$

$$\times \left\{S_{\alpha\beta,n}(E) + \frac{n\hbar\Omega_0}{2}\frac{\partial S_{\alpha\beta,n}(E)}{\partial E} + \hbar\Omega_0 A_{\alpha\beta,n}(E)\right\} = \delta_{\beta\gamma}\delta_{m0}.$$

Taking into account that the matrix $\hat{S}(t, E)$ is unitary and omitting the terms of order Ω_0^2, we get

$$\sum_{n=-\infty}^{\infty}\sum_{\alpha=1}^{N_r}\left\{S_{\alpha\beta,n}\left(n-\frac{n-m}{2}\right)\frac{\partial S^*_{\alpha\gamma,n-m}}{\partial E}+\frac{n}{2}\frac{\partial S_{\alpha\beta,n}}{\partial E}S^*_{\alpha\gamma,n-m}+\right.$$

$$\left.[S_{\alpha\beta,n}A^*_{\alpha\gamma,n-m}+A_{\alpha\beta,n}S^*_{\alpha\gamma,n-m}]\right\}=0.$$

Next we make the inverse Fourier transformation using the following properties,

$$n\,X_n = \frac{i}{\Omega_0}\left(\frac{\partial X}{\partial t}\right)_n, \quad n\,X^*_n = -\frac{i}{\Omega_0}\left(\frac{\partial X^*}{\partial t}\right)_{-n},$$

$$X^*_n = (X^*)_{-n}, \quad \sum_{n=-\infty}^{\infty} X_{-n}Y_{n-m} = (XY)_{-m},$$

(3.51)

and arrive at the following matrix equation,

$$\frac{i}{\Omega_0}\frac{\partial \hat{S}^\dagger}{\partial E}\frac{\partial \hat{S}}{\partial t} + \frac{i}{2\Omega_0}\left\{\frac{\partial^2 \hat{S}^\dagger}{\partial t\partial E}\hat{S} + \hat{S}^\dagger\frac{\partial^2 \hat{S}}{\partial t\partial E}\right\} + \hat{A}^\dagger \hat{S} + \hat{S}^\dagger \hat{A} = \hat{0}.$$

To simplify it we use the identity $\partial^2(\hat{S}^\dagger \hat{S})/\partial t\partial E = \hat{0}$, following from Eq. (3.47), which can be rewritten as follows

$$\frac{\partial^2 \hat{S}^\dagger}{\partial t\partial E}\hat{S} + \hat{S}^\dagger\frac{\partial^2 \hat{S}}{\partial t\partial E} = -\frac{\partial \hat{S}^\dagger}{\partial t}\frac{\partial \hat{S}}{\partial E} - \frac{\partial \hat{S}^\dagger}{\partial E}\frac{\partial \hat{S}}{\partial t}.$$

Then we arrive finally at the following equation (a consequence of the unitarity of scattering) for the matrix $\hat{A}$ [41]

$$\hbar\Omega_0\left[\hat{S}^\dagger(t,E)\,\hat{A}(t,E) + \hat{A}^\dagger(t,E)\,\hat{S}(t,E)\right] = \frac{1}{2}P\left\{\hat{S}^\dagger(t,E),\hat{S}(t,E)\right\},$$

(3.52)

where $P\{\hat{S}^\dagger,\hat{S}\}$ is the Poisson bracket with respect to energy and time,

$$P\left\{\hat{S}^\dagger,\hat{S}\right\} = i\hbar\left(\frac{\partial \hat{S}^\dagger}{\partial t}\frac{\partial \hat{S}}{\partial E} - \frac{\partial \hat{S}^\dagger}{\partial E}\frac{\partial \hat{S}}{\partial t}\right).$$

(3.53)

It is a self-adjoint and traceless matrix,

$$P\left\{\hat{S}^\dagger, \hat{S}\right\} = \left(P\left\{\hat{S}^\dagger, \hat{S}\right\}\right)^\dagger, \tag{3.54}$$

$$\text{tr}\left[P\left\{\hat{S}^\dagger, \hat{S}\right\}\right] \equiv \sum_{\alpha=1}^{N_r} P_{\alpha\alpha}\left\{\hat{S}^\dagger, \hat{S}\right\} = 0. \tag{3.55}$$

To prove Eq. (3.54) we use $\left(\hat{X}^\dagger \hat{Y}\right)^\dagger = \hat{Y}^\dagger \hat{X}$. To prove Eq. (3.55) we use unitarity, $\hat{S}^\dagger \hat{S} = \hat{S}\hat{S}^\dagger = \hat{I}$, its consequence, $\left(\partial \hat{S}^\dagger/\partial E\right)\hat{S} = -\hat{S}^\dagger \left(\partial S/\partial E\right)$, and a property of the trace, $\text{tr}\left[\hat{X}\hat{Y}\right] = \text{tr}\left[\hat{Y}\hat{X}\right]$:

$$\text{tr}[P] = i\hbar \text{tr}\left[\frac{\partial \hat{S}^\dagger}{\partial t}\frac{\partial \hat{S}}{\partial E} - \frac{\partial \hat{S}^\dagger}{\partial E}\hat{S}\hat{S}^\dagger\frac{\partial \hat{S}}{\partial t}\right] = i\hbar \text{tr}\left[\frac{\partial \hat{S}^\dagger}{\partial t}\frac{\partial \hat{S}}{\partial E} - \hat{S}^\dagger \frac{\partial \hat{S}}{\partial E}\frac{\partial \hat{S}^\dagger}{\partial t}\hat{S}\right]$$

$$= i\hbar \text{tr}\left[\frac{\partial \hat{S}^\dagger}{\partial t}\frac{\partial \hat{S}}{\partial E} - \frac{\partial \hat{S}^\dagger}{\partial t}\hat{S}\hat{S}^\dagger\frac{\partial \hat{S}}{\partial E}\right] = i\hbar \text{tr}\left[\frac{\partial \hat{S}^\dagger}{\partial t}\frac{\partial \hat{S}}{\partial E} - \frac{\partial \hat{S}^\dagger}{\partial t}\frac{\partial \hat{S}}{\partial E}\right] = 0.$$

Note if we start from Eq. (3.28b) then we arrive at the following equation [42]

$$\hbar \Omega_0 \left[\hat{A}(t,E)\hat{S}^\dagger(t,E) + \hat{S}(t,E)\hat{A}^\dagger(t,E)\right] = \frac{1}{2}P\left\{\hat{S}(t,E), \hat{S}^\dagger(t,E)\right\}, \tag{3.56}$$

which is equivalent to Eq. (3.52). If we multiply Eq. (3.52) by $\hat{S}$ on the left and by $\hat{S}^\dagger$ on the right we arrive at Eq. (3.56).

The symmetry conditions for the Floquet scattering matrix, Eq. (3.31), result in symmetry conditions for the matrix $\hat{A}(t,E)$. To derive these we proceed as follows. With parameters from Eq. (3.30) we have for the frozen scattering matrix, $\hat{S}(t,E;H,\{\varphi\}) = \hat{S}(-t,E;H,\{-\varphi\})$. Then from Eq. (3.52) we find, $\hat{A}(t,E;H,\{\varphi\}) = -\hat{A}(-t,E;H,\{-\varphi\})$. In terms of the Fourier coefficients these equations read, $\hat{S}_n(E;H,\{\varphi\}) = \hat{S}_{-n}(E;H,\{-\varphi\})$ and $\hat{A}_n(E;H,\{\varphi\}) = -\hat{A}_{-n}(E;H,\{-\varphi\})$. Finally substituting the sum of Eqs. (3.46) and (3.48) into Eq. (3.31) and taking into account the above relations between the Fourier coefficients we find the following symmetry condition [42]

$$A_{\alpha\beta}(t, E; H, \{\varphi\}) = -A_{\beta\alpha}(t, E; -H, \{\varphi\}). \tag{3.57}$$

The analogous condition for the frozen scattering matrix follows from Eq. (1.29),

$$S_{\alpha\beta}(t, E; H, \{\varphi\}) = S_{\beta\alpha}(t, E; -H, \{\varphi\}). \tag{3.58}$$

Let us consider the case with $H = 0$. The non-diagonal elements of the matrix $\hat{A}$ change sign under the reversal of incoming and outgoing channels, which is a striking difference to the behavior of the non-diagonal elements of the frozen scattering matrix. Therefore, we name the matrix $\hat{A}$ as *the anomalous scattering matrix*. Such a sign reversal results in different probabilities for direct, $\alpha \to \beta$, and reverse, $\beta \to \alpha$, transmission through the dynamic scatterer. The diagonal elements of the anomalous scattering matrix are zero (in the absence of a magnetic field). Therefore, the reflection amplitudes up to terms of order Ω_0 are defined entirely by the frozen scattering matrix. This last circumstance justifies our representation for the elements of the Floquet scattering matrix in Eq. (3.48).

3.5 Beyond the adiabatic approximation

In some simple cases the Floquet scattering matrix can be calculated analytically. Hence, it is convenient to turn to the mixed representation when the scattering matrix depends on energy and time.

3.5.1 *Scattering matrix in mixed energy-time representation*

Let us introduce the following scattering matrix, $\hat{S}_{in}(t, E)$ and $\hat{S}_{out}(E, t)$, in such a way that their Fourier coefficients are related to the Floquet scattering matrix elements as follows [45]

$$\hat{S}_F(E_n, E) = \hat{S}_{in,n}(E) \equiv \int_0^{\mathcal{T}} \frac{dt}{\mathcal{T}} e^{in\Omega_0 t} \hat{S}_{in}(t, E), \tag{3.59a}$$

$$\hat{S}_F(E, E_n) = \hat{S}_{out,-n}(E) \equiv \int_0^{\mathcal{T}} \frac{dt}{\mathcal{T}} e^{-in\Omega_0 t} \hat{S}_{out}(E, t). \quad (3.59b)$$

As we will see later in examples, the elements of the matrix $\hat{S}_{in}(t, E)$ are scattering amplitudes for particles incident with energy E and leaving the scattering region at time t. The dual matrix $\hat{S}_{out}(E, t)$ is composed of the scattering amplitudes for particles incident at time t and leaving the scatterer with energy E. Note this interpretation is consistent with the Heisenberg uncertainty principle. For instance, if the time when an electron leaves a scatterer is defined then its energy is not defined. In this case, in accordance with Eq. (3.59a), an electron's energy can be one of $E_n = E + n\hbar\Omega_0$. Similarly, if an initial time is defined then the initial energy is not. This energy can differ from the energy E of an electron leaving the scatterer. The probability, that an initial energy of an electron incident from the lead β and scattered into the lead α was $E_m = E + m\hbar\Omega_0$, is equal to $|S_{out,\alpha\beta,-m}(E)|^2$.

Substituting the definition for $\hat{S}_{in}$ into Eq. (3.28a) and the definition for $\hat{S}_{out}$ into Eq. (3.28b) and making the inverse transformation we get the following unitarity conditions [42, 45]

$$\int_0^{\mathcal{T}} \frac{dt}{\mathcal{T}} e^{im\Omega_0 t} \hat{S}^{\dagger}_{in}(t, E_m) \hat{S}_{in}(t, E) = \delta_{m,0} \hat{I}, \quad (3.60a)$$

$$\int_0^{\mathcal{T}} \frac{dt}{\mathcal{T}} e^{im\Omega_0 t} \hat{S}_{out}(E_m, t) \hat{S}^{\dagger}_{out}(E, t) = \delta_{m,0} \hat{I}. \quad (3.60b)$$

To the zeroth order in the adiabaticity parameter, $\varpi \to 0$, it follows from Eq. (3.46) that the matrices $\hat{S}_{in}$ and $\hat{S}_{out}$ are the same and they are equal to the frozen scattering matrix. Already in the first order in ϖ these matrices are different. From Eq. (3.48) we can find [42]

$$\hat{S}_{in}(t, E) = \hat{S}(t, E) + \frac{i\hbar}{2} \frac{\partial^2 \hat{S}(t, E)}{\partial t \partial E} + \hbar\Omega_0 \hat{A}(t, E) + \mathcal{O}(\varpi^2), \quad (3.61a)$$

$$\hat{S}_{out}(t, E) = \hat{S}(t, E) - \frac{i\hbar}{2} \frac{\partial^2 \hat{S}(t, E)}{\partial t \partial E} + \hbar\Omega_0 \hat{A}(t, E) + \mathcal{O}(\varpi^2), \quad (3.61b)$$

where $\mathcal{O}(\varpi^2)$ stands for the terms of order ϖ^2 or above.

Despite their difference, the matrices $\hat{S}_{in}$ and $\hat{S}_{out}$ are related due to micro-reversibility. From Eqs. (3.31) and (3.59) [45]

$$\hat{S}_{in}(t, E; H, \{\varphi\}) = \hat{S}_{out}^{T}(E, -t; -H, \{-\varphi\}). \qquad (3.62)$$

Moreover, from Eq. (3.59) one can find

$$\hat{S}_{in,n}(E) = \hat{S}_{out,n}(E_n), \qquad (3.63)$$

which in the time representation reads

$$\hat{S}_{in}(t, E) = \sum_{n=-\infty}^{\infty} \int_0^{\mathcal{T}} \frac{dt'}{\mathcal{T}} e^{in\Omega_0(t'-t)} \hat{S}_{out}(E_n, t'), \qquad (3.64a)$$

$$\hat{S}_{out}(E, t) = \sum_{n=-\infty}^{\infty} \int_0^{\mathcal{T}} \frac{dt'}{\mathcal{T}} e^{-in\Omega_0(t'-t)} \hat{S}_{in}(t', E_n). \qquad (3.64b)$$

For the sake of completeness we give the current in terms of $\hat{S}_{in}$. So we use Eq. (3.59a) in Eq. (3.38) and then in Eq. (3.39) and finally calculate [46]

$$I_\alpha(t) = \frac{e}{h} \int_0^\infty dE \sum_{\beta=1}^{N_r} \sum_{n=-\infty}^{\infty} \{f_\beta(E) - f_\alpha(E_n)\}$$

$$\times \int_0^{\mathcal{T}} \frac{dt'}{\mathcal{T}} e^{in\Omega_0(t-t')} S_{in,\alpha\beta}(t, E) S_{in,\alpha\beta}^*(t', E). \qquad (3.65)$$

We transform this equation to exclude the reference to the periodicity of the driving potential. Hence, we use the following correspondences

$$n\Omega_0 \to \omega,$$

$$\sum_{n=-\infty}^{\infty} \to \frac{\mathcal{T}}{2\pi} \int_{-\infty}^{\infty} d\omega, \qquad (3.66)$$

$$\int_0^{\mathcal{T}} dt' \, e^{in\Omega_0 t'} \to \int_{-\infty}^{\infty} dt' \, e^{i\omega t'},$$

which in fact means changing from the discrete Fourier transformation to the continuous Fourier transformation. After that the current reads

$$I_\alpha(t) = \frac{e}{h} \int dE \, \frac{1}{2\pi} \int_{-\infty}^{\infty} d\omega \sum_{\beta=1}^{N_r} [f_\beta(E) - f_\alpha(E + \hbar\omega)]$$

$$\times \int_{-\infty}^{\infty} dt' e^{i\omega(t-t')} S_{in,\alpha\beta}(t,E) S^*_{in,\alpha\beta}(t',E). \qquad (3.67)$$

Thus we derived an expression that can be used to calculate a time-dependent current in terms of the scattering matrix elements in the case of driving with an arbitrary (not necessarily periodic) dependence on time. In the particular case of a drive with period $\mathcal{T} = 2\pi/\Omega_0$ we have

$$\int_{-\infty}^{\infty} dt' e^{-i\omega t'} S^*_{in,\alpha\beta}(t',E) = \sum_{n=-\infty}^{\infty} 2\pi \delta(\omega - n\Omega_0) S^*_{in,\alpha\beta,n}(E). \qquad (3.68)$$

The use of this equation transforms Eq. (3.67) into Eq. (3.65) as expected.

Further we consider several simple examples and calculate analytically the elements of the scattering matrix $\hat{S}_{in}$.

3.5.2 Dynamic point-like potential

Let us consider a one-dimensional Schrödinger equation,

$$i\hbar \frac{\partial \Psi}{\partial t} = \left\{ -\frac{\hbar^2}{2m} \frac{\partial^2}{\partial x^2} + V(t,x) \right\} \Psi, \qquad (3.69)$$

with point-like potential $V(t, x)$ whose strength oscillates in time,[2]

$$V(t, x) = \delta(x) V(t), \quad V(t) = V_0 + 2V_1 \cos(\Omega_0 t + \varphi). \quad (3.70)$$

According to the Floquet theorem the solution to Eq. (3.69) with a potential periodic in time, Eq. (3.70), is of the following form,

$$\Psi(t, x) = e^{-i\frac{E}{\hbar}t} \sum_{n=-\infty}^{\infty} e^{-in\Omega_0 t} \psi_n(x), \quad (3.71)$$

where $\psi_n(x)$ is a general solution of the corresponding stationary problem. The potential is zero unless $x = 0$. Therefore, for the function $\psi_n(x \neq 0)$ we can take a general solution to the Schrödinger equation for a free particle,

$$\psi_n(x) = \begin{cases} a_n^{(-)} e^{ik_n x} + b_n^{(-)} e^{-ik_n x}, & x < 0, \\ a_n^{(+)} e^{ik_n x} + b_n^{(+)} e^{-ik_n x}, & x > 0, \end{cases} \quad (3.72)$$

with $k_n = \sqrt{2m(E + n\hbar\Omega_0)}/\hbar$.

To match the wave function on the left and on the right for $x = 0$ we use the following. At $x = 0$ the wave function should be continuous. For its derivative we integrate out Eq. (3.69) over an infinitesimal vicinity around the point $x = 0$. We find that the derivative has a jump at this point. Therefore, we have the following boundary conditions

$$\Psi(t, x = -0) = \Psi(t, x = +0),$$
$$\left.\frac{\partial \Psi(t, x)}{\partial x}\right|_{x=+0} - \left.\frac{\partial \Psi(t, x)}{\partial x}\right|_{x=-0} = \frac{2m}{\hbar^2} V(t) \Psi(t, x = 0), \quad (3.73)$$

which connect the unknown coefficients of the wave function, Eq. (3.72), at $x > 0$ and at $x < 0$.

Now we formulate a proper scattering problem, which in particular includes the boundary conditions at $x \to \pm\infty$. The coefficients $a_n^{(-)}$ and $b_n^{(+)}$

---

[2] This problem can also be solved by the continued fractions method [47]. We mention also Ref. [48] where the periodically driven two-dimensional point contact is considered.

in Eq. (3.72) correspond to incident waves, while the coefficients $a_n^{(+)}$ and $b_n^{(-)}$ correspond to outgoing (scattered) waves. So we can write

$$\psi_n(x) = \psi_n^{(in)}(x) + \psi_n^{(out)}(x), \tag{3.74}$$

where

$$\psi_n^{(in)}(x) = \begin{cases} a_n^{(-)} e^{ik_n x}, & x < 0, \\ b_n^{(+)} e^{-ik_n x}, & x > 0, \end{cases} \tag{3.75a}$$

and

$$\psi_n^{(out)}(x) = \begin{cases} b_n^{(-)} e^{-ik_n x}, & x < 0, \\ a_n^{(+)} e^{ik_n x}, & x > 0. \end{cases} \tag{3.75b}$$

Correspondingly the wave function, Eq. (3.71), can be written as the sum $\Psi(t,x) = \Psi^{(in)}(t,x) + \Psi^{(out)}(t,x)$. Note the coefficients $a_n^{(-)}$ and $b_n^{(+)}$ are defined by a given incident wave. In contrast the coefficients $a_n^{(+)}$ and $b_n^{(-)}$ must be calculated.

First we consider scattering of a wave with unit amplitude[3] corresponding to a particle with energy E incident from the left, Fig. 3.1,

$$\Psi_1^{(in)}(t,x) = e^{-i\frac{E}{\hbar}t} \begin{cases} e^{ikx}, & x < 0, \\ 0, & x > 0. \end{cases} \tag{3.76}$$

Comparing this equation with Eqs. (3.71) and (3.75a) we find, $a_{1,n}^{(-)} = \delta_{n0}$ and $b_{1,n}^{(+)} = 0$. To calculate the coefficients $a_{1,n}^{(+)}$, $b_{1,n}^{(-)}$ of the scattered wave $\Psi_1^{(out)}$ we use the boundary conditions, Eq. (3.73), and collect the coefficients having the same dependence on time, $\sim e^{-i\frac{E+n\hbar\Omega_0}{\hbar}t}$. As a result we arrive at the following set of linear equations, $n = 0, \pm 1, \pm 2, \ldots,$

$$\begin{cases} \delta_{n0} + b_{1,n}^{(-)} = a_{1,n}^{(+)}, \\ (k_n + ip_0) a_{1,n}^{(+)} = k\delta_{n0} - i\left(p_{+1} a_{1,n-1}^{(+)} + p_{-1} a_{1,n+1}^{(+)}\right), \end{cases} \tag{3.77}$$

where $p_0 = mV_0/\hbar^2$ and $p_{\pm 1} = mV_1 e^{\mp i\varphi}/\hbar^2$ are the Fourier coefficients for $p(t) = mV(t)/\hbar^2$.

---

[3]This wave is not normalized on a unit flux. Hence, there is a factor $\sqrt{k_n/k} \equiv \sqrt{v_n/v}$ in Eq. (3.78).

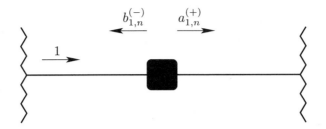

Fig. 3.1 Scattering of a wave with unit amplitude onto the point-like dynamic potential barrier. The arrows show propagation directions. The labels are amplitudes of the corresponding waves: 1 is an amplitude of an incoming wave, $b_{1,n}^{(-)}$ is an amplitude of a reflected wave, $a_{1,n}^{(+)}$ is an amplitude of a transmitted wave. Only a single (nth) component of the Floquet wave function for scattered electrons is shown.

The coefficients $b_{1,n}^{(-)}$ and $a_{1,n}^{(+)}$ define the corresponding Floquet scattering matrix elements for a point-like potential barrier, $\hat{S}_F^{(1)}(E_n, E)$, and, correspondingly, the elements of a matrix $\hat{S}_{in}^{(1)}(E)$,

$$S_{F,11}^{(1)}(E_n, E) = S_{in,11,n}^{(1)}(E) = \sqrt{\frac{k_n}{k}} b_{1,n}^{(-)}, \qquad (3.78a)$$

$$S_{F,21}^{(1)}(E_n, E) = S_{in,21,n}^{(1)}(E) = \sqrt{\frac{k_n}{k}} a_{1,n}^{(+)}. \qquad (3.78b)$$

Here the lower indices 1 and 2 correspond to the left ($x \to -\infty$) and right ($x \to +\infty$) reservoirs, respectively. The square root $\sqrt{k_n/k}$ appeared because the absolute value square of the scattering matrix element is defined as the ratio of the current of scattered particles, $\sim k_n \left|\psi_n^{(out)}\right|^2$, to the current of incident particles, $\sim k \left|\psi_n^{(in)}\right|^2$.

Substituting Eq. (3.78) into Eq. (3.77) we find

$$\begin{cases} \delta_{n0} + S_{in,11,n}^{(1)}(E) = S_{in,21,n}^{(1)}(E), \\[4pt] (k_n + ip_0) S_{in,21,n}^{(1)}(E) = k\delta_{n0} \\[4pt] \quad - ip_{+1}\sqrt{\dfrac{k_n}{k_{n-1}}} S_{in,21,n-1}^{(1)}(E) - ip_{-1}\sqrt{\dfrac{k_n}{k_{n+1}}} S_{in,21,n+1}^{(1)}(E). \end{cases} \qquad (3.79)$$

Let us solve this system of equations with accuracy to the first order in the parameter $\epsilon = \hbar\Omega_0/E$ introduced in Eq. (3.29). Notice in the problem under consideration the energy E is only a characteristic energy. Therefore, in this case the parameter ϵ coincides with the adiabaticity parameter, $\epsilon \sim \varpi$.

To the first order in ϵ we can approximate,

$$k_n = k + \frac{n\Omega_0}{v} + \mathcal{O}\left(\epsilon^2\right), \qquad \sqrt{\frac{k_n}{k_{n\mp 1}}} = 1 \pm \frac{\Omega_0}{2vk} + \mathcal{O}\left(\epsilon^2\right), \qquad (3.80)$$

where $v = \hbar k/m$ is the velocity of an electron with energy E. Using these expansions in Eq. (3.79) and omitting terms of order ϵ^2 we find after the inverse Fourier transformation

$$\begin{cases} 1 + S_{in,11}^{(1)}(t,E) = S_{in,21}^{(1)}(t,E), \\ \{k + ip(t)\} S_{in,21}^{(1)}(t,E) = k - \frac{i}{v}\frac{\partial S_{in,21}^{(1)}(t,E)}{\partial t} + \frac{1}{2vk}\frac{dp(t)}{dt} S_{in,21}^{(1)}(t,E). \end{cases}$$
(3.81)

Because these equations are derived to the first order in $\epsilon \sim \Omega_0$, we can solve them by iterating those terms that have a time derivative. Omitting such terms we get a zero-order solution, i.e., the elements of the frozen scattering matrix,

$$S_{11}^{(1)}(t,E) = \frac{-ip(t)}{k + ip(t)}, \qquad S_{12}^{(1)}(t,E) = \frac{k}{k + ip(t)}. \qquad (3.82)$$

Using this solution in Eq. (3.81) we calculate the elements $\hat{S}_{in}^{(1)}$ up to the first order in ϵ terms [42]

$$S_{in,11}^{(1)}(t,E) = \frac{-ip(t)}{k+ip(t)} - \frac{1}{2v}\frac{dp(t)}{dt}\frac{k-ip(t)}{[k+ip(t)]^3},$$

$$S_{in,21}^{(1)}(t,E) = \frac{k}{k+ip(t)} - \frac{1}{2v}\frac{dp(t)}{dt}\frac{k-ip(t)}{[k+ip(t)]^3}.$$
(3.83)

With Eq. (3.82) we show that

$$\frac{\partial^2 S^{(1)}_{11}(t,E)}{\partial t \partial E} = \frac{\partial^2 S^{(1)}_{21}(t,E)}{\partial t \partial E} = \frac{i}{\hbar v}\frac{dp(t)}{dt}\frac{k - ip(t)}{[k + ip(t)]^3}.$$

Therefore, Eq. (3.83) can be rewritten as

$$S^{(1)}_{in,11}(t,E) = S^{(1)}_{11}(t,E) + \frac{i\hbar}{2}\frac{\partial^2 S^{(1)}_{11}(t,E)}{\partial t \partial E},$$

$$S^{(1)}_{in,21}(t,E) = S^{(1)}_{21}(t,E) + \frac{i\hbar}{2}\frac{\partial^2 S^{(1)}_{21}(t,E)}{\partial t \partial E}. \tag{3.84}$$

Solving the same problem but with a wave incident from the right,

$$\Psi^{(in)}_2(t,x) = e^{-i\frac{E}{\hbar}t}\begin{cases} 0, & x < 0, \\ e^{-ikx}, & x > 0, \end{cases} \tag{3.85}$$

(or just using symmetry) we calculate,

$$S^{(1)}_{22}(t,E) = S^{(1)}_{11}(t,E), \quad S^{(1)}_{12}(t,E) = S^{(1)}_{21}(t,E),$$

$$S^{(1)}_{in,22}(t,E) = S^{(1)}_{in,11}(t,E), \, S^{(1)}_{in,12}(t,E) = S^{(1)}_{in,21}(t,E). \tag{3.86}$$

Thus using Eq. (3.84) we can write down the following relation between the scattering matrix $\hat{S}^{(1)}_{in}(t,E)$ and the frozen scattering matrix $\hat{S}(t,E)$:

$$\hat{S}^{(1)}_{in}(t,E) = \hat{S}^{(1)}(t,E) + \frac{i\hbar}{2}\frac{\partial^2 \hat{S}^{(1)}(t,E)}{\partial t \partial E}, \tag{3.87}$$

with

$$\hat{S}^{(1)}(t,E) = \frac{1}{k + ip(t)}\begin{pmatrix} -ip(t) & k \\ k & -ip(t) \end{pmatrix}. \tag{3.88}$$

Recall that equation (3.87) is derived with accuracy of order ϵ, which, in the case under consideration, is of the same order as the adiabaticity parameter ϖ. Comparing Eqs. (3.87) and (3.61a) we conclude that the anomalous scattering matrix is identically zero for a point-like scatterer,

$$\hat{A}^{(1)}(t,E) = 0. \tag{3.89}$$

Therefore, the dynamic point-like scatterer does not break the symmetry of scattering with respect to a spatial direction reversal inherent to stationary scattering. To break such a symmetry dynamicly it would be necessary to have a scatterer of a finite size that is able to keep an electron for a finite time [49–54] comparable with the period, $\mathcal{T} = 2\pi/\Omega_0$, of drive.

In conclusion we give relations between the coefficients of a scattered wave and the elements of the Floquet scattering matrix in the case with incident waves from both the left and the right,

$$\Psi^{(in)}(t,x) = e^{-i\frac{E}{\hbar}t} \begin{cases} a_0^{(-)} e^{ikx}, & x < 0, \\ b_0^{(+)} e^{-ikx}, & x > 0. \end{cases} \tag{3.90}$$

Because of the superposition principle if the incident wave is $\Psi^{(in)} = a_0^{(-)} \Psi_1^{(in)} + b_0^{(+)} \Psi_2^{(in)}$ then the scattered wave is $\Psi^{(out)} = a_0^{(-)} \Psi_1^{(out)} + b_0^{(+)} \Psi_2^{(out)}$. Using Eqs. (3.78) for the coefficients of $\Psi_1^{(out)}$ and the analogous relations between the coefficients of $\Psi_2^{(out)}$ and $S_{F,2j}^{(1)}(E_n, E)$, $j = 1, 2$, we find the coefficients of the scattered wave,

$$\Psi^{(out)}(t,x) = e^{-i\frac{E}{\hbar}t} \sum_{n=-\infty}^{\infty} e^{-in\Omega_0 t} \begin{cases} b_n^{(-)} e^{-ik_n x}, & x < 0, \\ a_n^{(+)} e^{ik_n x}, & x > 0, \end{cases} \tag{3.91}$$

as follows

$$b_n^{(-)} = \sqrt{\frac{k}{k_n}} S_{F,11}^{(1)}(E_n, E) a_0^{(-)} + \sqrt{\frac{k}{k_n}} S_{F,12}^{(1)}(E_n, E) b_0^{(+)}, \tag{3.92a}$$

$$a_n^{(+)} = \sqrt{\frac{k}{k_n}} S_{F,21}^{(1)}(E_n, E) a_0^{(-)} + \sqrt{\frac{k}{k_n}} S_{F,22}^{(1)}(E_n, E) b_0^{(+)}. \tag{3.92b}$$

Thus if the scattering matrix is known, then the solution to the boundary problem (3.73) with a wave function $\Psi(t,x) = \Psi^{(in)}(t,x) + \Psi^{(out)}(t,x)$, Eqs. (3.90) and (3.91), can be written down using the Floquet scattering matrix elements as given in Eq. (3.92). These equations can be written more compactly if we introduce a column vector, $\hat{\Psi}_0^{(in)}$, for the coefficients of the incident wave with energy E and a column vector, $\hat{\Psi}_n^{(out)}$, for the coefficients of the scattered wave with energy E_n,

$$\hat{\Psi}_0^{(in)} = \begin{pmatrix} a_0^{(-)} \\ b_0^{(+)} \end{pmatrix}, \quad \hat{\Psi}_n^{(out)} = \begin{pmatrix} b_n^{(-)} \\ a_0^{(+)} \end{pmatrix}. \quad (3.93)$$

Then equation (3.92) becomes,

$$\hat{\Psi}_n^{(out)} = \sqrt{\frac{k}{k_n}}\, \hat{S}_F\left(E_n, E\right) \hat{\Psi}_0^{(in)}. \quad (3.94)$$

In the case where the incident wave is also of the Floquet function type having sidebands with different energies E_m,

$$\Psi^{(in)}(t, x) = e^{-i\frac{E}{\hbar}t} \sum_{m=-\infty}^{\infty} e^{-im\Omega_0 t} \begin{cases} a_m^{(-)} e^{ik_m x}, & x < 0, \\ b_m^{(+)} e^{-ik_m x}, & x > 0, \end{cases} \quad (3.95)$$

we introduce corresponding column vectors,

$$\hat{\Psi}_m^{(in)} = \begin{pmatrix} a_m^{(-)} \\ b_m^{(+)} \end{pmatrix}, \quad (3.96)$$

and, using the superposition principle, generalize Eq. (3.94) as follows,

$$\hat{\Psi}_n^{(out)} = \sum_{m=-\infty}^{\infty} \sqrt{\frac{k_m}{k_n}}\, \hat{S}_F\left(E_n, E_m\right) \hat{\Psi}_m^{(in)}. \quad (3.97)$$

This equation is needed for a system comprising a set of point-like dynamic scatterers.

3.5.3 *Dynamic double-barrier potential*

Let the potential $V(t, x)$, in the Schrödinger equation (3.69), consist of two point-like potentials oscillating in time, $V_j(t)$, $j = L, R$, located at a

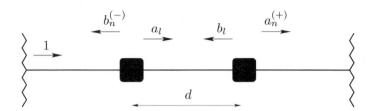

Fig. 3.2 Two dynamic point-like potentials separated by a ballistic wire of length d. The arrows show propagation directions. The labels are amplitudes of the corresponding waves, see Eqs. (3.99), (3.100), and (3.106).

distance d from each other, Fig. 3.2, and a uniform potential $U(t)$ oscillating in time between the first two,[4]

$$V(t,x) = V_L(t)\delta(x) + V_R(t)\delta(x-d) + U(t)\theta(x)\theta(d-x),$$

$$V_j(t) = V_{j,0} + 2V_{j,1}\cos(\Omega_0 t + \varphi_j), \qquad j = L, R, \qquad (3.98)$$

$$U(t) = 2U\cos(\Omega_0 t + \varphi_U),$$

where the Heaviside step function $\theta(x) = 1$ at $x > 0$ and $\theta(x) = 0$ at $x < 0$. Our aim is to calculate the Floquet scattering matrix $\hat{S}_F^{(2)}(E_n, E)$ for such a potential [45].

To calculate the elements $S_{F,11}^{(2)}(E_n, E)$ and $S_{F,21}^{(2)}(E_n, E)$ we consider the scattering problem for a particle with energy E incident from the left. Its wave function is

$$\Psi^{(in)}(t,x) = e^{-i\frac{E}{\hbar}t}\begin{cases} e^{ikx}, & x < 0, \\ 0, & x > 0. \end{cases} \qquad (3.99)$$

The scattered wave is of the Floquet function type,

$$\Psi^{(out)}(t,x) = e^{-i\frac{E}{\hbar}t}\sum_{n=-\infty}^{\infty}e^{-in\Omega_0 t}\begin{cases} b_n^{(-)}e^{-ik_n x}, & x < 0, \\ a_n^{(+)}e^{ik_n x}, & x > d, \end{cases} \qquad (3.100)$$

---

[4]The solution to the similar problem using the method of continued fractions is given in Ref. [40] within the scattering matrix approach and in Ref. [55] within the Green's function formalism.

where the coefficients $b_n^{(-)}$ and $a_n^{(+)}$ define the elements of the Floquet scattering matrix,

$$S_{F,11}^{(2)}(E_n, E) = S_{in,11,n}^{(2)}(E) = \sqrt{\frac{k_n}{k}}\, b_n^{(-)}, \qquad (3.101a)$$

$$S_{F,21}^{(2)}(E_n, E) = S_{in,21,n}^{(2)}(E) = \sqrt{\frac{k_n}{k}}\, a_n^{(+)}\, e^{ik_n d}. \qquad (3.101b)$$

Notice in the case of a finite-sized structure the transmission amplitude includes a factor with a corresponding propagation phase. In our case it is $e^{ik_n d}$.

The wave function inside the scattering region, $0 < x < d$, can also be represented as the Floquet function, Eq. (3.71). To find the corresponding functions $\psi_n(x)$ we take into account the following. In Section 3.1.3 we calculated the general solution to the Schrödinger equation with a uniform oscillating potential, Eq. (3.24). In the one-dimensional case for the potential $U(t)$, Eq. (3.98), it reads as follows,

$$\Psi_E(t,x) = e^{-i\left\{\frac{E}{\hbar}t + \frac{2U}{\hbar\Omega_0}\sin(\Omega_0 t + \varphi_U)\right\}}\left(a_E\, e^{ikx} + b_E\, e^{-ikx}\right), \qquad (3.102)$$

where a_E and b_E are constants (independent of t and x). This wave function corresponds to a particle with energy E and wave number $k = \sqrt{2mE}/\hbar$ in the region with a uniform potential $U(t)$ oscillating in time. We use $\Psi_E(t,x)$ as a basis for calculating the wave function at $0 < x < d$. It should be noted that when interacting with a potential $V_L(t)$ an incident electron can change its initial energy E and, correspondingly, its initial wave number k. In such a case an electron enters the region with potential $U(t)$ having energy $E_l = E + l\hbar\Omega_0$ and wave number k_l. Therefore, the most general solution within the region $0 < x < d$ is the following

$$\Psi^{(mid)}(t,x) = \sum_{l=-\infty}^{\infty} C_l \Psi_{E_l}(t,x). \qquad (3.103)$$

Next from Eq. (3.102) we expand the term,

$$\Upsilon(t) = e^{-i\frac{2U}{\hbar\Omega_0}\sin(\Omega_0 t + \varphi_U)}, \qquad (3.104)$$

into the Fourier series, $\Upsilon(t) = \sum_{q=-\infty}^{\infty} e^{-iq\Omega_0 t} \Upsilon_q$, with

$$\Upsilon_q = J_q\left(\frac{2U}{\hbar\Omega_0}\right) e^{-iq\varphi_U}, \qquad (3.105)$$

(J_q is the Bessel function of the first kind of the qth order). Then collecting together all the terms with the same dependence on time in Eq. (3.103) and introducing the following notation, $a_l = C_l\, a_{E_l}$ and $b_l = C_l\, b_{E_l}$, we finally get the required equation,

$$\Psi^{(mid)}(t,x) = e^{-i\frac{E}{\hbar}t} \sum_{n=-\infty}^{\infty} e^{-in\Omega_0 t}\, \psi_n(x), \qquad (3.106)$$

$$\psi_n(x) = \sum_{l=-\infty}^{\infty} \Upsilon_{n-l}\left(a_l\, e^{ik_l x} + b_l\, e^{-ik_l x}\right), \quad 0 < x < d,$$

which was suggested in Refs. [56, 57].

The sum of Eqs. (3.99), (3.100), and (3.106) determines the electron wave function,

$$\Psi(t,x) = \Psi^{(in)}(t,x) + \Psi^{(out)}(t,x) + \Psi^{(mid)}(t,x), \qquad (3.107)$$

at all points except at $x = 0$ and $x = d$. At these two points we should use boundary conditions similar to those given in Eq. (3.73):

$$\Psi(t, x=-0) = \Psi(t, x=+0),$$

$$\left.\frac{\partial \Psi(t,x)}{\partial x}\right|_{x=+0} - \left.\frac{\partial \Psi(t,x)}{\partial x}\right|_{x=-0} = \frac{2m}{\hbar^2} V_L(t)\Psi(t, x=0), \qquad (3.108)$$

$$\Psi(t, x=d-0) = \Psi(t, x=d+0),$$

$$\left.\frac{\partial \Psi(t,x)}{\partial x}\right|_{x=d+0} - \left.\frac{\partial \Psi(t,x)}{\partial x}\right|_{x=d-0} = \frac{2m}{\hbar^2} V_R(t)\Psi(t, x=d). \qquad (3.109)$$

Collecting terms having the same dependence on time we obtain an infinite system of equations for coefficients $b_n^{(-)}$, $a_n^{(+)}$, a_l, and b_l.

The same system of equations can be derived in another way with the help of scattering matrices for the relevant potentials. We designate as $\hat{L}_F$ the Floquet scattering matrix for a potential $V_L(t)$. Correspondingly, $\hat{R}_F$ is the Floquet scattering matrix for a potential $V_R(t)$. Further reasoning is quite analogous to what we used deriving Eq. (3.97) from the boundary conditions given in Eq. (3.73).

First we consider Eq. (3.108). Near $x = 0$ the wave function can be represented as follows, $\Psi(t,x) = \Psi_L^{(in)}(t,x) + \Psi_L^{(out)}(t,x)$, where $\Psi_L^{(in)}(t,x)$ corresponds to a wave incident to the barrier $V_L(t)$, while $\Psi_L^{(out)}(t,x)$ corresponds to a wave scattered by it. From Eqs. (3.99), (3.100), and (3.106) we find

$$\Psi_L^{(in)}(t,x) = e^{-i\frac{E}{\hbar}t} \sum_{n=-\infty}^{\infty} e^{-in\Omega_0 t} \begin{cases} \delta_{n0} e^{ikx}, & x < 0, \\ \sum_{l=-\infty}^{\infty} \Upsilon_{n-l} b_l e^{-ik_l x}, & x > 0, \end{cases} \quad (3.110)$$

$$\Psi_L^{(out)}(t,x) = e^{-i\frac{E}{\hbar}t} \sum_{n=-\infty}^{\infty} e^{-in\Omega_0 t} \begin{cases} b_n^{(-)} e^{-ik_n x}, & x < 0, \\ \sum_{l=-\infty}^{\infty} \Upsilon_{n-l} a_l e^{ik_l x}, & x > 0. \end{cases} \quad (3.111)$$

Collecting all the wave function amplitudes corresponding to the same energy E_n into the column vectors,

$$\hat{\Psi}_{Ln}^{(in)} = \begin{pmatrix} \delta_{n0} \\ \sum_{l=-\infty}^{\infty} \Upsilon_{n-l} b_l \end{pmatrix}, \quad \hat{\Psi}_{Ln}^{(out)} = \begin{pmatrix} b_n^{(-)} \\ \sum_{l=-\infty}^{\infty} \Upsilon_{n-l} a_l \end{pmatrix}, \quad (3.112)$$

and using Eq. (3.97) we obtain the following matrix equation

$$\begin{pmatrix} b_n^{(-)} \\ \sum_{l=-\infty}^{\infty} \Upsilon_{n-l} a_l \end{pmatrix} = \sum_{m=-\infty}^{\infty} \sqrt{\frac{k_m}{k_n}} \hat{L}_F(E_n, E_m) \begin{pmatrix} \delta_{m0} \\ \sum_{l=-\infty}^{\infty} \Upsilon_{m-l} b_l \end{pmatrix}, \quad (3.113)$$

which is completely equivalent to the boundary conditions given in Eq. (3.108) for the wave function given in Eq. (3.107).

The second pair of boundary conditions, Eq. (3.109), relate the coefficients of the wave function, Eq. (3.107), at $x = d$. Near this point the incident, $\Psi_R^{(in)}(t,x)$, and scattered, $\Psi_R^{(out)}(t,x)$, waves are

$$\Psi_R^{(in)}(t,x) = e^{-i\frac{E}{\hbar}t} \sum_{n=-\infty}^{\infty} e^{-in\Omega_0 t} \begin{cases} \sum_{l=-\infty}^{\infty} \Upsilon_{n-l}\, a_l\, e^{ik_l x}, & x < d, \\ 0, & x > d, \end{cases} \quad (3.114)$$

$$\Psi_R^{(out)}(t,x) = e^{-i\frac{E}{\hbar}t} \sum_{n=-\infty}^{\infty} e^{-in\Omega_0 t} \begin{cases} \sum_{l=-\infty}^{\infty} \Upsilon_{n-l}\, b_l\, e^{-ik_l x}, & x < d, \\ a_n^{(+)} e^{ik_n x}, & x > d. \end{cases} \quad (3.115)$$

The corresponding column vectors are

$$\hat{\Psi}_{Rn}^{(in)} = \begin{pmatrix} \sum_{l=-\infty}^{\infty} \Upsilon_{n-l}\, a_l\, e^{ik_l d} \\ 0 \end{pmatrix}, \quad \hat{\Psi}_{Rn}^{(out)} = \begin{pmatrix} \sum_{l=-\infty}^{\infty} \Upsilon_{n-l}\, b_l\, e^{-ik_l d} \\ a_n^{(+)} e^{ik_n d} \end{pmatrix}. \quad (3.116)$$

Applying Eq. (3.97) to the right point-like potential, we get the equation,

$$\begin{pmatrix} \sum_{l=-\infty}^{\infty} \Upsilon_{n-l}\, b_l\, e^{-ik_l d} \\ a_n^{(+)} e^{ik_n d} \end{pmatrix} = \sum_{m=-\infty}^{\infty} \sqrt{\frac{k_m}{k_n}}\, \hat{R}_F(E_n, E_m) \begin{pmatrix} \sum_{l=-\infty}^{\infty} \Upsilon_{m-l}\, a_l\, e^{ik_l d} \\ 0 \end{pmatrix}, \quad (3.117)$$

which is equivalent to Eq. (3.109) for the wave function given in Eq. (3.107).

Let us solve the system of equations (3.113) and (3.117) with accuracy of the zeroth order in the parameter $\epsilon = \hbar\Omega_0/E \ll 1$, Eq. (3.29). Notice, in contrast to the case with a point-like scatterer, when the adiabaticity parameter ϖ coincides with a parameter ϵ, in the case of a finite-sized scatterer, whose length d is much larger than the de Broglie wavelength, $\lambda_E = h/\sqrt{2mE}$, of an electron with energy E, the adiabaticity parameter is large compared to ϵ: $\varpi \sim \epsilon d/\lambda_E \gg \epsilon$. This fact allows us to analyze both the adiabatic, $\varpi \ll 1$, and non-adiabatic, $\varpi \gg 1$, regimes within the approach used.

So, to the zeroth order in ϵ we write

$$\frac{k_m}{k_n} = 1 + \mathcal{O}(\epsilon),$$

$$e^{\pm ik_l d} = e^{\pm ikd} e^{\pm il\Omega_0 \tau [1+\mathcal{O}(\epsilon)]},$$
(3.118)

where $\tau = L/v$ is the time of flight between the barriers for an electron with energy E. Further simplification is achieved using the following. As we showed earlier, the Floquet scattering matrix elements for a point-like barrier to the zeroth order in ϵ are the Fourier coefficients for the frozen scattering matrix, see Eqs. (3.78) and (3.82). Designating the frozen scattering matrices for the left and right barriers as $\hat{L}(t, E)$ and $\hat{R}(t, E)$, respectively, we have

$$\hat{X}_F(E_n, E_m) = \hat{X}_{n-m}(E) + \mathcal{O}(\epsilon), \qquad X = L, R.$$
(3.119)

Then using Eqs. (3.118), (3.119), and (3.101) we can rewrite the system of equations (3.113) and (3.117) in the following way,

$$\begin{pmatrix} S^{(2)}_{in,11,n}(E) \\ \sum_{l=-\infty}^{\infty} \Upsilon_{n-l} a_l \end{pmatrix} = \sum_{m=-\infty}^{\infty} \hat{L}_{n-m}(E) \begin{pmatrix} \delta_{m0} \\ \sum_{l=-\infty}^{\infty} \Upsilon_{m-l} b_l \end{pmatrix},$$

$$\begin{pmatrix} e^{-ikd} \sum_{l=-\infty}^{\infty} \Upsilon_{n-l} b_l e^{-il\Omega_0 \tau} \\ S^{(2)}_{in,21,n}(E) \end{pmatrix} = \sum_{m=-\infty}^{\infty} \hat{R}_{n-m}(E) \begin{pmatrix} e^{ikd} \sum_{l=-\infty}^{\infty} \Upsilon_{m-l} a_l e^{il\Omega_0 \tau} \\ 0 \end{pmatrix}.$$
(3.120)

Next we use the following trick. We assume that the quantities a_l and b_l are the Fourier coefficients for some functions $a(t) = a(t + \mathcal{T})$ and $b(t) = b(t + \mathcal{T})$ that are periodic in time. With these functions we can apply the inverse Fourier transformation to Eqs. (3.120) and calculate

$$\begin{pmatrix} S^{(2)}_{in,11}(t,E) \\ \Upsilon(t)\,a(t) \end{pmatrix} = \hat{L}(t,E) \begin{pmatrix} 1 \\ \Upsilon(t)\,b(t) \end{pmatrix},$$

$$\begin{pmatrix} e^{-ikd}\Upsilon(t)\,b(t+\tau) \\ S^{(2)}_{in,21}(t,E) \end{pmatrix} = \hat{R}(t,E) \begin{pmatrix} e^{ikd}\Upsilon(t)\,a(t-\tau) \\ 0 \end{pmatrix}, \tag{3.121}$$

where we took into account that the quantities $b_l\, e^{-il\Omega_0 \tau}$ and $a_l\, e^{il\Omega_0 \tau}$ are the Fourier coefficients for $b(t+\tau)$ and $a(t-\tau)$, respectively. It is easy to check. For instance,

$$[b(t+\tau)]_l = \int_0^{\mathcal{T}} \frac{dt}{\mathcal{T}} e^{il\Omega_0 t} b(t+\tau) = \int_0^{\mathcal{T}} \frac{dt'}{\mathcal{T}} e^{il\Omega_0(t'-\tau)} b(t') = b_l\, e^{-il\Omega_0 \tau}.$$

Note the system of equations (3.121) contains only four equations, while initially we had an infinite system of equations, Eq. (3.120). The first and the fourth equations in (3.121) define the quantities of interest, $S^{(2)}_{in,11}(t,E)$ and $S^{(2)}_{in,21}(t,E)$, while the second and the third equations allow us to calculate $a(t)$ and $b(t)$. Substituting the third equation into the second one we get the following (for brevity we omit E)

$$a(t) = \Upsilon^*(t)\,L_{21}(t) + e^{i2kL} L_{22}(t)\, R_{11}(t-\tau)\, a(t-2\tau). \tag{3.122}$$

In addition here we used $\Upsilon^{-1}(t) = \Upsilon^*(t)$, because $|\Upsilon(t)|^2 = 1$ for the function $\Upsilon(t)$ introduced in Eq. (3.104). Because the absolute value of the quantities in Eq. (3.122) is less than unity, we can write down the solution for this equation as the following series

$$a(t) = \sum_{q=0}^{\infty} e^{i2qkd} \lambda^{(q)}(t)\, \Upsilon^*(t-2q\tau)\, L_{21}(t-2q\tau), \tag{3.123}$$

$$\lambda^{(q>0)}(t) = \prod_{j=0}^{q-1} L_{22}(t-2j\tau)\, R_{11}(t-[2j+1]\tau),$$

$$\lambda^{(0)}(t) = 1.$$

This series can be solved if we consider the second term on the right hand side of Eq. (3.122) as a perturbation and sum up the terms in all the orders of the perturbation theory.

Using Eq. (3.123) in Eq. (3.121) we calculate $b(t)$ and then the Floquet scattering matrix elements [45]

$$S_{in,\alpha 1}^{(2)}(t,E) = \sum_{q=0}^{\infty} e^{i 2 q_{\alpha 1} k d} S_{\alpha 1}^{(q)}(t,E), \quad \alpha = 1, 2, \quad (3.124)$$

where $2q_{\alpha 1} = 2q + 1 - \delta_{\alpha 1}$ and

$$S_{\alpha 1}^{(q)}(t,E) = e^{-i\Phi_{q_{\alpha\beta}}} \sigma_{\alpha 1}^{(q)}(t,E), \quad (3.125)$$

$$\Phi_{q_{\alpha 1}} = \frac{1}{\hbar} \int_{t-2q_{\alpha 1}\tau}^{t} dt' \, U(t'). \quad (3.126)$$

$$\sigma_{11}^{(0)}(t) = L_{11}(t),$$
$$\sigma_{11}^{(q>0)}(t) = L_{12}(t) R_{11}(t-\tau) L_{21}(t-2q\tau) \lambda^{(q-1)}(t-2\tau), \quad (3.127)$$

$$\sigma_{21}^{(q)}(t) = R_{21}(t) L_{21}(t - [2q+1]\tau) \lambda^{(q)}(t-\tau). \quad (3.128)$$

For brevity in Eqs. (3.127) and (3.128) we do not show the argument E. Note the time-dependent phase factor in Eq. (3.125) can be written as $e^{-i\Phi_{q_{\alpha 1}}} = \Upsilon(t) \Upsilon^*(t - 2q_{\alpha 1}\tau)$.

Let us analyze Eq. (3.124). The scattering matrix element $\hat{S}_{in,\alpha 1}^{(2)}(t,E)$ is the sum of the partial amplitudes, $e^{i 2 q_{\alpha 1} k d} S_{in,\alpha 1}^{(q)}(t,E)$. Each amplitude corresponds to some path $\mathcal{L}_{\alpha 1}^{(q)}$ inside the scattering region. An electron with energy E enters the system through the lead 1, follows this path undergoing $2q_{\alpha 1} - 1$ reflections, and leaves the system through the lead α at a time moment t. The trajectory $\mathcal{L}_{\alpha 1}^{(q)}$ consists of $2q_{\alpha 1}$ segments of length d. The partial scattering amplitude $e^{i 2 q_{\alpha 1} k d} S_{in,\alpha 1}^{(q)}(t,E)$ is the product of a number of amplitudes $L_{\alpha\alpha}$ and $R_{\alpha\alpha}$, corresponding to an instantaneous reflection from the point-like barriers, amplitudes $L_{\alpha\neq\beta}$ and $R_{\alpha\neq\beta}$, corresponding to an instantaneous tunneling through the point-like barriers, and amplitudes $e^{i\left\{kd-\hbar^{-1}\int_{t_j-\tau}^{t_j}dt'U(t')\right\}}$, and corresponding to a propagation (starting at time $t_j - \tau$ and lasting a time period $\tau = d/v$) between the two barriers in a uniform oscillating potential $U(t)$. The times, $t_j = t - j\tau$, at which the instantaneous reflection/transmission amplitudes are calculated, are counted backwards along the path $\mathcal{L}_{\alpha 1}^{(q)}$ in a descending order starting from the moment t when the particle leaves the system through the left (for $\alpha = 1$) or right (for $\alpha = 2$) barrier.

Because $\mathcal{S}_{\alpha 1}^{(q)}(t, E)$ depends on scattering amplitudes calculated at different times, the scattering matrix $\hat{S}_{in}^{(2)}(t, E)$ is non-local in time, in contrast to the (local in time) frozen scattering matrix $\hat{S}(t, E)$. This non-locality arises as a consequence of a finite (minimal) time τ spent by an electron inside the scattering region. If the period $\mathcal{T}$ becomes as small as τ the system enters a non-adiabatic scattering regime. Therefore, a natural adiabaticity parameter for the system under consideration is the product $\varpi_0 = \Omega_0 \tau/(2\pi)$.

To calculate the remaining elements $S_{F,\alpha 2}^{(2)}(E_n, E)$, $\alpha = 1, 2$, and, correspondingly, $S_{in,\alpha 2,n}^{(2)}$, we have to consider the scattering of an electron with energy E incident from the right. Then the corresponding elements of the scattering matrix $\hat{S}_{in}^{(2)}(t, E)$ are given by equations analogous to Eqs. (3.124)–(3.126) with

$$\sigma_{12}^{(q)} = L_{12}(t)\, R_{12}(t - [2q+1]\tau)\, \rho^{(q)}(t - \tau), \tag{3.129}$$

$$\sigma_{22}^{(0)} = R_{22}(t),$$

$$\sigma_{22}^{(q>0)} = R_{21}(t)\, L_{22}(t-\tau)\, R_{12}(t - 2q\tau)\, \rho^{(q-1)}(t - 2\tau). \tag{3.130}$$

Here the quantity $\rho^{(q)}(t)$ is

$$\rho^{(q>0)}(t) = \prod_{j=0}^{q-1} R_{11}(t - 2j\tau)\, L_{22}(t - [2j+1]\tau),$$

$$\rho^{(0)} = 1. \tag{3.131}$$

Thus, we have calculated the scattering matrix,

$$\hat{S}_{in}^{(2)}(t, E) = \sum_{q=0}^{\infty} e^{i2q\alpha_1 kd}\hat{S}^{(q)}(t, E), \qquad (3.132)$$

allowing a description of the transport through the dynamic double-barrier in adiabatic as in non-adiabatic regimes.

3.5.3.1 *Adiabatic approximation*

Let us consider the limit, $\varpi \to 0$, and calculate the anomalous scattering matrix, see Eq. (3.48), for the double-barrier structure. We denote it by $\hat{A}^{(2)}(t, E)$. Recall that this matrix is responsible for the chiral asymmetry of scattering at slow driving.

To the zeroth order in ϖ the matrix $\hat{S}_{in}^{(2)}(t, E)$ coincides with the frozen scattering matrix, which we denote as $\hat{S}^{(2)}(t, E)$ for the double-barrier under consideration. To calculate it we use Eq. (3.132) where we ignore any change in the quantities during a time period τ. Then in Eqs. (3.123), (3.127)–(3.131) all the quantities are calculated at a time t, while equation (3.126) (for $\beta = 1$) and an analogous one for $\beta = 2$, becomes

$$\Phi_{q_{\alpha\beta}} \approx U(t)\tau\hbar^{-1}(2q + 1 - \delta_{\alpha\beta}).$$

As a result we get

$$S_{\alpha\beta}^{(2)}(t, E) = \sum_{q=0}^{\infty} \bar{S}_{\alpha\beta}^{(q)}(t, E),$$

$$\bar{S}_{\alpha\beta}^{(q)}(t, E) = e^{i(kd - U(t)\tau/\hbar)(2q+1-\delta_{\alpha\beta})}\bar{\sigma}_{\alpha\beta}^{(q)}(t, E). \qquad (3.133)$$

Here the elements of the matrix $\hat{\bar{\sigma}}^{(q)}(t, E)$ are given in Eqs. (3.127)–(3.130) where we put $\tau = 0$.

To calculate the matrix $\hat{A}^{(2)}(t, E)$ we calculate $\hat{S}_{in}^{(2)}(t, E)$ in the first order in ϖ. To this end we expand the right hand side of Eq. (3.132) up to the linear in τ terms. Then we use Eq. (3.133) for the frozen scattering matrix and Eq. (3.61a) to extract the anomalous scattering matrix. Calculating the time and energy derivatives we take into account the following. The frozen matrix $\hat{S}^{(2)}$ depends on time via the potential $U(t)$ and the matrices

$\hat{L}(t)$ and $\hat{R}(t)$. The energy dependence of $\hat{S}^{(2)}$, within the approximations used, Eqs. (3.118) and (3.119), is defined by the phase factor e^{2iqkd} only.[5] Then after simple algebra we find

$$\hbar\Omega A^{(2)}_{\alpha\beta}(t,E) = \sum_{q=0}^{\infty} \bar{S}^{(q)}_{\alpha\beta}(t,E) \mathcal{A}^{(q)}_{\alpha\beta}(t,\mu), \qquad (3.134\text{a})$$

where

$$\mathcal{A}^{(q)}_{11} = \tau_0 q \frac{\partial}{\partial t} \ln\left(\frac{L_{12}}{L_{21}}\right), \qquad (3.134\text{b})$$

$$\mathcal{A}^{(q)}_{21} = -\frac{\tau_0(2q+1)}{2} \frac{\partial}{\partial t} \ln\left(\frac{L_{21}}{R_{21}}\right) - \frac{\tau_0 q}{2} \frac{\partial}{\partial t} \ln\left(\frac{R_{11}}{L_{22}}\right), \qquad (3.134\text{c})$$

$$\mathcal{A}^{(q)}_{12} = -\frac{\tau_0(2q+1)}{2} \frac{\partial}{\partial t} \ln\left(\frac{R_{12}}{L_{12}}\right) - \frac{\tau_0 q}{2} \frac{\partial}{\partial t} \ln\left(\frac{L_{22}}{R_{11}}\right), \qquad (3.134\text{d})$$

$$\mathcal{A}^{(q)}_{22} = \tau_0 q \frac{\partial}{\partial t} \ln\left(\frac{R_{21}}{R_{12}}\right). \qquad (3.134\text{e})$$

The equations above show that the anomalous scattering matrix $\hat{A}^{(2)}$ possesses symmetry properties with respect to interchange of lead indices, which are different from those of the frozen scattering [42]. The symmetry of the $\hat{A}^{(2)}$ matrix depends on differences between the matrix elements of the $\hat{L}$ and $\hat{R}$ matrices. The main point is that the symmetry of the anomalous scattering matrix is fundamentally different from that of the frozen scattering matrix.

3.5.4 Unitarity and the sum over trajectories

The scattering matrix elements for any structure comprising point-like scatterers connected via ballistic segments can be represented as the sum over trajectories similar to Eq. (3.132).[6] On the other hand, as we saw, the use of the unitarity conditions allows us to simplify the calculations. Therefore, it seems to be useful to formulate the unitarity conditions directly in

---

[5]The energy dependence of the scattering matrices $\hat{L}$ and $\hat{R}$ results in corrections of order ϵ, which we ignore.

[6]The representation of the scattering amplitude as the sum over trajectories is well known from the semiclassical approach, see, e.g., Refs. [58, 59]. For the quantum pumping problem it was used in Refs. [60, 61].

terms of the partial scattering amplitudes, $S^{(q)}_{\alpha\beta}(t, E)$, corresponding to the propagation of an electron along one of trajectories.

So, we substitute Eq. (3.132) into Eq. (3.28b) and use the inverse Fourier transformation. Then we use the expansion given in Eq. (3.118) and get

$$\sum_{q=0}^{\infty} \hat{S}^{(q)}(t, E)\hat{S}^{(q)\dagger}(t, E)$$

$$+ \sum_{p=0}^{\infty}\sum_{s=1}^{\infty} e^{-2iskd} \hat{S}^{(p)}(t, E)\hat{S}^{(p+s)\dagger}(t + 2\tau s, E) \qquad (3.135)$$

$$+ \sum_{q=0}^{\infty}\sum_{s=1}^{\infty} e^{2iskd} \hat{S}^{(q+s)}(t, E)\hat{S}^{(q)\dagger}(t - 2s\tau, E) = \hat{I}.$$

This identity should hold at any energy E.

Note within the approximation used the quantities $\hat{S}^{(q)}$ should be kept as energy independent on the scale over which the phase kd changes by 2π. In such a case Eq. (3.135) can be considered as the Fourier expansion for the unit matrix $\hat{I}$ in the basis of plane waves, e^{2ilkd}, $l = 0, \pm 1, \pm 2, \ldots$ Expanding the right hand side of Eq. (3.135) into this basis and calculating the corresponding Fourier coefficients we arrive at the following equations [45]

$$\sum_{q=0}^{\infty} \hat{S}^{(q,\tau)}(t, E)\hat{S}^{(q,\tau)\dagger}(t, E) = \hat{I}, \qquad (3.136a)$$

$$\sum_{p=0}^{\infty} \hat{S}^{(p,\tau)}(t, E)\hat{S}^{(p+s,\tau)\dagger}(t + 2\tau s, E) = \hat{0}, \qquad (3.136b)$$

$$\sum_{q=0}^{\infty} \hat{S}^{(q+s,\tau)}(t, E)\hat{S}^{(q,\tau)\dagger}(t - 2\tau s, E) = \hat{0}, \qquad (3.136c)$$

where $\hat{0}$ is a zero matrix.

We stress that compared to Eq. (3.60a) the equations given above are less general, because they rely essentially on the expansion (3.132), where the matrices $\hat{S}^{(q)}$ are energy independent over the scale of order $\hbar\Omega_0$.

3.5.5 Current and the sum over trajectories

Let us use Eq. (3.65) and calculate the current generated by the dynamic double-barrier structure connected to the reservoirs having the same potential, $\mu_\alpha = \mu$, and temperature, $T_\alpha = T$, hence $f_\alpha(E) = f_0(E)$, $\alpha = 1, 2$.

We substitute Eq. (3.132) into Eq. (3.65) and simplify it. We assume that both the energy quantum $\hbar\Omega_0$ and the temperature are small compared to the Fermi energy,

$$\hbar\Omega_0, \, k_B T \ll \mu. \tag{3.137}$$

Then to integrate over energy in Eq. (3.65) we use the following expansion, $kd \approx k_\mu d + (E - \mu)/(\hbar\tau_\mu^{-1})$, where the lower index μ indicates that the corresponding quantity is evaluated at the Fermi energy. Within this accuracy we can treat the matrices $\hat{S}^{(q)}$ as energy independent over the relevant energy window and evaluate them at $E = \mu$. The latter simplification is correct because the elements of the scattering matrices $\hat{L}$ and $\hat{R}$ defining the elements of the matrix $\hat{S}^{(q)}$ change significantly only if the energy $E \sim \mu$ changes by a quantity of order μ. Therefore, they can be kept as constant while integrating over energy over the window of the order of $\max(\hbar\Omega_0, k_B T) \ll \mu$.

Using the simplifications introduced above we can integrate over energy in Eq. (3.65) and represent the time-dependent current,

$$I_\alpha(t) = I_\alpha^{(d)}(t) + I_\alpha^{(nd)}(t), \tag{3.138a}$$

as the sum of diagonal, $I_\alpha^{(d)}(t)$, and non-diagonal, $I_\alpha^{(nd)}(t)$, contributions [45, 62].

3.5.5.1 Temperature-independent contribution to generated current

The diagonal part comprises contributions of different dynamic scattering channels, which can be labeled by the index q dependent on the number of reflections (equal to $2q - \delta_{\alpha\beta}$ for $q > 0$) experienced by an electron propagating through the system [45]

$$I_\alpha^{(d)}(t) = \sum_{q=0}^{\infty} J_\alpha^{(q)}(t), \tag{3.138b}$$

where the contribution of the qth dynamic scattering channel is

$$J_\alpha^{(q)}(t) = -i\frac{e}{2\pi}\left(\hat{S}^{(q)}(t,\mu)\frac{\partial \hat{S}^{(q)\dagger}(t,\mu)}{\partial t}\right)_{\alpha\alpha}. \qquad (3.138c)$$

Apparently the contribution $I_\alpha^{(d)}(t)$ is independent of the temperature.[7]

3.5.5.2 Contribution to generated current dependent on temperature

The non-diagonal contribution to the current is the sum of temperature-dependent non-diagonal contributions in dynamic scattering channels [45]

$$I_\alpha^{(nd)}(t) = \sum_{p=0}^{\infty}\sum_{\substack{q=0\\q\neq p}}^{\infty} e^{i2(p-q)k_\mu d}\,\eta\left(\frac{[p-q]T}{T^*}\right) J_\alpha^{(p,q)}(t), \qquad (3.138d)$$

with

$$J_\alpha^{(p,q)}(t) = -i\frac{e}{2\pi}\left(\hat{S}^{(p)}(t,\mu)\frac{\hat{S}^{(q)\dagger}(t,\mu) - \hat{S}^{(q)\dagger}(t - 2\tau_\mu[p-q],\mu)}{2\tau_\mu[p-q]}\right)_{\alpha\alpha}. \qquad (3.138e)$$

Here $\eta(x) = x/\sinh(x)$, where $x = |p-q|T/T^*$ and $k_B T^* = \hbar/(2\pi\tau_\mu)$.

The factor $\eta\left(|p-q|T/T^*\right)$ describes the effect of averaging over the energies of incident electrons within the temperature widening of the edge of the Fermi distribution function. The time of flight, $\tau_\mu = d/v_\mu$, (for an electron with Fermi energy) between the barriers plays a twofold role. On one hand, it separates adiabatic, $\mathcal{T} \gg \tau_\mu$, and non-adiabatic, $\mathcal{T} \leq \tau_\mu$, regimes. On the other hand, it defines the crossover temperature, T^*, separating low-temperature and high-temperature regimes. At low temperatures, $T \ll T^*$, the factor $\eta = 1$. While at relatively high temperatures, $T \gg T^*$, this factor is small, $\eta\left(|p-q|T/T^*\right) \approx 2|p-q|(T/T^*)e^{-|p-q|T/T^*}$. Therefore, at high temperatures the non-diagonal current, $I_\alpha^{(nd)}(t)$, is exponentially suppressed. Note the temperature effect we are discussing here

---

[7]The direct current generated by the dynamic scatterer at different temperatures was calculated numerically in Refs. [63, 64].

is due to averaging over the energy of incident electrons[8] and has nothing to do with inelastic (or other) processes destroying the phase coherence.

The unitarity conditions, Eq. (3.136), allows us to simplify $I_\alpha^{(nd)}(t)$ and to show that both the diagonal contribution and the non-diagonal contribution are real. So, taking a time derivative of Eq. (3.136a) we conclude that Eq. (3.138b) is real. Note each term $J_\alpha^{(q)}(t)$ in Eq. (3.138b) in general is not real, only their sum is necessarily real. Therefore, the interpretation of the quantity $J_\alpha^{(q)}(t)$ as a contribution of the qth dynamic scattering channel into the current $I_\alpha^{(d)}(t)$ is correct only in the case when $J_\alpha^{(q)}(t)$ is real.

To show that Eq. (3.138d) is real we first simplify it. From Eqs. (3.136b) and (3.136c) it follows that the product of the scattering matrix elements corresponding to electrons leaving the scatterer at different times, t and $t - 2\tau_\mu[p-q]$, drops out from Eq. (3.138d). Then the non-diagonal contribution is reduced to

$$I_\alpha^{(nd)}(t) = \frac{e}{2\pi\tau_\mu} \Im \sum_{s=1}^{\infty} e^{i2sk_\mu d} \frac{\eta\left(\frac{sT}{T^*}\right)}{s} \mathcal{C}_\alpha^{(s)}(t,\mu),$$

(3.139)

$$\mathcal{C}_\alpha^{(s)}(t,\mu) = \sum_{q=0}^{\infty} \left(\hat{S}^{(q+s)}(t,\mu)\hat{S}^{(q)\dagger}(t,\mu)\right)_{\alpha\alpha}.$$

Here the quantity $\mathcal{C}_\alpha^{(s)}$ is the sum of the interference contributions from all the pairs of photon-assisted amplitudes corresponding to trajectories with the same length difference $2sd$. Note this length difference enters the phase factor $e^{i2sk_\mu d}$. All these amplitudes correspond to electrons leaving the scatterer at the time t when the current $I_\alpha^{(nd)}(t)$ is calculated.

3.5.5.3 Nature of two contributions to generated current

The two parts, $I_\alpha^{(d)}(t)$ and $I_\alpha^{(nd)}(t)$, of a generated current result from different processes that leads to different temperature dependencies. The first part, $I_\alpha^{(d)}(t)$, is the sum of the contributions, $J_{\alpha\beta}^{(q)}$, arising from different electron paths $\mathcal{L}_{\alpha\beta}^{(q)}$ inside the system. These paths differ by incoming (β) and outgoing (α) leads, and by the index q counting the number of reflections inside the system. Therefore, one can consider the contribution $J_{\alpha\beta}^{(q)}$ as due

---

[8]The temperature T^* is known as the crossover temperature in the persistent current problem [29, 65], and it appeared in the problem of stationary transport in ballistic mesoscopic structures with interference [66–68].

to photon-assisted interference processes taking place within the same spatial path $\mathcal{L}_{\alpha\beta}^{(q)}$. Each such path is characterized by a delay time $2q_{\alpha\beta}\tau$, i.e., the difference of times when an electron leaves and enters the system. If this time is not small compared with the driving period, $\mathcal{T}$, then dynamic effects become important for an electron scattering off the system. Therefore, one can consider the path $\mathcal{L}_{\alpha\beta}^{(q)}$ as an effective *dynamic scattering channel*. Then we interpret $J_\alpha^{(q)}$ as arising due to intra-channel photon-assisted interference processes. Because all the quantum-mechanical amplitudes corresponding to such processes are multiplied by the same dynamic factor $e^{2iq_{\alpha\beta}kd}$, the corresponding probability is independent of energy. Consequently the integration of the energy becomes trivial.

In contrast, the second part, $I_\alpha^{(nd)}(t)$, due to interference between different paths (i.e., due to inter-channel interference) is defined as the sum of terms oscillating in energy. Consequently it vanishes at high temperatures.

From Eq. (3.138) it follows that when increasing either the temperature or the drive frequency the different dynamic scattering channels contribute independently to the generated current, $I_\alpha(t) \approx I_\alpha^{(d)}(t)$. With regard to the temperature such a conclusion is evident because $I_\alpha^{(d)}(t)$ is temperature-independent while $I_\alpha^{(nd)}(t)$ is exponentially suppressed at $T > T^*$. With regard to the frequency this follows from the observation that the ratio $I_\alpha^{(d)}(t)/I_\alpha^{(nd)}(t)$ behaves as $\Omega\tau_0$. Therefore, at $\Omega \to \infty$ the contribution $I_\alpha^{(d)}(t)$ dominates.

We emphasize that the current $I_\alpha^{(d)}(t)$ cannot be considered as a classical part of the generated current $I_\alpha(t)$. This part is due to interference, therefore, it has a quantum-mechanical nature. However, it is due to interference taking pace within the same spatial trajectory, therefore this current is insensitive to energy averaging and, correspondingly, it is temperature-independent.

Chapter 4

Direct current generated by the dynamic scatterer

A current generated by a dynamic scatterer [43] has a DC component under some conditions [69]. In other words, an excitation, which is periodic in time, of a mesoscopic scatterer can result in a direct current even in the absence of a bias between the reservoirs. This effect is called *the quantum pump effect* [70–114].[1] A dynamic mesoscopic scatterer generating a direct current is called *a quantum pump* [116–119].

Under specific conditions [74, 81, 120–124] the quantum pump can transfer the same integer charge in each cycle [125–128]. Then we call this *a quantized emission regime*. Also, there is a pump generating a quantized alternating current, which works in both the adiabatic as well as the non-adiabatic regimes [129]. Note also devices (generating a quantized direct current) based on the strong Coulomb interaction [130], which can be extremely accurate [131]. However, they require the fine adjustment of several parameters and they operate at much smaller, MHz frequencies compared to the pumps in the quantized emission regime mentioned above, which work at GHz frequencies.

In some systems[2] quantized pumping is topologically protected [70, 133–139].[3] The equivalence between the topological description and the scattering description of the pumped charge is established in Refs. [141, 142].

---

[1]Note also the pump effect in classical stochastic systems. For a review, see Ref. [115].
[2]For a bosonic system, see Ref. [132].
[3]This is correct for adiabatic pumping. With increasing frequency the pumping charge decreases [140].

4.1 Steady particle flow

The existence of a direct current in the system means that there is a steady particle flow in the leads connecting the scatterer to the reservoirs. To characterize the intensity of such a flow in some direction [from the scatterer to the reservoir, the upper index (in), or back, the upper index (out)] it is convenient to use a *distribution function*, $f_\alpha^{(in/out)}(E)$, which defines how many particles with energy within the interval dE near E in unit time pass through a cross-section of a lead α. After integrating over energy we get the total flow in some direction in the given lead. The direct current in a lead is defined as the difference between particle flows directed from the scatterer to the reservoir and those directed back times an electron charge. Charge conservation requires the sum of direct currents flowing in all the leads to be equal zero.

4.1.1 *Distribution function*

Since we assume that the reservoirs are in equilibrium, the electrons moving in leads from the reservoirs to the scatterer, the incident electrons, are described by the Fermi distribution function, $f_\alpha(E)$, where the index $\alpha = 1, \ldots, N_r$, numbers the reservoirs. The distribution function $f_\alpha(E)$ depends on both the chemical potential, μ_α, and the temperature, T_α, of the corresponding reservoir. Later in this chapter we assume the chemical potentials and temperatures, hence the distribution functions, to be the same for all the reservoirs,

$$\mu_\alpha = \mu_0, \quad T_\alpha = T_0, \quad \alpha = 1 \ldots, N_r,$$

$$f_\alpha(E) = f_0(E).$$

(4.1)

In contrast, electrons scattered by a dynamic sample are not in equilibrium. Therefore, they are characterized by the non-equilibrium distribution function. Let us show that the distribution function for scattered electrons is different from the Fermi distribution function.

The single-particle distribution function, $f_\alpha^{(out)}(E)$, for electrons scattered into the lead α and moving out of the scatterer, is defined as [88]

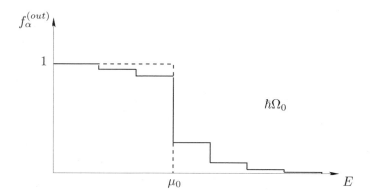

Fig. 4.1 The non-equilibrium distribution function, $f_\alpha^{(out)}(E)$, for electrons scattered into the lead α at zero temperature. The step width is $\hbar\Omega_0$. The Fermi distribution function at zero temperature is shown by the dashed line.

$$\left\langle \hat{b}_\alpha^\dagger(E) \hat{b}_\beta(E') \right\rangle = \delta_{\alpha\beta}\delta(E-E') f_\alpha^{(out)}(E), \quad (4.2)$$

$$f_\alpha^{(out)}(E) = \sum_{n=-\infty}^{\infty} \sum_{\beta=1}^{N_r} \left| S_{F,\alpha\beta}(E,E_n) \right|^2 f_\beta(E_n).$$

Using this definition we rewrite Eq. (3.40) for a direct current, $I_{\alpha,0}$, generated by a dynamic scatterer

$$I_{\alpha,0} = \frac{e}{h} \int_0^\infty dE \left\{ f_\alpha^{(out)}(E) - f_\alpha(E) \right\}. \quad (4.3)$$

From this equation it follows directly that the direct current exists in the case where the distribution function for scattered electrons is different from the one for incoming electrons.

In the case of a dynamic scatterer, even if Eq. (4.1) is fulfilled, the distribution function $f_\alpha^{(out)}(E)$, Eq. (4.2), differs from the Fermi distribution function, $f_0(E)$. Let us illustrate it in the case of zero temperatures, Fig. 4.1. In this case for each energy E the sum over n in Eq. (4.2) is restricted by those n for which $E_n \equiv E + n\hbar\Omega_0 \leq \mu_0$. Therefore, we can write

$$f_\alpha^{(out)}(E) = \sum_{n=-\infty}^{\left[\frac{\mu-E}{\hbar\Omega_0}\right]} \sum_{\beta=1}^{N_r} |S_{F,\alpha\beta}(E,E_n)|^2 = \begin{cases} < 1, E < \mu_0, \\ > 0, E > \mu_0, \end{cases} \quad (4.4)$$

where $[X]$ is the integer part of X. The above equation reaches unity only if the upper limit in the sum over n approaches infinity. This follows directly from the unitarity of the Floquet scattering matrix, see Eq. (3.28b).

Note the distribution function $f_\alpha^{(out)}(E)$ is different from the equilibrium one only at energies near the Fermi level, $E \approx \mu_0$. For energies far from μ_0 the distribution function for scattered electrons is almost in equilibrium:

$$f_\alpha^{(out)}(E) \approx \begin{cases} 1, E \ll \mu_0, \\ 0, E \gg \mu_0. \end{cases} \quad (4.5)$$

Therefore, we conclude: The dynamic scatterer forces an electron system out of equilibrium. This is, perhaps, the most prominent difference between the dynamic scatterer and the stationary one.

4.1.2 Adiabatic regime: Current linear in the pump frequency

Let us analyze a direct current in the limit of a small pump frequency, see Eq. (3.49). This is the so-called *adiabatic regime* of current generation. In this case it is convenient to use Eq. (3.42). With Eq. (4.1) we can write

$$I_{\alpha,0} = \frac{e}{h} \int_0^\infty dE \sum_{n=-\infty}^\infty \{f_0(E) - f_0(E_n)\} \sum_{\beta=1}^{N_r} |S_{F,\alpha\beta}(E_n,E)|^2. \quad (4.6)$$

Expanding the difference of the Fermi functions up to linear in $\hbar\Omega_0$ terms and using the zero-order adiabatic approximation for the Floquet scattering matrix, see Eq. (3.46a), we calculate

$$I_{\alpha,0} = \frac{e\Omega_0}{2\pi} \int_0^\infty dE \left(-\frac{\partial f_0}{\partial E}\right) \sum_{\beta=1}^{N_r} \sum_{n=1}^\infty n \left\{|S_{\alpha\beta,n}(E)|^2 - |S_{\alpha\beta,-n}(E)|^2\right\}, \quad (4.7)$$

where the lower index n indicates the Fourier coefficient for the corresponding frozen scattering matrix element $S_{\alpha\beta}$.

It follows from the equation given above that the current $I_{\alpha,0}$ can be non-zero if the Fourier coefficients corresponding to the positive, $n > 0$ (emission), and negative, $n < 0$ (absorption), harmonics are different. After an inverse Fourier transformation this condition reads

$$\hat{S}(t, E) \neq \hat{S}(-t, E). \tag{4.8}$$

Therefore, *the broken time-reversal symmetry of the frozen scattering matrix is a necessary condition for direct current generation by the dynamic mesoscopic scatterer in the adiabatic regime.*

In fact we are dealing with a *dynamic* break of the time-reversal symmetry by the parameters of a scatterer, $p_i(t)$, varying under the action of the external perturbations that are periodic in time. For instance, in the case of two parameters varying with the same frequency but shifted in phase,

$$p_1(t) = p_{1,0} + p_{1,1} \cos(\Omega_0 t),$$
$$p_2(t) = p_{2,0} + p_{2,1} \cos(\Omega_0 t + \varphi), \tag{4.9}$$

the time-reversal symmetry is broken. To show this we note that in this case the time reversal, $t \to -t$, is equivalent to a phase reversal, $\varphi \to -\varphi$, which, at $\varphi \neq 0, 2\pi$, changes the parameter set for the frozen scattering matrix. As a result we arrive at Eq. (4.8).

Performing an inverse Fourier transformation on Eq. (4.7) we get a more compact expression for an adiabatic direct current [69, 79, 95]

$$I_{\alpha,0} = -i\frac{e}{2\pi} \int_0^\infty dE \left(-\frac{\partial f_0(E)}{\partial E}\right) \int_0^{\mathcal{T}} \frac{dt}{\mathcal{T}} \left(\hat{S}(E,t) \frac{\partial \hat{S}^\dagger(E,t)}{\partial t}\right)_{\alpha\alpha}. \tag{4.10}$$

To show that this equation above is real we use the unitarity property of the scattering matrix, $\hat{S}\hat{S}^\dagger = \hat{I}$. Then we find that all of the diagonal elements $\left(\hat{S}d\hat{S}^\dagger\right)_{\alpha\alpha} = -\left(d\hat{S}\hat{S}^\dagger\right)_{\alpha\alpha}$ are imaginary, hence Eq. (4.10) is real.

Let us show that Eq. (4.10) conserves charge. In the case of direct currents the charge conservation law (the charge continuity equation) reads

$$\sum_{\alpha=1}^{N_r} I_{\alpha,0} = 0. \tag{4.11}$$

Following Ref. [95] we use the Birman–Krein formula (see, e.g., Ref. [53]),

$$d \ln\left(\det \hat{S}\right) = -\mathrm{tr}\left(\hat{S} d \hat{S}^\dagger\right). \tag{4.12}$$

Summing over α in Eq. (4.10) and using the identity (4.12), we find

$$\sum_{\alpha \sim 1}^{N_r} I_{\alpha,0} \sim \int_0^{\mathcal{T}} dt\, \mathrm{tr}\left(\hat{S} \frac{\partial \hat{S}^\dagger}{\partial t}\right) = -\int_0^{\mathcal{T}} dt\, \frac{d}{dt} \ln\left(\det \hat{S}\right)$$

$$= \ln\left(\det \hat{S}(0)\right) - \ln\left(\det \hat{S}(\mathcal{T})\right) = 0,$$

where in the last equality we have used the periodicity of the frozen scattering matrix.

In the particular case of a scatterer with two leads when the scattering matrix is given by Eq. (1.63), with phases γ, θ, ϕ and the reflection coefficient R all being functions periodic in time, the direct current generated, Eq. (4.10), is ($I_0 \equiv I_{1,0} = -I_{2,0}$)

$$I_0 = \frac{e}{4\pi} \int_0^\infty dE \left(-\frac{\partial f_0(E)}{\partial E}\right) \int_0^{\mathcal{T}} \frac{dt}{\mathcal{T}} \left\{ R(t) \frac{\partial \theta(t)}{\partial t} + T(t) \frac{\partial \phi(t)}{\partial t} \right\}. \tag{4.13}$$

As one can see the current generated depends essentially on the phases of the scattering matrix elements. This fact emphasizes once more the quantum-mechanical nature of a current generated by dynamic scatterer. Note that without a magnetic field $\phi \equiv 0$.

Notice equation (4.10) defines a current at zero as well as at finite temperatures. From the formal point of view the expansion in powers of Ω_0 we used in Eq. (4.6) is valid only at $\hbar\Omega_0 \ll k_B T_0$. However, one can show that Eq. (4.10) is valid in the opposite case, $\hbar\Omega_0 \gtrsim k_B T_0$, also. To do so, we note that at zero temperature the integration over energy in each term in the sum over n in Eq. (4.6) is restricted by the interval of order $\sim |n|\hbar\Omega_0$ near the Fermi energy μ_0. At the same time the adiabatic approximation, Eq. (3.46a), is valid under the condition given by Eq. (3.49). This condition allows us to keep the frozen scattering matrix, $\hat{S}(t,E)$, as energy independent within the energy interval mentioned above and to calculate it at $E = \mu_0$. The remaining integral over energy in Eq. (4.6) becomes trivial. It gives $n\hbar\Omega_0$. As a result the first order in the pump frequency expression for a direct generated current reads ($\hbar\Omega_0 \gtrsim k_B T_0$)

$$I_{\alpha,0} = -i\frac{e}{2\pi}\int_0^{\mathcal{T}}\frac{dt}{\mathcal{T}}\left(\hat{S}(t,\mu)\frac{\partial \hat{S}^\dagger(t,\mu)}{\partial t}\right)_{\alpha\alpha}. \quad (4.14)$$

The same equation can be obtained from Eq. (4.10) in the zero temperature limit (formally at $T_0 = 0$) when $-\partial f_0/\partial E = \delta(E-\mu)$.

Equation (4.14) admits an elegant geometrical formulation of the necessary condition for the existence of a direct current generated in the adiabatic regime, see Ref. [69]. Let us consider a space of parameters p_i of the frozen scattering matrix, see Fig. 4.2, for the two varying parameters, p_1 and p_2. Take a point, A(t), in this space with coordinates $p_i(t)$. During the period, $0 < t < \mathcal{T}$, the point A(t) follows a closed trajectory $\mathcal{L}$. We denote $\hat{S} \equiv \hat{S}(\{p_i(t)\}, \mu)$, where $\{p_i(t)\}$ is a set of all the parameters, and rewrite Eq. (4.14) as [79]

$$I_{\alpha,0} = -i\frac{e\Omega_0}{4\pi^2}\oint_{\mathcal{L}}\left(\hat{S}d\hat{S}^\dagger\right)_{\alpha\alpha} \quad (4.15)$$

where the linear dependence on Ω_0 is explicit.

For the sake of simplicity we consider the case with only two parameters, $p_1(t)$ and $p_2(t)$, which vary with small amplitudes, $p_{i,1} \ll p_{i,0}$, $i = 1, 2$, see Eq. (4.9). Then we can write,

$$d\hat{S}^\dagger = \frac{\partial \hat{S}^\dagger}{\partial p_1}dp_1 + \frac{\partial \hat{S}^\dagger}{\partial p_2}dp_2.$$

Using the Green theorem in Eq. (4.15),

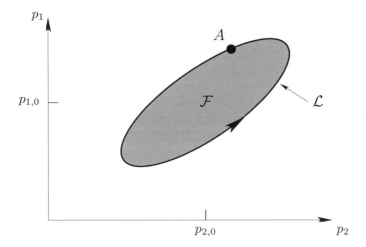

Fig. 4.2 The parameter space for the frozen scattering matrix in the case of two varying parameters, p_1 and p_2. During one period the point $A(t)$ with coordinates $(p_1(t), p_2(t))$ follows a trajectory $\mathcal{L}$. $\mathcal{F}$ stands for the surface area. The arrow indicates the direction of movement for $\varphi > 0$, see Eq. (4.9).

$$\iint_{\mathcal{F}} \left\{ \frac{\partial}{\partial p_1} \left(\hat{S} \frac{\partial \hat{S}^{\dagger}}{\partial p_2} \right) - \frac{\partial}{\partial p_2} \left(\hat{S} \frac{\partial \hat{S}^{\dagger}}{\partial p_1} \right) \right\}_{\alpha\alpha} dp_1 dp_2$$

$$= \oint_{\mathcal{L}} \hat{S} \frac{\partial \hat{S}^{\dagger}}{\partial p_1} dp_1 + \hat{S} \frac{\partial \hat{S}^{\dagger}}{\partial p_2} dp_2,$$

we finally arrive at the following [69]:

$$I_{\alpha,0} = \mathcal{F} \frac{e\Omega_0}{2\pi^2} \operatorname{Im} \left(\frac{\partial \hat{S}}{\partial p_1} \frac{\partial \hat{S}^{\dagger}}{\partial p_2} \bigg|_{p_i = p_{i,0}} \right)_{\alpha\alpha}, \quad (4.16)$$

where $\mathcal{F} = \pi p_{1,1} p_{2,1} \sin(\varphi)$ is an area of the surface (in the present case it is an ellipse) enclosed by the curve $\mathcal{L}$. The value of $\mathcal{F}$ is positive if the point A moves counterclockwise, as shown in Fig. 4.2. In Eq. (4.16) we also took into account the following. If the parameters vary with small amplitudes then to the leading order we can keep the derivatives of the scattering matrix elements constant in the surface integral and calculate them at $p_i = p_{i,0}$.

So, *if the area $\mathcal{F}$ encircled by the representing point $A(t)$ in the parameter space of a scattering matrix during a period is non-zero, then in the* general case[4] the direct current generated in the adiabatic regime is non-zero.

In the small amplitude limit the current is proportional to the area $\mathcal{F}$, that is the current is a quadratic form of the parameter amplitudes. Therefore, the pump effect is an essentially non-linear effect. Notice, it follows from Eq. (4.16) that the value and even the direction of a direct current can be changed simply by varying the phase difference φ between the parameters $p_1(t)$ and $p_2(t)$. This was shown experimentally in Ref. [116].

Equation (4.16) illustrates also an already mentioned relation between the existence of a direct current and broken time-reversal symmetry. Such a relation is clearly seen from the following. Under time reversal the direction of motion of a point A reverses. Therefore, the oriented surface $\mathcal{F}$ changes sign.

It should be noted that there are frozen scattering matrix derivatives in Eq. (4.16). They are not connected directly to the driving. However, at some particular values of the parameters, $p_{i,0}$, these derivatives (or either of them) can vanish. That results in the vanishing of the direct current. Therefore, the pump effect depends not only on the parameters of the dynamic influence but also on the stationary characteristics of the scatterer. More precisely, the direct current arises only in the case of a spatially asymmetric scatterer. To show this we use Eq. (3.43), which under the conditions of Eq. (4.1) reads

$$I_{\alpha,0} = \frac{e}{h} \int_0^\infty dE\, f_0(E) \sum_{n=-\infty}^{\infty} \sum_{\beta=1}^{N_r} \left\{ \left|S_{F,\alpha\beta}(E_n,E)\right|^2 - \left|S_{F,\beta\alpha}(E_n,E)\right|^2 \right\}. \tag{4.17}$$

One can see the direct current is non-zero if the photon-assisted probability for scattering from the lead β to the lead α is different from the probability for scattering in the reversed direction.

So, *the necessary condition for the appearance of a quantum pump effect is a spatial-inversion asymmetry of the scatterer.*

The use of Eq. (4.17) in the adiabatic regime, $\hbar\Omega_0 \ll \delta E$, allows us to represent a generated current as the sum of contributions due to electrons with different energies and to introduce *the spectral density of generated currents*, $dI_\alpha(t,E)/dE$, which we will need to analyze a quantum pump

---

[4]The current can be zero if the derivatives of the scattering matrix elements are zero. In addition in the large amplitude regime when the integrand is not constant and changes sign, in some cases the current is zero even if the area is not zero. Then it is natural to speak about accidental current nullifying.

under an external bias. However, it follows from Eq. (4.14) that without a bias and at a zero temperature the current can be expressed in terms of quantities characterizing the scattering of electrons with Fermi energy only.

Using Eq. (3.50) we find the square of the absolute value of a Floquet scattering matrix element up to linear in Ω_0 terms:

$$|S_{F,\alpha\beta}(E_n, E)|^2 \approx |S_{\alpha\beta,n}(E)|^2 + \frac{n\hbar\Omega_0}{2} \frac{\partial |S_{\alpha\beta,n}(E)|^2}{\partial E}$$
$$+ 2\hbar\Omega_0 \mathrm{Re}\left[S^*_{\alpha\beta,n}(E) A_{\alpha\beta,n}(E)\right]. \quad (4.18)$$

Also we use $\sum_n \sum_\beta |S_{F,\beta\alpha}(E_n, E)|^2 = 1$. Substituting these equations in Eq. (4.17), taking into account that $\sum_n \sum_\beta |S_{\alpha\beta,n}(E)|^2 = 1$, performing the inverse Fourier transformation, and using the identity (3.56), we finally calculate a direct current that is linear in the pump frequency, Ω_0 [95]

$$I_{\alpha,0} = \int_0^{\mathcal{T}} \frac{dt}{\mathcal{T}} \int_0^\infty dE\, f_0(E) \frac{dI_\alpha(t, E)}{dE}, \quad (4.19)$$

where the current spectral density, dI_α/dE, is related to the diagonal element of the following matrix Poisson brackets,

$$\frac{dI_\alpha(t, E)}{dE} = \frac{e}{h} P\left\{\hat{S}, \hat{S}^\dagger\right\}_{\alpha\alpha} \equiv i\frac{e}{2\pi}\left(\frac{\partial \hat{S}}{\partial t}\frac{\partial \hat{S}^\dagger}{\partial E} - \frac{\partial \hat{S}}{\partial E}\frac{\partial \hat{S}^\dagger}{\partial t}\right)_{\alpha\alpha}. \quad (4.20)$$

This quantity is subject to the conservation law at each energy and at any time:

$$\sum_{\alpha=1}^{N_r} \frac{dI_\alpha}{dE} = \frac{e}{h} \sum_{\alpha=1}^{N_r} P\left\{\hat{S}, \hat{S}^\dagger\right\}_{\alpha\alpha} = 0. \quad (4.21)$$

This equation is a direct consequence of the identity (3.55).

We stress that both equations, (4.10) and (4.19), define the same quantity, $I_{\alpha,0}$. The difference is in the way they are written. Substituting Eq. (4.20) into Eq. (4.19) and integrating the first term by parts over time t, and both terms by parts over the energy E we arrive at Eq. (4.10).

Thus we see that the dynamic scatterer is in principle different from the stationary one even in the case of slowly varying parameters. The difference is due to the existence of currents with spectral density $dI_\alpha(t,E)/dE$ generated in the leads connecting the scatterer and the reservoirs.

4.1.3 Current quadratic in the pump frequency

If the phase difference φ of the pumping parameters, see Eq. (4.9), is zero then the current that is linear in frequency, Eq. (4.16), vanishes. In particular such a current is absent if only one parameter of the scattering matrix varies in time. However, even in this case the dynamic scatterer can generate a direct current that is quadratic in the pump frequency, $I_{\alpha,0} \sim \Omega_0^2$. The direct current proportional to Ω_0^n for $n > 1$ is usually called *non-adiabatic*.

To calculate the direct current that is quadratic in Ω_0 we substitute Eq. (4.18) into Eq. (4.6) and expand the difference of the Fermi functions up to terms proportional to Ω_0^2. After simple algebra we find the direct current,

$$I_{\alpha,0} = \frac{e}{2\pi} \int_0^\infty dE \left(-\frac{\partial f_0}{\partial E}\right) \int_0^\mathcal{T} \frac{dt}{\mathcal{T}} \operatorname{Im} \left\{ \hat{S} \frac{\partial \hat{S}^\dagger}{\partial t} + 2\hbar\Omega_0 \hat{A} \frac{\partial \hat{S}^\dagger}{\partial t} \right\}_{\alpha\alpha}.$$

(4.22)

If the frozen scattering matrix is time-reversal invariant, $\hat{S}(t) = \hat{S}(-t)$, then the first term in the curly brackets in Eq. (4.22), linear in the pump frequency, does not contribute to the current. In this case the contribution that is quadratic in Ω_0 is dominant,

$$I_{\alpha,0}^{(2)} = \frac{e\hbar\Omega_0}{\pi} \int_0^\infty dE \left(-\frac{\partial f_0}{\partial E}\right) \int_0^\mathcal{T} \frac{dt}{\mathcal{T}} \operatorname{Im} \left\{ \hat{A} \frac{\partial \hat{S}^\dagger}{\partial t} \right\}_{\alpha\alpha}.$$

(4.23)

Earlier we showed that the current that is linear in the pump frequency is subject to the conservation law, Eq. (4.11). Since the current, $I_{\alpha,0}^{(2)}$, is also a direct current, it should satisfy the same conservation law. Therefore, we have

$$\int_0^\mathcal{T} \frac{dt}{\mathcal{T}} \operatorname{Im} \operatorname{tr} \left(\hat{A}(t,E) \frac{\partial \hat{S}^\dagger(t,E)}{\partial t} \right) = 0.$$

(4.24)

This equation, fulfilled at any energy E, puts an additional constraint onto the anomalous scattering matrix $\hat{A}$.

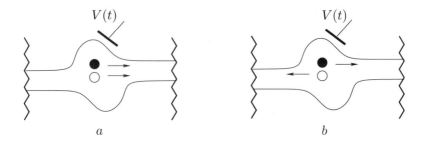

Fig. 4.3 The creation and scattering of quasi-electron-hole pairs in a dynamic sample with two leads. Under the action of a potential that is periodic in time, $V(t) = V(t+\mathcal{T})$, an electron can absorb one or several energy quanta, $\hbar\Omega_0$. As a result it jumps from the occupied level to the non-occupied one, which can be viewed as the creation of a quasi-electron-hole pair. A quasi-electron (a dark circle) and a hole (a light circle) can leave the scattering region through the same lead (a) or through different leads (b). In the latter case a net charge is transferred from one reservoir to the other.

4.2 Quantum pump effect

The direct current generated by the mesoscopic dynamic scatterer is due to asymmetric redistribution of (equal) electron flows incident to the scatterer from the reservoirs. It does not require any source (or drain) of charge inside the scattering region. Before we outline a physical mechanism responsible for such an asymmetry, we give simple arguments to illustrate the possibility of generating a direct current without a bias.

4.2.1 Quasi-particle picture of direct current generation

The possibility of the appearance of a direct current is more clear if one goes from the real particle picture to the quasi-particle picture [88]. The particle with energy above the Fermi level μ_0 we will call *a quasi-electron*, while an empty state with energy below μ_0 we will call *a hole*.

For the sake of simplicity we assume all the reservoirs are at zero temperature (and have the same chemical potential). Then the quasi-particles are absent in equilibrium. Therefore, there is zero quasi-particle flow incident to the scatterer. On the other hand the dynamic scatterer acts as a source of quasi-electron-hole pairs moving from the scatterer to the reservoirs, Fig. 4.3 [88, 143–145]. The quasi-electron-hole pair is created when a (real) electron absorbs one, $n = 1$, or several, $n > 1$, energy quanta, $\hbar\Omega_0$, while interacting with a dynamic scatterer. During this process an electron empties the state with energy $E < \mu_0$ (a hole is created) and it occupies the state with energy $E_n = E + n\hbar\Omega_0 > \mu_0$ (a quasi-electron is created). We emphasize that the created pair is charge neutral.[5] However if a quasi-electron and a hole leave the scattering region through the different leads, see Fig. 4.3(b), then a current pulse is generated between the corresponding reservoirs. The sum of the currents in the different leads is obviously zero. On the other hand, if a quasi-electron and a hole are both scattered into the same lead, then a current does not appear at all, see Fig. 4.3(a).

From this picture it is clear that the appearance of a direct current is a consequence of broken quasi-electron-hole symmetry. Otherwise the number of quasi-electrons and holes scattered to the same lead would be equal on average, hence the current averaged over a long time (a direct current) would not arise.

4.2.2 Interference mechanism of direct current generation

As we already mentioned, within the real particle picture the appearance of a direct current is due to asymmetry in the scattering of electrons from one lead to another and back, see Eq. (4.17). The physical mechanism leading to such an asymmetry stems from the interference of photon-assisted scattering amplitudes [69, 75, 86, 88, 146].

To show this we follow Ref. [147] and consider a one-dimensional scatterer comprising two potentials, $V_1(t) = 2V\cos(\Omega_0 t + \varphi_1)$ and $V_2(t) = 2V\cos(\Omega_0 t + \varphi_2)$, oscillating with the same amplitude and located at a distance L from each other, Fig. 4.4. For simplicity we assume both potentials oscillate with a small amplitude. Let an electron with energy E fall upon the scatterer. Since for small amplitude oscillations only single-photon processes are relevant [51, 56, 57], there are three different outcomes:

(i) The electron does not interact with the potentials, hence it does not change its energy. In this case the electron leaves the scattering region with energy $E^{(out)}$ equal to its initial energy, $E^{(out)} = E$.

(ii) The electron absorbs one energy quantum: $E^{(out)} = E + \hbar\Omega_0$.

(iii) The electron emits one energy quantum: $E^{(out)} = E - \hbar\Omega_0$.

Since these possibilities correspond to different final states, which differ in final energy $E^{(out)}$, the total transmission probability T is the sum of the probabilities for these three processes,

$$T = T^{(0)}(E, E) + T^{(+)}(E + \hbar\Omega_0, E) + T^{(-)}(E - \hbar\Omega_0, E), \quad (4.25)$$

---

[5]The charge of a filled Fermi sea is treated as a reference point.

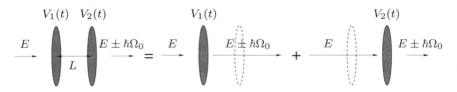

Fig. 4.4 Photon-assisted scattering amplitudes for an electron propagating through a dynamic double-barrier. An electron can absorb (or emit) an energy quantum $\hbar\Omega_0$ interacting either with a potential $V_1(t)$ or with a potential $V_2(t)$. Therefore, the photon-induced scattering amplitude is a sum of two terms.

where the first argument is a final energy while the second one is an initial energy. The probability $T^{(0)}$, like the probability for scattering by the stationary scatterer, does not depend on the direction of propagation. In contrast both probabilities $T^{(+)}$ and $T^{(-)}$ depend on it. Therefore, we concentrate on these latter probabilities.

First we calculate $T^{(+)}$. Note there are two possibilities for passing through the scatterer and absorbing an energy quantum, see Fig. 4.4. The first possibility is to absorb an energy quantum when interacting with the potential $V_1(t)$. And the second possibility is to absorb an energy quantum interacting with $V_2(t)$. Since in both these cases the final state is the same, the corresponding amplitudes (not probabilities!) should be added. Denoting corresponding the amplitudes as $\mathcal{A}^{(j,+)}$, $j = 1, 2$, we calculate the probability,

$$T^{(+)} = \left| \mathcal{A}^{(1,+)} + \mathcal{A}^{(2,+)} \right|^2. \qquad (4.26)$$

Each of the amplitudes $\mathcal{A}^{(j,+)}$ can be represented as the product of two terms, the amplitude, $\mathcal{A}^{(free)}(E) = e^{ikL}$, of free propagation from one potential barrier to the other[6] and the photon-assisted amplitude, $\mathcal{A}_j^{(+)}$, describing the absorption of an energy quantum $\hbar\Omega_0$ during an interaction with the potential V_j. The amplitude $\mathcal{A}_j^{(+)}$ is proportional to the Fourier coefficient for $V_j(t)$: $\mathcal{A}_j^{(+)} = \kappa V e^{-i\varphi_j}$, where κ is a proportionality constant.

We consider separately two cases. First, when an electron is incident from the side of the potential V_1 and, second, when an electron is incident from the side of the potential V_2. We will label the corresponding probabilities with the help of lower indices $\rightarrow$ and $\leftarrow$, respectively. Our aim is to show that

---

[6]The effect of dephasing on the pumped current was addressed in [62, 146, 148].

$$T_{\rightarrow}^{(+)} \neq T_{\leftarrow}^{(+)}, \tag{4.27}$$

see also Refs. [12, 71, 149, 150].

Calculating $T_{\rightarrow}^{(+)}$ we take into account that an electron first meets the potential $V_1(t)$ and only then, after a distance L, does it reach the potential $V_2(t)$. Therefore, if an electron absorbs energy near V_1 then it propagates between the potential barriers with an enhanced energy, $E_+ = E + \hbar\Omega$. The corresponding amplitude is $\mathcal{A}_{\rightarrow}^{(1,+)} = \mathcal{A}_1^{(+)} \mathcal{A}^{(free)}(E_+)$. In contrast, if an electron absorbs energy near the potential $V_2(t)$ then it propagates between the barriers with initial energy E. The corresponding amplitude is $\mathcal{A}_{\rightarrow}^{(2,+)} = \mathcal{A}^{(free)}(E)\mathcal{A}_2^{(+)}$. If $\hbar\Omega_0 \ll E$ we can expand the phase of the amplitude $\mathcal{A}^{(free)}(E_+)$ up to linear in Ω_0 terms: $k(E_+)L \approx kL + \Omega_0\tau$, where $k = k(E)$ and $\tau = Lm/(\hbar k)$ is the time for free propagation from the barrier V_1 to the barrier V_2. So we write

$$\mathcal{A}_{\rightarrow}^{(1,+)} = \kappa V e^{-i\varphi_1} e^{i(kL + \Omega_0\tau)}, \quad \mathcal{A}_{\rightarrow}^{(2,+)} = e^{ikL} \kappa V e^{-i\varphi_2}. \tag{4.28}$$

Substituting these amplitudes into Eq. (4.26) we calculate

$$T_{\rightarrow}^{(+)} = 2\kappa^2 V^2 \left\{ 1 + \cos\left(\varphi_1 - \varphi_2 - \Omega_0\tau\right) \right\}. \tag{4.29}$$

Now we calculate the probability $T_{\leftarrow}^{(+)}$. Going from the right to the left an electron first meets V_2 and only then does it meet V_1. By analogy with the calculations above we find

$$\mathcal{A}_{\leftarrow}^{(1,+)} = e^{ikL} \kappa V e^{-i\varphi_1}, \quad \mathcal{A}_{\leftarrow}^{(2,+)} = \kappa V e^{-i\varphi_2} e^{i(kL + \Omega_0\tau)}, \tag{4.30}$$

and correspondingly

$$T_{\leftarrow}^{(+)} = 2\kappa^2 V^2 \left\{ 1 + \cos\left(\varphi_1 - \varphi_2 + \Omega_0\tau\right) \right\}. \tag{4.31}$$

Comparing Eqs. (4.29) and (4.31) we see that indeed the probability depends on the propagation direction, as in Eq. (4.27). The directional asymmetry of scattering can be characterized via the difference, $\Delta T^{(+)} = T_{\rightarrow}^{(+)} - T_{\leftarrow}^{(+)}$, which is equal to

$$\Delta T^{(+)} = 4\kappa^2 V^2 \sin\left(\Delta\varphi\right) \sin\left(\Omega_0 \tau\right), \tag{4.32}$$

where $\Delta\varphi = \varphi_1 - \varphi_2$.

The probability of propagation with emission of an energy quantum $\hbar\Omega_0$ is characterized by the same asymmetry, $\Delta T^{(-)} = \Delta T^{(+)}$, for our simple model. Therefore, if the same electron flows with intensity I_0 fall upon the scatterer from both sides, then the asymmetric redistribution of scattered electrons results in a direct current, $I_{dc} = I_0 \left(\Delta T^{(+)} + \Delta T^{(-)}\right) = 2I_0 \Delta T^{(+)}$. This current depends on two phase factors. On one hand, it depends on the phase difference, $\Delta\varphi$, between the potentials $V_1(t)$ and $V_2(t)$. On the other hand, the current depends on an additional contribution to the dynamic phase, $\Omega_0\tau = \Omega_0 L/v$ (where $v = \hbar k/m$ is the electron velocity), due to the energy change during scattering. The first factor breaks the time-reversal invariance allowing the existence of a direct current in the system where there would be no current in the stationary regime. While the second factor characterizes the system as spatially asymmetric (comprising two different potentials at a distance L). It is interesting to note that in the case under consideration spatial-inversion symmetry is broken only if $\varphi_1 \neq \varphi_2$, therefore, one can speak about the *dynamicly* broken spatial symmetry. It follows from Eq. (4.32) that the breaking of only one of the two symmetries, either spatial-inversion or time-reversal, is not enough for direct current generation.

4.3 Single-parameter adiabatic direct current generation

Accordingly to Brouwer's arguments [69][7] in order to generate a direct current in the adiabatic regime it is necessary to have at least two parameters which are varied out of phase. The variation of a single parameter can result in a direct current that is at least quadratic in frequency, see Section 4.1.3. This conclusion is confirmed by both experiment [151, 152] and theory [40, 64, 86, 153–158][8]. However, in Refs. [165–167] it was shown theoretically that under the slow rotation of a potential it is possible to

---

[7] See Fig. 4.2 and related discussion in the text.
[8] Note the papers [127, 159–164] where a direct current that is linear in pumping frequency was generated by varying only a single parameter in a strongly non-adiabatic regime.

generate a direct current that is linear in the rotation frequency[9]. If the rotation angle is treated as a parameter then this is clearly an example of a single-parameter adiabatic direct current generation. It is natural to call this device *a quantum Archimedes screw*. Below we give simple arguments showing that in structures with a cyclic coordinate single-parameter direct current generation is a rule rather than an exception.

Let the scattering matrix be dependent on a single parameter varying in time, $\hat{S}(t) = \hat{S}[p(t)]$. In the case when the system returns periodically to its initial state we have two possibilities: (i) The parameter p is a periodic function of time, $p(t) = p(t+\mathcal{T})$, or (ii) the parameter p is an angle, i.e., the scattering matrix depends periodically on p, see, e.g., Ref. [79], $S \sim e^{ip}$. In the latter case the parameter space can be rolled up into a cylinder (with $0 \leq p < 2\pi$) and the parameter p can be an increasing function of time, for example, $p \sim t$.

If the parameter p is small, then the adiabatic time-dependent current $I_\alpha(t)$, Eq. (5.13), can be linearized,

$$I_\alpha(t) = e\, C_{\alpha\alpha}(0)\, \frac{\partial p}{\partial t}, \qquad (4.33)$$

where the constant,

$$C_{\alpha\alpha}(p) = -\frac{i}{2\pi} \left(\hat{S}\, \frac{\partial \hat{S}^\dagger}{\partial p} \right)_{\alpha\alpha},$$

is calculated at $p = 0$. In case (i) the current is periodic in time without a DC component. While in case (ii) the current can have a DC component if $p \sim t$ and $C_{\alpha\alpha}(0) \neq 0$. Therefore, only a topologically non-trivial parameter space allows single-parameter adiabatic pumping.

This conclusion remains valid at large p also, when C becomes a function of p. In case (i) we can expand $C(p)$ into a Taylor series in powers of p. Each term of this series results only in an alternating current. In case (ii) we expand $C(p)$ into a Fourier series. Again all the terms but the zero mode produce alternating currents. In contrast, the zero mode results in a direct current (if $p \sim t$). Therefore, if the diagonal element α of the matrix

---

[9]Note also that a uniformly translating potential can generate a direct current [168, 169]. For a slow translation the current is proportional to the speed. If we treat a spatial coordinate as a parameter then this is also an example of a single-parameter adiabatic direct current generation. Here the current results from the classical drag effect, i.e., the momentum transfer from the moving potential to the electron system is primary. In contrast, in the quantum pump effect the energy transfer is primary.

$\hat{C} = \hat{S} \partial \hat{S}^\dagger / \partial p$ has a constant term (a zero mode) in the Fourier expansion in a cyclic coordinate p, then by varying p with a constant speed, $p = \Omega_0 t$, one can generate a direct current $I_\alpha \sim \Omega_0$ [9].

Chapter 5

Alternating current generated by the dynamic scatterer

In contrast to a direct current, which exists only under special conditions, an alternating current is generated only if the properties of the scatterer change periodically in time. As we will see below, there are several physical processes responsible for the appearance of alternating currents. First of all, a redistribution of incident electrons among the outgoing channels is considered to be an intrinsic property of a dynamic scatterer needed to generate a current. Second, alternating currents result from a possible periodic change of a charge localized on the scatterer. And finally, the potential difference between the electronic reservoirs can also lead to the appearance of a current. We emphasize that even DC bias can lead to alternating currents since the conductance of a dynamic scatterer changes in time.

5.1 Adiabatic alternating current

Let us calculate the time-dependent current $I_\alpha(t)$, Eq. (3.39), flowing through the dynamic scatterer in the adiabatic regime, $\varpi = \hbar\Omega_0/\delta E \to 0$. To this end we transform Eq. (3.37b) for the Fourier harmonics of a current as follows. First, in the term having a factor $f_\beta(E_n)$ we make the following replacements: $E_n \to E$ and $n \to -n$. Then use an expansion (3.50) and calculate the product:

$$S^*_{F,\alpha\beta}(E_n, E) S_{F,\alpha\beta}(E_{l+n}, E) = S^*_{\alpha\beta,n} S_{\alpha\beta,l+n} + \hbar\Omega_0 \left\{ \frac{n}{2} \frac{\partial S^*_{\alpha\beta,n}}{\partial E} S_{\alpha\beta,l+n} \right.$$
$$\left. + \frac{(n+l)}{2} \frac{\partial S_{\alpha\beta,n+l}}{\partial E} S^*_{\alpha\beta,n} + \left(S^*_{\alpha\beta,n} A_{\alpha\beta,l+n} + A^*_{\alpha\beta,n} S_{\alpha\beta,n+l} \right) \right\} + \mathcal{O}\left(\varpi^2\right).$$

After that we sum over n,

$$\sum_{n=-\infty}^{\infty} S_{F,\alpha\beta}^*(E_n, E) S_{F,\alpha\beta}(E_{l+n}, E) = \left(|S_{\alpha\beta}|^2\right)_l +$$

$$\frac{i\hbar}{2}\left(-\frac{\partial^2 S_{\alpha\beta}^*}{\partial t \partial E}S_{\alpha\beta} + \frac{\partial^2 S_{\alpha\beta}}{\partial t \partial E}S_{\alpha\beta}^*\right)_l + \hbar\Omega_0\left(S_{\alpha\beta}^* A_{\alpha\beta} + A_{\alpha\beta}^* S_{\alpha\beta}\right)_l + \mathcal{O}(\varpi^2),$$

where on the right hand side (RHS) of the equation above the lower index l denotes the Fourier harmonics for the corresponding quantity. Then we get the following equation for a current linear in the pump frequency as the sum of three terms,

$$I_\alpha(t) = I_\alpha^{(V)}(t) + I_\alpha^{(Q)}(t) + I_\alpha^{(gen)}(t). \tag{5.1}$$

The first term,

$$I_\alpha^{(V)}(t) = \frac{e}{h}\int_0^\infty dE \sum_{\beta=1}^{N_r} |S_{\alpha\beta}(t, E)|^2 \{f_\beta(E) - f_\alpha(E)\}, \tag{5.2}$$

is non-zero if the chemical potentials (or the temperatures) are different for different reservoirs. From the unitarity condition, Eq. (3.47), it follows that the quantity $I_\alpha^{(V)}(t)$ is subject to the conservation law,

$$\sum_{\alpha=1}^{N_r} I_\alpha^{(V)}(t) = 0, \tag{5.3}$$

the same as for a direct current, see Eq. (1.48). This fact justifies the separation of $I_\alpha^{(V)}(t)$ from the total current and allows us to relate it to the potential (or temperature) difference between the reservoirs.

The second term in Eq. (5.1),

$$I_\alpha^{(Q)}(t) = -e\frac{\partial}{\partial t}\int_0^\infty dE \sum_{\beta=1}^{N_r} f_\beta(E)\frac{dN_{\alpha\beta}(t, E)}{dE}, \tag{5.4}$$

is that part of the current due to the variation in time of a charge $Q(t)$ of a scatterer. In this equation we have introduced the frozen *partial density of states* (DOS),

$$\frac{dN_{\alpha\beta}(t,E)}{dE} = \frac{i}{4\pi}\left\{S_{\alpha\beta}(t,E)\frac{\partial S^*_{\alpha\beta}(t,E)}{\partial E} - \frac{\partial S_{\alpha\beta}(t,E)}{\partial E}S^*_{\alpha\beta}(t,E)\right\}, \tag{5.5}$$

which is expressed in terms of the elements of the frozen scattering matrix $\hat{S}(t,E)$ in the same way that the partial DOS of a stationary scatterer is expressed in terms of the stationary scattering matrix elements, see Ref. [43]. The notion of the partial DOS is important for understanding the low-frequency properties of mesoscopic conductors [9, 43, 170–172].

Summing the currents $I_\alpha^{(Q)}$ in all of the leads we arrive at the charge conservation law,

$$\sum_{\alpha=1}^{N_r} I_\alpha^{(Q)}(t) + \frac{\partial Q(t)}{\partial t} = 0, \tag{5.6}$$

where the charge localized on the scatterer is

$$Q(t) = e\int_0^\infty dE \sum_{\alpha=1}^{N_r}\sum_{\beta=1}^{N_r} f_\beta(E)\frac{dN_{\alpha\beta}(E,t)}{dE}. \tag{5.7}$$

Strictly speaking the total current I_α should enter Eq. (5.6). However, it follows from Eqs. (5.3) and (5.10) that neither $I_\alpha^{(V)}(t)$ nor $I_\alpha^{(gen)}(t)$ contribute to the equation under consideration. This allows us to interpret $I_\alpha^{(Q)}(t)$ as a current due to a variation of a scatterer charge.

We see that $I_\alpha^{(V)}$ and $I_\alpha^{(Q)}$ can be explained as characteristics (conductance and DOS) that are inherent to a stationary scatterer. In contrast the third contribution, a current generated by the dynamic scatterer in the lead α,

$$I_\alpha^{(gen)}(t) = \int_0^\infty dE \sum_{\beta=1}^{N_r} f_\beta(E)\frac{dI_{\alpha\beta}(t,E)}{dE}, \tag{5.8}$$

requires a quantity absent in the stationary case [41],

$$\frac{dI_{\alpha\beta}}{dE} = \frac{e}{h}\left(2\hbar\Omega_0 \text{Re}\left[S^*_{\alpha\beta}A_{\alpha\beta}\right] + \frac{1}{2}P\left\{S_{\alpha\beta},S^*_{\alpha\beta}\right\}\right). \quad (5.9)$$

This is *a partial spectral current density* for the flow generated by the dynamic scatterer from the reservoir β into the reservoir α.

The generated current, $I_\alpha^{(gen)}(t)$, is subject to the conservation law,

$$\sum_{\alpha=1}^{N_r} I_\alpha^{(gen)}(t) = 0, \quad (5.10)$$

which directly follows from the property of the partial spectral current density,

$$\sum_{\alpha=1}^{N_r} \frac{dI_{\alpha\beta}(t,E)}{dE} = 0. \quad (5.11)$$

This condition tells us that there is no internal source of charge (see Section 4.2): The scatterer takes a current $dI_{\alpha\beta}(E)/dE$ incoming from the lead β and pushes it into the lead α. The Fermi function $f_\beta(E)$ in Eq. (5.8) shows us how populated this stream is.

To prove the identity (5.11) we use the diagonal element of the matrix expression (3.52),

$$4\hbar\Omega_0 \sum_{\alpha=1}^{N_r} \text{Re}\left\{S^*_{\alpha\beta}A_{\alpha\beta}\right\} = P\left\{\hat{S}^\dagger,\hat{S}\right\}_{\beta\beta}, \quad (5.12)$$

and find

$$\frac{2h}{e}\sum_{\alpha=1}^{N_r}\frac{dI_{\alpha\beta}}{dE} = 4\hbar\Omega_0\sum_{\alpha=1}^{N_r}\text{Re}\left\{S^*_{\alpha\beta}A_{\alpha\beta}\right\} + \sum_{\alpha=1}^{N_r}P\left\{S_{\alpha\beta},S^*_{\alpha\beta}\right\}$$

$$= P\left\{\hat{S}^\dagger,\hat{S}\right\}_{\beta\beta} - P\left\{\hat{S}^\dagger,\hat{S}\right\}_{\beta\beta} = 0.$$

If we sum the quantity $dI_{\alpha\beta}/dE$ over all the incoming scattering channels (the index β) then we get the spectral current density generated into the lead α,

$$\frac{dI_\alpha}{dE} = \sum_{\beta=1}^{N_r} \frac{dI_{\alpha\beta}}{dE} = \frac{e}{2h}\left(4\hbar\Omega_0 \sum_{\beta=1}^{N_r} \mathrm{Re}\left\{S^*_{\alpha\beta}A_{\alpha\beta}\right\} + \sum_{\beta=1}^{N_r} P\left\{S_{\alpha\beta},S^*_{\alpha\beta}\right\}\right)$$

$$= \frac{e}{2h}\left(P\left\{\hat{S},\hat{S}^\dagger\right\}_{\alpha\alpha} + P\left\{\hat{S},\hat{S}^\dagger\right\}_{\alpha\alpha}\right) = \frac{e}{h}P\left\{\hat{S},\hat{S}^\dagger\right\}_{\alpha\alpha},$$

that coincides with Eq. (4.20).

The generated current $I_\alpha^{(gen)}(t)$ is essentially related to the anomalous scattering matrix, $\hat{A}(t,E)$, violating the symmetry of scattering with respect to a movement direction reversal, compare with Eqs. (3.57) and (3.58). Note for the point-like scatterer $\hat{A} = \hat{0}$, see Eq. (3.89), and also $P\left\{S_{\alpha\beta}, S^*_{\alpha\beta}\right\} = 0$, which directly follows from Eq. (3.88), hence $I_\alpha^{(gen)} = 0$. Therefore, with no external bias (when $I_\alpha^{(V)} = 0$) the current of a dynamic point-like scatterer is only due to the variation of its charge, $I_\alpha(t) = I_\alpha^{(Q)}(t)$. For an arbitrary dynamic scatterer having reservoirs with the same potentials and temperatures, $f_\alpha(E) = f_0(E)$, $\forall \alpha$, the current $I_\alpha(t) = I_\alpha^{(Q)}(t) + I_\alpha^{(gen)}(t)$ is

$$I_\alpha(t) = -\frac{ie}{2\pi} \int_0^\infty dE \left(-\frac{\partial f_0(E)}{\partial E}\right)\left(\hat{S}(t,E)\frac{\partial \hat{S}^\dagger(t,E)}{\partial t}\right)_{\alpha\alpha}, \quad (5.13)$$

which is nothing but a generalization of the Büttiker–Thomas–Prêtre formula [43].

5.2 External AC bias

Now we calculate a current flowing through the dynamic mesoscopic scatterer if the reservoirs are biased with a voltage that is periodic in time $V_{\alpha\beta}(t) = V_{\alpha\beta}(t+\mathcal{T}) \equiv V_\alpha(t) - V_\beta(t)$. This case is special since the alternating currents due to the bias, $V_{\alpha\beta}(t)$, interfere with the currents, $I_\alpha^{(gen)}(t)$, generated by the scatterer itself. As a result there arises an additional, so-called *interference*, contribution to the current.

So, let the potentials applied to the reservoirs vary in time with the same frequency as the parameters of scatterer:

$$V_\alpha(t) = V_\alpha \cos\left(\Omega_0 t + \phi_\alpha\right), \quad \alpha = 1, \ldots, N_r. \quad (5.14)$$

Due to the approach for phase-coherent transport phenomena of Refs. [173, 174] the potential that is periodic in time $V_\alpha(t)$ of an electron reservoir can be treated as spatially uniform and it is accounted for in the phase of the wave function of electrons incident from the reservoir to the scatterer. At the same time the chemical potential μ_α entering the Fermi distribution function $f_\alpha(E)$ is constant and independent of $V_\alpha(t)$.

As we know, the Schrödinger equation with a spatially uniform potential $V_\alpha(t)$,

$$i\hbar \frac{\partial \Psi_\alpha}{\partial t} = H_{0,\alpha}\Psi_\alpha + eV_\alpha(t)\Psi_\alpha, \quad (5.15)$$

can be integrated in time. Then the electron wave function can be written as follows, see Section 3.1.3,

$$\Psi_\alpha = \Psi_{0,\alpha} e^{-i\hbar^{-1} \int_{-\infty}^{t} dt' eV_\alpha(t')}, \quad (5.16)$$

where $\Psi_{0,\alpha}$ is a solution to Eq. (5.15) with $V_\alpha(t) = 0$. Such a solution corresponding to energy E, is

$$\Psi_{0E,\alpha} = e^{-i\frac{E}{\hbar}t} \psi_{E,\alpha}(\vec{r}). \quad (5.17)$$

With potential $V_\alpha(t)$, Eq. (5.14), the wave function, Eq. (5.16), corresponding to energy E, becomes ($eV_\alpha > 0$):

$$\Psi_{E,\alpha} = e^{-i\frac{E}{\hbar}t} \bar{\psi}_{E,\alpha}(\vec{r}) \sum_{n=-\infty}^{\infty} e^{-in\phi_\alpha} J_n\left(\frac{eV_\alpha}{\hbar\Omega_0}\right) e^{-in\Omega_0 t}, \quad (5.18)$$

where we have used the following Fourier series:

$$e^{-iX \sin(\Omega_0 t + \phi_\alpha)} = \sum_{n=-\infty}^{\infty} J_n(X) e^{-in(\Omega_0 t + \phi_\alpha)}, \quad (5.19)$$

and have included the constant $C = e^{ieV_\alpha/(\hbar\Omega_0)\,\sin(\Omega_0 t' + \phi_\alpha)}|_{t'=-\infty}$ from Eq. (5.16) in the function $\bar{\psi}_{E,\alpha}(\vec{r}) = C\psi_{E,\alpha}(\vec{r})$.

The wave function $\Psi_{E,\alpha}$ is of the Floquet function type, see Eqs. (3.22) and (3.27). Note the spatial part $\bar{\psi}_{E,\alpha}$ depends on the Floquet energy E but does not depend on the sub-band number n. Therefore, the Floquet wave function is normalized exactly as the stationary wave function $\psi_{E,\alpha}$:

$$\int d^3r \, |\Psi_{E,\alpha}|^2 = \int d^3r \, |\psi_{E,\alpha}|^2 . \tag{5.20}$$

Indeed, using the following property of Bessel functions

$$\sum_{n=-\infty}^{\infty} J_n(X) J_{n+q}(X) = \delta_{q0} , \tag{5.21}$$

we find from Eq. (5.18),

$$|\Psi_{E,\alpha}|^2 = |\psi_{E,\alpha}|^2 \sum_{n=-\infty}^{\infty} \sum_{m=-\infty}^{\infty} e^{-i(n-m)\phi_\alpha} e^{-i(n-m)\Omega_0 t} J_n(X) J_m(X)$$

$$= |\psi_{E,\alpha}|^2 \sum_{q=-\infty}^{\infty} e^{iq\phi_\alpha} e^{iq\Omega_0 t} \sum_{n=-\infty}^{\infty} J_n(X) J_{n+q}(X) = |\psi_{E,\alpha}|^2 .$$

Here we denoted $X = eV_\alpha/(\hbar\Omega_0)$, introduced $q = m - n$, and took into account $|C|^2 = 1$.

The state with wave function $\Psi_{E,\alpha}$ can be occupied at most by one electron. The measurement of an electron energy in the state $\Psi_{E,\alpha}$ can result in any of the values $E_n = E + n\hbar\Omega_0$ with probability $J_n^2(eV_\alpha/\hbar\Omega_0)$. However, the mean energy, $E[\Psi_{E,\alpha}]$, is equal to the energy E of the corresponding stationary state, $\Psi_{0E,\alpha}$:

$$E[\Psi_{E,\alpha}] = \sum_{n=-\infty}^{\infty} E_n J_n^2 = E \sum_{n=-\infty}^{\infty} J_n^2 + \hbar\Omega_0 \sum_{n=1}^{\infty} n \left(J_n^2 - J_{-n}^2\right) = E .$$

Therefore, the distribution function reflecting the occupation of the states $\Psi_{E,\alpha}$ is the Fermi distribution function dependent on the Floquet energy E.

5.2.1 Second quantization operators for incident and scattered electrons

Let us introduce the creation and annihilation operators, $\hat{a}'^{\dagger}_{\alpha}(E)$ and $\hat{a}'_{\alpha}(E)$, for electrons in the Floquet state $\Psi_{E,\alpha}$. They are anti-commuting, Eq. (1.30). The quantum-statistical average of the following product

$$\left\langle \hat{a}'^{\dagger}_{\alpha}(E)\, \hat{a}'_{\beta}(E') \right\rangle = \delta_{\alpha\beta}\, \delta(E - E')\, f_{\alpha}(E) \qquad (5.22)$$

is expressed through the Fermi distribution function $f_{\alpha}(E)$ dependent on the Floquet energy E.

Strictly speaking we should consider scattering of the whole Floquet state, $\Psi_{E,\alpha}$, incident on the mesoscopic sample. However, considering Eq. (3.29) and if the amplitude of the oscillating potential is small,

$$eV_{\alpha} \ll E, \qquad (5.23)$$

then the scattering of any sub-band of the Floquet state is independent of the scattering of other sub-bands. Therefore, following the approach of Ref. [174] we, as before, consider the scattering of an electron that is in a state with fixed energy.

We suppose that the potential $V_{\alpha}(t)$ is present in the reservoir α but it is absent in the lead α connecting the reservoir and the scatterer. Then an electron in the lead is described by the wave function with fixed energy. For an incident electron in the lead α this wave function is $\Psi^{(in)}_{\alpha} = e^{-iEt/\hbar}\, \psi^{(in)}_{\alpha}$, where $\psi^{(in)}_{\alpha}$ is given by Eq. (1.33). Notice there are a number of Floquet states $\Psi_{E',\alpha}$, Eq. (5.18), having a sub-band with energy E in the reservoir α. For such states the Floquet energy E' should be different from E by the integer number of energy quanta $\hbar\Omega_0$. For instance, if $E' = E + n\hbar\Omega_0$ then the sub-band E'_{-n} has an energy E since

$$E'_{-n} = E' - n\hbar\Omega_0 = E + n\hbar\Omega_0 - n\hbar\Omega_0 = E.$$

All such Floquet states contribute to the wave function, $\Psi^{(in)}_{\alpha}$, of an electron in the lead. Therefore, the operators $\hat{a}^{\dagger}_{\alpha}(E)/\hat{a}_{\alpha}(E)$ creating/annihilating an electron in the state $\Psi^{(in)}_{\alpha}$ in the lead α can be expressed in terms of the operators $\hat{a}^\dagger_\alpha(E_n)/\hat{a}_\alpha(E_m)$ creating/annihilating an electron in the reservoir α, as follows:

$$\hat{a}_\alpha(E) = \sum_{m=-\infty}^{\infty} e^{-im\phi_\alpha} J_m\left(\frac{eV_\alpha}{\hbar\Omega_0}\right) \hat{a}'_\alpha(E - m\hbar\Omega_0),$$

$$\hat{a}^\dagger_\alpha(E) = \sum_{n=-\infty}^{\infty} e^{in\phi_\alpha} J_n\left(\frac{eV_\alpha}{\hbar\Omega_0}\right) \hat{a}'^\dagger_\alpha(E - n\hbar\Omega_0).$$

(5.24)

The spatial parts of the corresponding wave functions, Eq. (5.18), are assumed to be the same where the reservoir is connected to the lead (an adiabatic connection condition). Therefore, they do not enter the equations given above.

The operators $\hat{a}'$ are for electrons in a reservoir. They are anti-commuting by definition, see Eq. (1.30). Let us show that the operators $\hat{a}$, Eq. (5.24), for electrons in a lead are also anti-commuting. Using Eq. (5.21) we calculate

$$\{\hat{a}^\dagger_\alpha(E), \hat{a}_\beta(E')\} = \sum_{n=-\infty}^{\infty} \sum_{m=-\infty}^{\infty} e^{i\phi_\alpha n} e^{-i\phi_\beta m} J_n\left(\frac{eV_\alpha}{\hbar\Omega_0}\right) J_m\left(\frac{eV_\beta}{\hbar\Omega_0}\right)$$
$$\times \left\{\hat{a}'^\dagger_\alpha(E - n\hbar\Omega_0), \hat{a}'_\beta(E' - m\hbar\Omega_0)\right\}$$

$$= \delta_{\alpha\beta} \sum_{l=-\infty}^{\infty} e^{i\phi_\alpha l} \delta(E - E' - l\hbar\Omega_0) \sum_{n=-\infty}^{\infty} J_n\left(\frac{eV_\alpha}{\hbar\Omega_0}\right) J_{n-l}\left(\frac{eV_\alpha}{\hbar\Omega_0}\right)$$

$$= \delta_{\alpha\beta} \sum_{l=-\infty}^{\infty} e^{i\phi_\alpha l} \delta(E - E' - l\hbar\Omega_0) \delta_{l0} = \delta_{\alpha\beta} \delta(E - E'),$$

where we introduced $l = n - m$.

Next we calculate the distribution function, $\tilde{f}_\alpha(E) = \langle \hat{a}^\dagger_\alpha(E) \hat{a}_\alpha(E) \rangle$, for electrons in the lead α

$$\tilde{f}_\alpha(E) = \sum_{n=-\infty}^{\infty} J_n^2\left(\frac{eV_\alpha}{\hbar\Omega_0}\right) f_\alpha(E - n\hbar\Omega_0). \quad (5.25)$$

This distribution function is non-equilibrium, that is due to changing conditions (the oscillating potential vanishes in the lead) there are no relaxation processes present. Despite the non-equilibrium state, the electrons in the lead incident to the scatterer carry a current $I_\alpha^{(in)}$, which is independent of the oscillating potential $V_\alpha(t)$. This current is time independent and it coincides with a current of equilibrium particles

$$I_\alpha^{(in)} = -\frac{e}{h}\int_0^\infty dE\, \tilde{f}_\alpha(E) = -\frac{e}{h}\int_0^\infty dE \sum_{n=-\infty}^{\infty} J_n^2\left(\frac{eV_\alpha}{\hbar\Omega_0}\right) f_\alpha(E - n\hbar\Omega_0)$$

(5.26)

$$= -\frac{e}{h}\int_0^\infty dE\, f_\alpha(E) \sum_{n=-\infty}^{\infty} J_n^2\left(\frac{eV_\alpha}{\hbar\Omega_0}\right) = -\frac{e}{h}\int_0^\infty dE\, f_\alpha(E).$$

In the second line of this equation we made a shift $E \to E + n\hbar\Omega_0$ under the integral over energy. As always, we use a wideband approximation, i.e., we assume that only electrons with energy $E \sim \mu$ are relevant for transport. Therefore, we can relax what is happening at $E \approx 0$, where, strictly speaking, the decomposition given in Eq. (5.24) fails.

As the next step we need to express the creation/annihilation operators, $\hat{b}_\alpha/\hat{b}_\alpha^\dagger$, for electrons scattered into the lead α in terms of operators, $\hat{a}'_\beta/\hat{a}'^\dagger_\beta$, for electrons in reservoirs. The relation between the operators $\hat{b}_\alpha$ for scattered electrons and the operators $\hat{a}_\beta$ for incident electrons is given in Eq. (3.32). Then using Eq. (5.24) we finally get

$$\hat{b}_\alpha(E) = \sum_{\delta=1}^{N_r} \sum_{n'=-\infty}^{\infty} \sum_{p'=-\infty}^{\infty} S_{F,\alpha\delta}(E, E_{n'}) e^{-i(n'+p')\phi_\delta} J_{n'+p'}\left(\frac{eV_\delta}{\hbar\Omega_0}\right) \hat{a}'_\delta(E_{-p'}),$$

(5.27)

$$\hat{b}_\alpha^\dagger(E) = \sum_{\gamma=1}^{N_r} \sum_{n=-\infty}^{\infty} \sum_{p=-\infty}^{\infty} S^*_{F,\alpha\gamma}(E, E_n) e^{i(n+p)\phi_\gamma} J_{n+p}\left(\frac{eV_\gamma}{\hbar\Omega_0}\right) \hat{a}'^\dagger_\gamma(E_{-p}).$$

These operators, as expected for fermionic operators, are anti-commuting. To show this we write

$$\left\{\hat{b}_\alpha^\dagger(E), \hat{b}_\beta(E')\right\} = \sum_{\gamma=1}^{N_r} \sum_{\delta=1}^{N_r} \sum_{n=-\infty}^{\infty} \sum_{p=-\infty}^{\infty} \sum_{n'=-\infty}^{\infty} \sum_{p'=-\infty}^{\infty} e^{i(n+p)\phi_\gamma} e^{-i(n'+p')\phi_\delta}$$

$$\times J_{n+p}\left(\frac{eV_\gamma}{\hbar\Omega_0}\right) J_{n'+p'}\left(\frac{eV_\delta}{\hbar\Omega_0}\right) S^*_{F,\alpha\gamma}(E, E_n) S_{F,\beta\delta}(E', E_{n'})$$

$$\times \left\{\hat{a}'^\dagger_\gamma(E - p\hbar\Omega_0), \hat{a}'_\delta(E' - p'\hbar\Omega_0)\right\}.$$

Then we take into account that

$$\left\{\hat{a}'^{\dagger}_{\gamma}\left(E-p\hbar\Omega_0\right),\hat{a}'^{\dagger}_{\delta}\left(E'-p'\hbar\Omega_0\right)\right\} = \delta_{\gamma\delta}\,\delta\left(E-E'+(p'-p)\hbar\Omega_0\right),$$

and proceed as follows. With the help of $\delta_{\gamma\delta}$ we sum over δ. Because of the Dirac delta function we write $E' = E + (p'-p)\hbar\Omega_0 \equiv E_{p'-p}$ instead of E'. Then we introduce $m = p' - p$ instead of p', $k = n - n' - m$ instead of n', and $q = n + p$ instead of p. After that we calculate:

$$\left\{\hat{b}^{\dagger}_{\alpha}(E),\hat{b}_{\beta}(E')\right\} = \sum_{\gamma=1}^{N_r}\sum_{m=-\infty}^{\infty}\delta\left(E-E'+m\hbar\Omega_0\right)\sum_{n=-\infty}^{\infty}\sum_{k=-\infty}^{\infty}S^*_{F,\alpha\gamma}(E,E_n)$$

$$\times e^{ik\phi_\gamma}S_{F,\beta\gamma}(E_m,E_{n-k})\sum_{q=-\infty}^{\infty}J_q\left(\frac{eV_\gamma}{\hbar\Omega_0}\right)J_{q+k}\left(\frac{eV_\gamma}{\hbar\Omega_0}\right).$$

Using Eq. (5.21) for Bessel functions we simplify the above equation as follows:

$$\left\{\hat{b}^{\dagger}_{\alpha}(E),\hat{b}_{\beta}(E')\right\} = \sum_{\gamma=1}^{N_r}\sum_{m=-\infty}^{\infty}\delta\left(E-E'+m\hbar\Omega_0\right)$$

$$\times \sum_{n=-\infty}^{\infty}S^*_{F,\alpha\gamma}(E,E_n)\,S_{F,\beta\gamma}(E_m,E_n).$$

Finally we take into account the unitarity of the Floquet scattering matrix, Eq. (3.28b), and find the required anti-commutation relation for operators of scattered electrons:

$$\left\{\hat{b}^{\dagger}_{\alpha}(E),\hat{b}_{\beta}(E')\right\} = \delta(E-E')\,\delta_{\alpha\beta}. \qquad (5.28)$$

For the sake of completeness we give a distribution function, $f^{(out)}_\alpha(E) = \left\langle\hat{b}^{\dagger}_{\alpha}(E)\hat{b}_{\alpha}(E)\right\rangle$, for electrons scattered into the lead α [41]:

$$f^{(out)}_\alpha(E) = \sum_{\gamma=1}^{N_r}\sum_{n=-\infty}^{\infty}\sum_{n'=-\infty}^{\infty}S^*_{\alpha\gamma}(E,E_n)\,S_{\alpha\gamma}(E,E_{n'})\,e^{i(n-n')\phi_\gamma}$$

$$(5.29)$$

$$\times \sum_{p=-\infty}^{\infty}J_{n+p}\left(\frac{eV_\gamma}{\hbar\Omega_0}\right)J_{n'+p}\left(\frac{eV_\gamma}{\hbar\Omega_0}\right)f_\gamma(E-p\hbar\Omega_0).$$

Note this equation is real. To show this one can calculate a complex conjugate quantity. Then after an irrelevant replacement, $n \leftrightarrow n'$, we arrive at the initial equation.

5.2.2 Alternating current

Substituting Eqs. (5.24) and (5.27) into Eq. (3.34) and taking into account Eq. (3.33a) we arrive at the current operator $\hat{I}_\alpha(t)$. Further, with Eq. (5.22) we average quantum statistically over the equilibrium state of the reservoirs and find the following equation for the time-dependent current, $I_\alpha(t) = \langle \hat{I}_\alpha(t) \rangle$:

$$I_\alpha(t) = \sum_{l=-\infty}^{\infty} e^{-il\Omega_0 t} I_{\alpha,l}, \qquad (5.30a)$$

$$I_{\alpha,l} = \frac{e}{h} \int_0^\infty dE \left\{ \sum_{\gamma=1}^{N_r} \sum_{n=-\infty}^{\infty} \sum_{n'=-\infty}^{\infty} e^{i(n-n'-l)\phi_\gamma} S^*_{\alpha\gamma}(E, E_n) S_{\alpha\gamma}(E_l, E_{n'+l}) \right.$$

$$\left. \times \sum_{p=-\infty}^{\infty} J_{n+p}\left(\frac{eV_\gamma}{\hbar\Omega_0}\right) J_{n'+l+p}\left(\frac{eV_\gamma}{\hbar\Omega_0}\right) f_\gamma(E - p\hbar\Omega_0) - \delta_{l0} f_\alpha(E) \right\}. \qquad (5.30b)$$

Let us transform this equation so that it is a difference of Fermi functions. To this end we use Eqs. (3.28) and (5.21) and find the following expression for the Fourier harmonics of the current:

$$I_{\alpha,l} = \frac{e}{h} \int_0^\infty dE \sum_{\gamma=1}^{N_r} \sum_{p=-\infty}^{\infty} \{f_\gamma(E - p\hbar\Omega_0) - f_\alpha(E)\} \sum_{n=-\infty}^{\infty} \sum_{n'=-\infty}^{\infty} e^{i(n-n'-l)\phi_\gamma}$$

$$\times S^*_{\alpha\gamma}(E, E_n) S_{\alpha\gamma}(E_l, E_{n'+l}) J_{n+p}\left(\frac{eV_\gamma}{\hbar\Omega_0}\right) J_{n'+l+p}\left(\frac{eV_\gamma}{\hbar\Omega_0}\right). \qquad (5.31)$$

This equation is convenient to use in the adiabatic regime, when we can expand the Fermi function difference in powers of Ω_0.

5.2.3 Direct current

A more compact equation can be obtained for a time-independent part of the current, $l = 0$. First of all we express the Floquet scattering matrix in terms of the scattering matrix $\hat{S}_{out}(E,t)$, see Eq. (3.59b):

$$S_{\alpha\gamma}(E, E_{n'}) = S_{out,\alpha\gamma,-n'}(E), \quad S^*_{\alpha\gamma}(E, E_n) = S^*_{out,\alpha\gamma,-n}(E).$$

Then using the series (5.19) we express the Bessel functions in terms of the Fourier coefficients for some exponential function dependent on an oscillating potential, $V_\gamma(t)$:

$$J_{n'+p}\left(\frac{eV_\gamma}{\hbar\Omega_0}\right) = e^{i(n'+p)\phi_\gamma}\left(e^{-i\hbar^{-1}\int_{-\infty}^{t}dt'eV_\gamma(t')}\right)_{n'+p}.$$

Note the lower limit in the time integral is irrelevant since it does not affect the value of the Fourier coefficient. Using the equation above in Eq. (5.31) and summing over n and n' with the help of the following property of the Fourier coefficients,

$$\sum_{n'=-\infty}^{\infty} A_{-n'} B_{p+n'} = (AB)_p \quad \sum_{n=-\infty}^{\infty} (A_{-n})^* (B^*)_{-p-n} = (A^*B^*)_{-p},$$
(5.32)

we finally calculate the direct current in the lead α [45]:

$$I_{\alpha,0} = \frac{e}{h}\int_0^\infty dE \sum_{\gamma=1}^{N_r} \sum_{p=-\infty}^{\infty} \{f_\beta(E - p\hbar\Omega_0) - f_\alpha(E)\}$$
(5.33)
$$\times \left|\left(e^{-i\hbar^{-1}\int_{-\infty}^{t}dt'eV_\gamma(t')}S_{out,\alpha\gamma}(E,t)\right)_p\right|^2.$$

As we see the reservoir's oscillating potential can be taken into account as an additional phase factor in the corresponding scattering matrix elements. Since, from Eq. (4.13), the phase of the scattering matrix elements defines a generated current, we can guess that the presence of oscillating potentials at the reservoirs modifies the generated current.

5.2.4 Adiabatic direct current

To clarify the effect of the potentials $V_\beta(t)$ on the direct current $I_{\alpha,0}$ we consider an adiabatic regime, $\varpi \ll 1$, and restrict ourselves to terms linear in the oscillating potentials,

$$|eV_\beta| \ll \hbar\Omega_0 \ll \delta E, \quad \forall \beta, \tag{5.34}$$

where δE is a characteristic energy introduced for Eq. (3.49). We assume also no bias conditions, Eq. (4.1).

Let us expand the difference of the Fermi functions in Eq. (5.33) in powers of the pump frequency:

$$f_0\left(E - p\hbar\Omega_0\right) - f_0\left(E\right) \approx \left(-\frac{\partial f_0}{\partial E}\right) p\hbar\Omega_0 + \frac{p^2\left(\hbar\Omega_0\right)^2}{2}\frac{\partial^2 f_0}{\partial E^2}. \tag{5.35}$$

Here we need to keep terms quadratic in Ω_0. They are necessary since the phase factors dependent on $V_\gamma(t')$ result in a factor Ω_0^{-1}.

We substitute Eq. (5.35) into Eq. (5.33) and sum over p. Then we take into account the adiabatic expansion, Eq. (3.61b), for the scattering matrix $\hat{S}_{out}$ and keep only terms linear in both $V_\gamma(t)$ and Ω_0. For brevity we introduce $\Upsilon_\gamma(t) = \exp\left\{-\frac{i}{\hbar}\int\limits_{-\infty}^{t} dt' eV_\gamma(t')\right\}$. So, the term linear in Ω_0 in Eq. (5.35) results in the following:

$$\hbar \sum_{n=-\infty}^{\infty} \Omega_0 p \left|(\Upsilon_\gamma S_{out,\alpha\gamma})_p\right|^2 = -i\hbar \int\limits_0^{\mathcal{T}} \frac{dt}{\mathcal{T}} \Upsilon_\gamma S_{out,\alpha\gamma} \frac{\partial}{\partial t}\left(\Upsilon_\gamma^* S_{out,\alpha\gamma}^*\right)$$

$$= \int\limits_0^{\mathcal{T}} \frac{dt}{\mathcal{T}} eV_\gamma(t) \left\{|S_{\alpha\gamma}|^2 - \frac{i\hbar}{2}\left(\frac{\partial^2 S_{\alpha\gamma}}{\partial t \partial E} S_{\alpha\gamma}^* - S_{\alpha\gamma}\frac{\partial^2 S_{\alpha\gamma}^*}{\partial t \partial E}\right)\right.$$

$$\left. + 2\hbar\Omega_0 \operatorname{Re}\left[S_{\alpha\gamma}^* A_{\alpha\gamma}\right]\right\} - i\hbar \int\limits_0^{\mathcal{T}} \frac{dt}{\mathcal{T}} S_{\alpha\gamma} \frac{\partial S_{\alpha\gamma}^*}{\partial t} + \mathcal{O}\left(\Omega_0^2\right).$$

The term that is quadratic in the pump frequency in Eq. (5.35) is

$$\sum_{n=-\infty}^{\infty} \Omega_0^2 p^2 \left|(\Upsilon_\gamma S_{out,\alpha\gamma})_p\right|^2 = \int\limits_0^{\mathcal{T}} \frac{dt}{\mathcal{T}} \frac{\partial}{\partial t}\left(\Upsilon_\gamma S_{out,\alpha\gamma}\right) \frac{\partial}{\partial t}\left(\Upsilon_\gamma^* S_{out,\alpha\gamma}^*\right)$$

$$= \frac{i}{\hbar} \int\limits_0^{\mathcal{T}} \frac{dt}{\mathcal{T}} eV_\gamma(t) \left\{\frac{\partial S_{\alpha\gamma}}{\partial t} S_{\alpha\gamma}^* - S_{\alpha\gamma}\frac{\partial S_{\alpha\gamma}^*}{\partial t}\right\} + \mathcal{O}\left(\Omega_0^2, V_\gamma^2\right).$$

The last equation applies to the current, Eq. (5.33), with factor $\partial^2 f_0/\partial E^2$. We integrate over energy by parts and calculate

$$\int_0^\infty dE \frac{\partial^2 f_0(E)}{\partial E^2} \left\{ \frac{\partial S_{\alpha\gamma}}{\partial t} S_{\alpha\gamma}^* - S_{\alpha\gamma} \frac{\partial S_{\alpha\gamma}^*}{\partial t} \right\} = \int_0^\infty dE \left(-\frac{\partial f_0(E)}{\partial E} \right)$$

$$\times \left\{ \frac{\partial^2 S_{\alpha\gamma}}{\partial t \partial E} S_{\alpha\gamma}^* - S_{\alpha\gamma} \frac{\partial^2 S_{\alpha\gamma}^*}{\partial t \partial E} + \frac{\partial S_{\alpha\gamma}}{\partial t} \frac{\partial S_{\alpha\gamma}^*}{\partial E} - \frac{\partial S_{\alpha\gamma}}{\partial E} \frac{\partial S_{\alpha\gamma}^*}{\partial t} \right\},$$

where we used $\partial f_0/\partial E|_{E=\infty} = 0$ and $\partial f_0/\partial E|_{E=0} = 0$. Note the latter is valid at $k_B T \ll \mu$.

With the above transformations we can represent a direct current, $I_{\alpha,0}$, as the sum of three terms linear in both Ω_0 and V_γ [41]:

$$I_{\alpha,0} = I_{\alpha,0}^{(pump)} + I_{\alpha,0}^{(rect)} + I_{\alpha,0}^{(int)}. \tag{5.36a}$$

Here the current $I_{\alpha,0}^{(pump)}$, generated by the dynamic scatterer in the absence of an oscillating bias, is given by Eq. (4.10). The next term,

$$I_{\alpha,0}^{(rect)} = \frac{e^2}{h} \int_0^\infty dE \left(-\frac{\partial f_0(E)}{\partial E} \right) \int_0^{\mathcal{T}} \frac{dt}{\mathcal{T}} \sum_{\gamma=1}^{N_r} V_\gamma(t) |S_{\alpha\gamma}(E,t)|^2 \tag{5.36b}$$

is a rectification current. It is due to the rectification of the alternating currents, produced by the time-dependent potentials $V_\gamma(t)$, in the time-dependent conductance. The coexistence of rectified and generated currents has been investigated theoretically [60, 83, 175–177] and experimentally [151, 152].

And, finally, the last term in Eq. (5.36a), an interference contribution,

$$I_{\alpha,0}^{(int)} = \frac{e^2}{h} \int_0^\infty dE \left(-\frac{\partial f_0}{\partial E} \right) \int_0^{\mathcal{T}} \frac{dt}{\mathcal{T}} \sum_{\gamma=1}^{N_r} V_\gamma(t)$$

$$\times \left(2\hbar\Omega_0 \text{Re}\left[S_{\alpha\gamma}^* A_{\alpha\gamma} \right] + \frac{1}{2} P\{ S_{\alpha\gamma} S_{\alpha\gamma}^* \} \right), \tag{5.36c}$$

is due to a mutual influence (an interference) between the currents generated by the scatterer and the currents due to an AC bias. This part of the current shares features with both the generated current (it is proportional to Ω_0) and the rectified current (it is proportional to V_γ).

Physically the splitting of Eq. (5.36a) into three parts is justified by the fact that each part separately is subject to the conservation law, Eq. (4.11):

$$\sum_{\alpha=1}^{N_r} I_\alpha^{(x)} = 0, \quad x = pump,\ rect,\ int. \tag{5.37}$$

Let us analyze the conditions necessary for the existence of each of the mentioned contributions. As we have already seen, see Section 4.1.2, the current $I_{\alpha,0}^{(pump)}$ is absent if the frozen scattering matrix is time-reversal invariant:

$$\hat{S}(t, E) = \hat{S}(-t, E). \tag{5.38}$$

The rectified current, $I_{\alpha,0}^{(rect)}$, depends in fact on the potential difference, $\Delta V_{\gamma\alpha}(t) = V_\gamma(t) - V_\alpha(t)$, and it vanishes if the potentials of all the reservoirs are the same,

$$V_\gamma(t) = V(t), \quad \forall \gamma. \tag{5.39}$$

To show this we use the unitarity of the scattering matrix, see Eq. (3.47), and find $\sum_{\gamma=1}^{N_r} |S_{\alpha\gamma}(t, E)|^2 = 1$. Moreover, since the potentials are periodic we have $\int_0^{\mathcal{T}} dt\, V_\alpha(t) = 0$. Using these two conditions we calculate

$$\int_0^{\mathcal{T}} \frac{dt}{\mathcal{T}} \sum_{\gamma=1}^{N_r} V_\alpha(t) |S_{\alpha\gamma}(t, E)|^2 = 0.$$

And finally subtracting the identity above from Eq. (5.36b), we find the required equation,

$$I_\alpha^{(rect)} = \int_0^{\mathcal{T}} \frac{dt}{\mathcal{T}} \sum_{\gamma=1}^{N_r} G_{\alpha\gamma}(t) \left\{ V_\gamma(t) - V_\alpha(t) \right\}, \tag{5.40}$$

where the frozen conductance matrix elements,

$$G_{\alpha\gamma}(t) = G_0 \int_0^\infty dE \left(-\frac{\partial f_0(E)}{\partial E} \right) |S_{\alpha\gamma}(t, E)|^2 \tag{5.41}$$

are defined in the same way as in the stationary case, see Eq. (1.54).

In contrast, the last contribution, $I_{\alpha,0}^{(int)}$, is present even if both Eqs. (5.38) and (5.39) are fulfilled, and neither the pumped nor the rectified currents exist. To show this we first use Eq. (5.39) and rewrite Eq. (5.36c) as follows:

$$I_{\alpha,0}^{(int)} = \frac{e^2}{h} \int_0^\infty dE \left(-\frac{\partial f_0}{\partial E}\right) \int_0^\mathcal{T} \frac{dt}{\mathcal{T}} V(t) P\left\{\hat{S}(t,E), \hat{S}^\dagger(t,E)\right\}_{\alpha\alpha}. \quad (5.42)$$

Here we have summed over γ using the following identity

$$4\hbar\Omega_0 \sum_{\gamma=1}^{N_r} \text{Re}\left\{S_{\alpha\gamma}^* A_{\alpha\gamma}\right\} = P\left\{\hat{S}, \hat{S}^\dagger\right\}_{\alpha\alpha}. \quad (5.43)$$

To prove this equation we multiply the matrix equation (3.52) from the left by $\hat{S}$ and from the right by $\hat{S}^\dagger$ and take its diagonal element.

Under the conditions given in Eq. (5.38) the pumped current is zero, while the interference contribution, Eq. (5.42), can survive. The current $I_{\alpha,0}^{(int)}$, Eq. (5.42), is not zero if the potential $V(t)$ is shifted in phase with respect to the parameters that vary in time $p_i(t)$ of the scatterer. Therefore, to analyze the ability of the entire system, i.e., the scatterer plus reservoirs, to generate a direct current, $I_{\alpha,0} \neq 0$, it is necessary to take into account phases of all the time-dependent quantities, as parameters of the scatterer as possible time-dependent potentials at the reservoirs.

Chapter 6

Noise generated by the dynamic scatterer

The current correlation function $P_{\alpha\beta}(t_1, t_2)$, is defined in Eqs. (2.30) and (2.39) in the time and frequency domains, respectively. Such a defined correlator is called *a symmetrized correlator*. It satisfies the following symmetries,

$$P_{\alpha\beta}(t_1, t_2) = P_{\beta\alpha}(t_2, t_1), \qquad (6.1a)$$

$$P_{\alpha\beta}(\omega_1, \omega_2) = P_{\beta\alpha}(\omega_2, \omega_1), \qquad (6.1b)$$

which are a direct consequence of the fact that the currents, measured in leads α and β, symmetrically enter the correlator.

6.1 Spectral noise power

If currents are generated by a periodic dynamic scatterer, then the correlation function can be represented as follows (compare with Eq. (2.33), valid in the case of a stationary scatterer) [44]:

$$P_{\alpha\beta}(\omega_1, \omega_2) = \sum_{l=-\infty}^{\infty} 2\pi\delta(\omega_1 + \omega_2 - l\Omega_0) \mathcal{P}_{\alpha\beta,l}(\omega_1, \omega_2), \qquad (6.2a)$$

where the spectral power $\mathcal{P}_{\alpha\beta,l}(\omega_1, \omega_2)$ is expressed in terms of the Floquet scattering matrix elements as

$$\mathcal{P}_{\alpha\beta,l}(\omega_1, \omega_2) = \frac{e^2}{h} \int_0^\infty dE \left\{ \delta_{\alpha\beta}\, \delta_{l0}\, F_{\alpha\alpha}(E, E + \hbar\omega_1) \right. \qquad (6.2b)$$

$$- \sum_{n=-\infty}^{\infty} F_{\alpha\alpha}(E, E+\hbar\omega_1) S_{F,\beta\alpha}^*(E_n+\hbar\omega_1, E+\hbar\omega_1) S_{F,\beta\alpha}(E_{n+l}, E)$$

$$- \sum_{n=-\infty}^{\infty} F_{\beta\beta}(E, E+\hbar\omega_2) S_{F,\alpha\beta}^*(E_n+\hbar\omega_2, E+\hbar\omega_2) S_{F,\alpha\beta}(E_{n+l}, E)$$

$$+ \sum_{\gamma=1}^{N_r} \sum_{\delta=1}^{N_r} \sum_{n=-\infty}^{\infty} \sum_{m=-\infty}^{\infty} \sum_{p=-\infty}^{\infty} F_{\gamma\delta}(E_{l+n}, E_m + \hbar\omega_1) S_{F,\beta\gamma}(E_{l+p}, E_{l+n})$$

$$\times S_{F,\alpha\gamma}^*(E, E_{l+n}) S_{F,\alpha\delta}(E+\hbar\omega_1, E_m+\hbar\omega_1) S_{F,\beta\delta}^*(E_p+\hbar\omega_1, E_m+\hbar\omega_1)\Big\}.$$

The quantity $F_{\alpha\beta}$, being a combination of the Fermi functions, is defined in Eq. (2.46). To derive the equation above we proceed as in Section 2.2.2 but instead of Eq. (1.39) now we use Eq. (3.32) to relate the operators $\hat{b}_\alpha$ for scattered electrons to the operators $\hat{a}_\beta$ for incident electrons.

First of all we represent $P_{\alpha\beta}(\omega_1, \omega_2)$ as the sum of four quantities $P_{\alpha\beta}^{(i,j)}(\omega_1, \omega_2)$, $i,j = in, out$ as Eq. (2.43). For instance, $P_{\alpha\beta}^{(in,out)}(\omega_1, \omega_2)$ is a correlation function for a current of incident electrons in the lead α and a current of scattered electrons in the lead β. So, the spectral power is

$$\mathcal{P}_{\alpha\beta,l}(\omega_1,\omega_2) = \sum_{i,j=in,out} \mathcal{P}_{\alpha\beta,l}^{(i,j)}(\omega_1,\omega_2). \qquad (6.3)$$

Since incident electrons still do not interact with the scatterer then the part of the correlator related to incident currents is the same in the dynamic as in the stationary case. Therefore, for $P_{\alpha\beta}^{(in,in)}$ we can use Eq. (2.45) and write

$$\mathcal{P}_{\alpha\beta,l}^{(in,in)}(\omega_1,\omega_2) = \delta_{\alpha\beta}\,\delta_{l0}\,\frac{e^2}{h}\int_0^\infty dE\, F_{\alpha\alpha}(E, E+\hbar\omega_1). \qquad (6.4)$$

Next we calculate $P_{\alpha\beta}^{(in,out)}$:

$$P_{\alpha\beta}^{(in,out)}(\omega_1,\omega_2) = e^2 \int_0^\infty dE_1 \int_0^\infty dE_2 \Big\{$$

$$\left\langle \hat{a}_\alpha^\dagger (E_1) \hat{a}_\alpha (E_1 + \hbar\omega_1) \right\rangle \left\langle \hat{b}_\beta^\dagger (E_2) \hat{b}_\beta (E_2 + \hbar\omega_2) \right\rangle$$

$$-\frac{1}{2}\left\langle \hat{a}_\alpha^\dagger (E_1) \hat{a}_\alpha (E_1 + \hbar\omega_1) \hat{b}_\beta^\dagger (E_2) \hat{b}_\beta (E_2 + \hbar\omega_2) \right\rangle$$

$$-\frac{1}{2}\left\langle \hat{b}_\beta^\dagger (E_2) \hat{b}_\beta (E_2 + \hbar\omega_2) \hat{a}_\alpha^\dagger (E_1) \hat{a}_\alpha (E_1 + \hbar\omega_1) \right\rangle \Big\}.$$

(6.5)

According to Wick's theorem (see, e.g., Ref. [30]) the mean of the product of four operators is the sum of products of two pair means. For instance:

$$\left\langle \hat{a}_\alpha^\dagger (E) \hat{a}_\alpha (E_1 + \hbar\omega_1) \hat{b}_\beta^\dagger (E_2) \hat{b}_\beta (E_2 + \hbar\omega_2) \right\rangle =$$

$$\left\langle \hat{a}_\alpha^\dagger (E_1) \hat{a}_\alpha (E_1 + \hbar\omega_1) \right\rangle \left\langle \hat{b}_\beta^\dagger (E_2) \hat{b}_\beta (E_2 + \hbar\omega_2) \right\rangle$$

$$+ \left\langle \hat{a}_\alpha^\dagger (E_1) \hat{b}_\beta (E_2 + \hbar\omega_2) \right\rangle \left\langle \hat{a}_\alpha (E_1 + \hbar\omega_1) \hat{b}_\beta^\dagger (E_2) \right\rangle.$$

We can use Wick's theorem since the operators $\hat{a}_\alpha$ correspond to particles in macroscopic reservoirs and the operators $\hat{b}_\beta$ are the linear combination of $\hat{a}_\alpha$. The first term on the right hand side (RHS) of the equation above does not contribute to the correlator, since it is compensated exactly by the corresponding product of currents [the first term on the RHS of Eq. (6.5)]. Therefore, only those pair means are relevant that comprise particle operators from both current operators $\hat{I}_\alpha^{(in)}$ and $\hat{I}_\beta^{(out)}$ simultaneously. To calculate such pair means we use Eq. (3.32). In particular:

$$\left\langle \hat{a}_\alpha^\dagger (E_1) \hat{b}_\beta (E_2 + \hbar\omega_2) \right\rangle = \sum_{\gamma=1}^{N_r} \sum_{m=-\infty}^{\infty} S_{F,\beta\gamma} (E_2 + \hbar\omega_2, E_2 + \hbar[\omega_2 + m\Omega_0])$$

$$\times \left\langle \hat{a}_\alpha^\dagger (E_1) \hat{a}_\gamma (E_2 + \hbar[\omega_2 + m\Omega_0]) \right\rangle$$

$$= \sum_{\gamma=1}^{N_r} \sum_{m=-\infty}^{\infty} S_{F,\beta\gamma} (E_2 + \hbar\omega_2, E_2 + \hbar[\omega_2 + m\Omega_0])$$

$$\times \delta_{\alpha\gamma} \delta (E_1 - E_2 - \hbar[\omega_2 + m\Omega_0]) f_\alpha (E_1)$$

$$= \sum_{m=-\infty}^{\infty} S_{F,\beta\alpha} (E_2 + \hbar\omega_2, E_2 + \hbar[\omega_2 + m\Omega_0])$$

$$\times \delta (E_1 - E_2 - \hbar[\omega_2 + m\Omega_0]) f_\alpha (E_1).$$

By analogy we calculate all the other pair means that appear in Eq. (6.5):

$$\left\langle \hat{a}_\alpha \left(E_1 + \hbar\omega_1\right) \hat{b}_\beta^\dagger \left(E_2\right) \right\rangle = \sum_{n=-\infty}^{\infty} S_{F,\beta\alpha}^* \left(E_2, E_2 + n\hbar\Omega_0\right)$$

$$\times \delta\left(E_1 + \hbar\omega_1 - E_2 - n\hbar\Omega_0\right) \left[1 - f_\alpha\left(E_1 + \hbar\omega_1\right)\right],$$

$$\left\langle \hat{b}_\beta^\dagger \left(E_2\right) \hat{a}_\alpha \left(E_1 + \hbar\omega_1\right) \right\rangle = \sum_{n=-\infty}^{\infty} S_{F,\beta\alpha}^* \left(E_2, E_2 + n\hbar\Omega_0\right)$$

$$\times \delta\left(E_1 + \hbar\omega_1 - E_2 - n\hbar\Omega_0\right) f_\alpha\left(E_1 + \hbar\omega_1\right),$$

$$\left\langle \hat{b}_\beta \left(E_2 + \hbar\omega_2\right) \hat{a}_\alpha^\dagger \left(E_1\right) \right\rangle = \sum_{m=-\infty}^{\infty} S_{F,\beta\alpha} \left(E_2 + \hbar\omega_2, E_2 + \hbar[\omega_2 + m\Omega_0]\right)$$

$$\times \delta\left(E_1 - E_2 - \hbar[\omega_2 + m\Omega_0]\right)\left[1 - f_\alpha\left(E_1\right)\right].$$

Substituting these pair means into Eq. (6.5) we arrive at the sum of two terms. Then using the Dirac delta function to integrate over energy, say over E_2, we get the following for each of these terms:

$$\int_0^\infty dE_2 \frac{1}{2} S_{F,\beta\alpha}\left(E_2 + \hbar\omega_2, E_{2,m} + \hbar\omega_2\right) \delta\left(E_1 - E_{2,m} - \hbar\omega_2\right)$$

$$\times f_\alpha\left(E_1\right) S_{F,\beta\alpha}^* \left(E_2, E_{2,n}\right) \delta\left(E_1 + \hbar\omega_1 - E_{2,n}\right)\left[1 - f_\alpha\left(E_1 + \hbar\omega_1\right)\right]$$

$$= \frac{1}{2\hbar} \delta\left(\omega_1 + \omega_2 + (m-n)\Omega_0\right) f_\alpha\left(E_1\right) \left[1 - f_\alpha\left(E_1 + \hbar\omega_1\right)\right]$$

$$\times S_{F,\beta\alpha}^*\left(E_{1,-n} + \hbar\omega_1, E_1 + \hbar\omega_1\right) S_{F,\beta\alpha}\left(E_{1,-m}, E_1\right),$$

$$\int_0^\infty dE_2 \frac{1}{2} S_{F,\beta\alpha}^* \left(E_2, E_{2,n}\right) \delta\left(E_1 + \hbar\omega_1 - E_{2,n}\right) f_\alpha\left(E_1 + \hbar\omega_1\right)$$

$$\times S_{F,\beta\alpha}\left(E_2 + \hbar\omega_2, E_{2,m} + \hbar\omega_2\right) \delta\left(E_1 - E_{2,m} - \hbar\omega_2\right)\left[1 - f_\alpha\left(E_1\right)\right]$$

$$= \frac{1}{2\hbar} \delta\left(\omega_1 + \omega_2 + (m-n)\Omega_0\right) f_\alpha\left(E_1 + \hbar\omega_1\right)\left[1 - f_\alpha\left(E_1\right)\right]$$

$$\times S^*_{F,\beta\alpha}\left(E_{1,-n} + \hbar\omega_1, E_1 + \hbar\omega_1\right) S_{F,\beta\alpha}\left(E_{1,-m}, E_1\right),$$

where $E_{i,k} = E_i + k\hbar\Omega_0$, $i = 1, 2$. Using the equations above in Eq. (6.5), introducing $l = n - m$ instead of m, and replacing $n \to -n$ and $E_1 \to E$, we finally find:

$$P^{(in,out)}_{\alpha\beta}(\omega_1, \omega_2) = \sum_{l=-\infty}^{\infty} 2\pi\delta\left(\omega_1 + \omega_2 - l\Omega_0\right) \mathcal{P}^{(in,out)}_{\alpha\beta,l}(\omega_1, \omega_2), \quad (6.6a)$$

with

$$\mathcal{P}^{(in,out)}_{\alpha\beta}(\omega_1, \omega_2) = -\frac{e^2}{h} \int_0^\infty dE \sum_{n=-\infty}^{\infty} F_{\alpha\alpha}(E, E + \hbar\omega_1) \qquad (6.6b)$$

$$\times S^*_{F,\beta\alpha}\left(E_n + \hbar\omega_1, E + \hbar\omega_1\right) S_{F,\beta\alpha}\left(E_{n+l}, E\right).$$

In the same way we find:

$$P^{(out,in)}_{\alpha\beta}(\omega_1, \omega_2) = e^2 \int_0^\infty dE_1 \int_0^\infty dE_2 \Big\{ \qquad (6.7)$$

$$\left\langle \hat{b}^\dagger_\alpha(E_1) \hat{b}_\alpha(E_1 + \hbar\omega_1) \right\rangle \left\langle \hat{a}^\dagger_\beta(E_2) \hat{a}_\beta(E_2 + \hbar\omega_2) \right\rangle$$

$$-\frac{1}{2} \left\langle \hat{b}^\dagger_\alpha(E_1) \hat{b}_\alpha(E_1 + \hbar\omega_1) \hat{a}^\dagger_\beta(E_2) \hat{a}_\beta(E_2 + \hbar\omega_2) \right\rangle$$

$$-\frac{1}{2} \left\langle \hat{a}^\dagger_\beta(E_2) \hat{a}_\beta(E_2 + \hbar\omega_2) \hat{b}^\dagger_\alpha(E_1) \hat{b}_\alpha(E_1 + \hbar\omega_1) \right\rangle \Big\}.$$

Comparing this to Eq. (6.5) we see that $P^{(out,in)}_{\alpha\beta}(\omega_1, \omega_2)$ can be calculated from Eq. (6.6) after the following replacements: $\alpha \leftrightarrow \beta$, $E_1 \leftrightarrow E_2$, and $\omega_1 \leftrightarrow \omega_2$. As a result we get for the spectral power (replace $E_2 \to E$):

$$\mathcal{P}^{(out,in)}_{\alpha\beta}(\omega_1, \omega_2) = -\frac{e^2}{h} \int_0^\infty dE \sum_{n=-\infty}^{\infty} F_{\beta\beta}(E, E + \hbar\omega_2)$$

$$(6.8)$$

$$\times S^*_{F,\alpha\beta}\left(E_n + \hbar\omega_2, E + \hbar\omega_2\right) S_{F,\alpha\beta}\left(E_{n+l}, E\right).$$

Then we calculate the last contribution:

$$P_{\alpha\beta}^{(out,out)}(\omega_1,\omega_2) = \frac{e^2}{2} \int_0^\infty dE_1 \int_0^\infty dE_2 \Big\{$$

$$\left\langle \hat{b}_\alpha^\dagger(E_1)\hat{b}_\beta(E_2+\hbar\omega_2)\right\rangle \left\langle \hat{b}_\alpha(E_1+\hbar\omega_1)\hat{b}_\beta^\dagger(E_2)\right\rangle \qquad (6.9)$$

$$+ \left\langle \hat{b}_\beta^\dagger(E_2)\hat{b}_\alpha(E_1+\hbar\omega_1)\right\rangle \left\langle \hat{b}_\beta(E_2+\hbar\omega_2)\hat{b}_\alpha^\dagger(E_1)\right\rangle \Big\},$$

where we have already expressed the mean for four operators in terms of pair means. The first of these is

$$\left\langle \hat{b}_\alpha^\dagger(E_1)\hat{b}_\beta(E_2+\hbar\omega_2)\right\rangle = \sum_{\gamma=1}^{N_r}\sum_{r=-\infty}^{\infty}\sum_{\delta=1}^{N_r}\sum_{s=-\infty}^{\infty} \left\langle \hat{a}_\gamma^\dagger(E_{1,r})\hat{a}_\delta(E_{2,s}+\hbar\omega_2)\right\rangle$$

$$\times S_{F,\alpha\gamma}^*(E_1,E_{1,r}) S_{F,\beta\delta}(E_2+\hbar\omega_2,E_{2,s}+\hbar\omega_2) = \sum_{\gamma=1}^{N_r}\sum_{r=-\infty}^{\infty}\sum_{s=-\infty}^{\infty} f_\gamma(E_{1,r})$$

$$\times \delta(E_{1,r}-E_{2,s}-\hbar\omega_2) S_{F,\alpha\gamma}^*(E_1,E_{1,r}) S_{F,\beta\gamma}(E_2+\hbar\omega_2,E_{2,s}+\hbar\omega_2),$$

and, correspondingly, the second one is equal to

$$\left\langle \hat{b}_\alpha(E_1+\hbar\omega_1)\hat{b}_\beta^\dagger(E_2)\right\rangle = \sum_{\delta=1}^{N_r}\sum_{m=-\infty}^{\infty}\sum_{\gamma=1}^{N_r}\sum_{q=-\infty}^{\infty} \left\langle \hat{a}_\delta(E_{1,m}+\hbar\omega_1)\hat{a}_\gamma^\dagger(E_{2,q})\right\rangle$$

$$\times S_{F,\alpha\delta}(E_1+\hbar\omega_1,E_{1,m}+\hbar\omega_1) S_{F,\beta\gamma}^*(E_2,E_{2,q})$$

$$= \sum_{\delta=1}^{N_r}\sum_{m=-\infty}^{\infty}\sum_{q=-\infty}^{\infty} [1-f_\delta(E_{1,m}+\hbar\omega_1)]\,\delta(E_{1,m}+\hbar\omega_1-E_{2,q})$$

$$\times S_{F,\alpha\delta}(E_1+\hbar\omega_1,E_{1,m}+\hbar\omega_1) S_{F,\beta\delta}^*(E_2,E_{2,q}).$$

Integrating the product of these means over E_2 we get

$$\int_0^\infty dE_2 \, \delta\left(E_{1,r} - E_{2,s} - \hbar\omega_2\right) \delta\left(E_{1,m} + \hbar\omega_1 - E_{2,q}\right) f_\gamma\left(E_{1,r}\right)$$

$$\times \left[1 - f_\delta\left(E_{1,m} + \hbar\omega\right)\right] S^*_{F,\alpha\delta}\left(E_1, E_{1,r}\right) S_{F,\beta\gamma}\left(E_2 + \hbar\omega_2, E_{2,s} + \hbar\omega_2\right)$$

$$\times S_{F,\alpha\delta}\left(E_1 + \hbar\omega_1, E_{1,m} + \hbar\omega_1\right) S^*_{F,\beta\delta}\left(E_2, E_{2,q}\right)$$

$$= \frac{1}{\hbar} \delta\left(\omega_1 + \omega_2 - [r + q - s - m]\Omega_0\right) f_\gamma\left(E_{1,r}\right) \left[1 - f_\delta\left(E_{1,m} + \hbar\omega_1\right)\right]$$

$$\times S^*_{F,\alpha\gamma}\left(E_1, E_{1,r}\right) S_{F,\beta\gamma}\left(E_{1,r-s}, E_{1,r}\right)$$

$$\times S_{F,\alpha\delta}\left(E_1 + \hbar\omega_1, E_{1,m} + \hbar\omega_1\right) S^*_{F,\beta\delta}\left(E_{1,m-q} + \hbar\omega_1, E_{1,m} + \hbar\omega_1\right)$$

$$= \frac{1}{\hbar} \delta\left(\omega_1 + \omega_2 - l\Omega_0\right) f_\gamma\left(E_{1,l+n}\right) \left[1 - f_\delta\left(E_{1,m} + \hbar\omega_1\right)\right]$$

$$\times S^*_{F,\alpha\gamma}\left(E_1, E_{1,l+n}\right) S_{F,\beta\gamma}\left(E_{1,l+p}, E_{1,l+n}\right)$$

$$\times S_{F,\alpha\delta}\left(E_1 + \hbar\omega_1, E_{1,m} + \hbar\omega_1\right) S^*_{F,\beta\delta}\left(E_{1,p} + \hbar\omega_1, E_{1,m} + \hbar\omega_1\right),$$

where at the end we introduced new indices: $p = m - q$ (instead of q), $n = s + m - q$ (instead of s), and $l = r - s + q - m$ (instead of r).

Comparing the first and the second terms on the RHS of Eq. (6.9) one can see that the latter one results in the same expression as above but with $f_\gamma(E_{l+n})\left[1 - f_\delta(E_m + \hbar\omega)\right]$ being replaced by $f_\delta(E_m + \hbar\omega)\left[1 - f_\gamma(E_{l+n})\right]$. Therefore, finally, equation (6.9) results in

$$P^{(out,out)}_{\alpha\beta}(\omega_1, \omega_2) = \sum_{l=-\infty}^{\infty} 2\pi\delta(\omega_1 + \omega_2 - l\Omega_0) \mathcal{P}^{(out,out)}_{\alpha\beta,l}(\omega_1, \omega_2),$$

(6.10a)

where

$$\mathcal{P}_{\alpha\beta}^{(out,out)}(\omega_1,\omega_2) = \frac{e^2}{h} \int_0^\infty dE \sum_{\gamma=1}^{N_r} \sum_{\delta=1}^{N_r} \sum_{n=-\infty}^{\infty} \sum_{m=-\infty}^{\infty} \sum_{p=-\infty}^{\infty}$$

$$\times F_{\gamma\delta}(E_{l+n}, E_m + \hbar\omega_1) S_{F,\alpha\gamma}^*(E, E_{l+n}) S_{F,\beta\gamma}(E_{l+p}, E_{l+n}) \quad (6.10b)$$

$$\times S_{F,\alpha\delta}(E + \hbar\omega_1, E_m + \hbar\omega_1) S_{F,\beta\delta}^*(E_p + \hbar\omega_1, E_m + \hbar\omega_1).$$

Summing Eqs. (6.4), (6.6b), (6.8), and (6.10b) we get the result given in Eq. (6.2b).

6.2 Zero frequency spectral noise power

The quantity $\mathcal{P}_{\alpha\beta}(0) \equiv \mathcal{P}_{\alpha\beta,0}(0,0)$, referred to as symmetrized *noise*, characterizes the mean square of current fluctuations (at $\alpha = \beta$) or a symmetrized current cross-correlator (at $\alpha \neq \beta$), averaged over a long time period. It can be written as

$$\mathcal{P}_{\alpha\beta}(0) = \frac{1}{2} \int_0^{\mathcal{T}} \frac{dt}{\mathcal{T}} \int_{-\infty}^{\infty} d\tau \left\langle \Delta\hat{I}_\alpha(t) \Delta\hat{I}_\beta(t+\tau) + \Delta\hat{I}_\beta(t+\tau) \Delta\hat{I}_\alpha(t) \right\rangle.$$

(6.11)

The noise expression in terms of the Floquet scattering matrix elements is given in Eq. (6.2b) at $l = 0$ and $\omega_1 = \omega_2 = 0$.

From Eq. (6.1b) it follows that the noise value does not change when the lead indices interchange,

$$\mathcal{P}_{\alpha\beta}(0) = \mathcal{P}_{\beta\alpha}(0). \quad (6.12)$$

This is another reason why this quantity is called symmetrized noise.

Like its stationary counterpart the quantity $\mathcal{P}_{\alpha\beta}(0)$ can be represented as the sum of thermal noise and shot noise, see Eq. (2.60). Thermal noise, $\mathcal{P}_{\alpha\beta}^{(th)}$, is due to fluctuations of quantum state occupations of electrons incident from reservoirs with a non-zero temperature. While shot noise, $\mathcal{P}_{\alpha\beta}^{(sh)}$, is due to fluctuations of quantum state occupations of scattered electrons: If an electron is scattered, say into the lead α, then in this contact the instantaneous current is larger than the average current, while in the other contacts, $\beta \neq \alpha$, the instantaneous current is zero, i.e., it is smaller than the corresponding average current.

Let us calculate the noise when all the reservoirs have the same chemical potential and temperature,

$$\mu_\alpha = \mu, \quad T_\alpha = T. \tag{6.13}$$

Hence the distribution functions for the electrons in the reservoirs are the same,

$$f_\alpha(E) = f_0(E). \tag{6.14}$$

Then from Eq. (6.2b) at $l = 0$, $\omega_1 = \omega_2 = 0$ it follows (see also Section 2.2.4) that $\mathcal{P}_{\alpha\beta}(0) = \mathcal{P}_{\alpha\beta}^{(th)} + \mathcal{P}_{\alpha\beta}^{(sh)}$ [178], where

$$\mathcal{P}_{\alpha\beta}^{(th)} = \frac{e^2}{h} \int_0^\infty dE\, f_0(E)[1 - f_0(E)] \left\{ \delta_{\alpha\beta} \left(1 + \sum_{n=-\infty}^\infty \sum_{\gamma=1}^{N_r} |S_{F,\alpha\gamma}(E_n, E)|^2 \right) \right.$$

$$\left. - \sum_{n=-\infty}^\infty \left(|S_{F,\alpha\beta}(E_n, E)|^2 + |S_{F,\beta\alpha}(E_n, E)|^2 \right) \right\}, \tag{6.15}$$

$$\mathcal{P}_{\alpha\beta}^{(sh)} = \frac{e^2}{h} \int_0^\infty dE \sum_{\gamma=1}^{N_r} \sum_{\delta=1}^{N_r} \sum_{n=-\infty}^\infty \sum_{m=-\infty}^\infty \sum_{p=-\infty}^\infty \frac{[f_0(E_n) - f_0(E_m)]^2}{2}$$

$$\times S_{F,\alpha\gamma}^*(E, E_n) S_{F,\alpha\delta}(E, E_m) S_{F,\beta\delta}^*(E_p, E_m) S_{F,\beta\gamma}(E_p, E_n). \tag{6.16}$$

It is clear that the thermal noise vanishes at zero temperature, since in that case $f_0(E)[1 - f_0(E)] = \theta(\mu - E)\theta(E - \mu) \equiv 0$. In contrast, the shot noise does exist at an arbitrary temperature. However, it vanishes in the equilibrium system, i.e., if the scatterer is stationary. In that case $\hat{S}_F(E_p, E) = \delta_{p0} \hat{S}(E)$, hence there are only terms with $n = 0$, $m = 0$, and $p = 0$ in Eq. (6.16). For these terms the difference of the Fermi functions is zero.

As we showed in Section (2.2.4.1) the unitarity of scattering results in conservation laws, Eq. (2.63), for stationary noise. The noise due to a dynamic scatterer is also subject to conservation laws. Moreover, thermal noise and shot noise satisfy them separately:

$$\sum_{\beta=1}^{N_r} \mathcal{P}_{\alpha\beta}^{(th)} = 0, \quad \sum_{\alpha=1}^{N_r} \mathcal{P}_{\alpha\beta}^{(th)} = 0, \tag{6.17a}$$

$$\sum_{\beta=1}^{N_r} \mathcal{P}_{\alpha\beta}^{(sh)} = 0, \quad \sum_{\alpha=1}^{N_r} \mathcal{P}_{\alpha\beta}^{(sh)} = 0, \tag{6.17b}$$

which follow directly from Eqs. (6.15) and (6.16) if one uses, in addition, the unitarity condition, Eq. (3.28). Note to prove the second equality in Eq. (6.17b) it is necessary to make the following replacement in Eq. (6.16): $E \to E - p\hbar\Omega_0$, $n \to n - p$, and $m \to m - p$.

Let us analyze the sign of the zero-frequency noise power. The cross-correlator $\mathcal{P}_{\alpha\neq\beta}$ is negative in the stationary case, see Eq. (2.64b). It is negative in the dynamic case too:

$$\mathcal{P}_{\alpha\neq\beta}^{(th)} \leq 0, \quad \mathcal{P}_{\alpha\neq\beta}^{(sh)} \leq 0. \tag{6.18}$$

For thermal noise this follows directly from Eq. (6.15):

$$\mathcal{P}_{\alpha\neq\beta}^{(th)} = -\frac{e^2}{h} \int_0^\infty dE\, f_0(E)\,[1 - f_0(E)]$$

$$\times \sum_{n=-\infty}^{\infty} \left(|S_{F,\alpha\beta}(E_n, E)|^2 + |S_{F,\beta\alpha}(E_n, E)|^2 \right) \leq 0.$$

To check this rule for shot noise, let us rewrite Eq. (6.16) for $\alpha \neq \beta$ as

$$\mathcal{P}_{\alpha\neq\beta}^{(sh)} = -\frac{e^2}{h} \int_0^\infty dE \sum_{p=-\infty}^{\infty}$$

$$\left| \sum_{n=-\infty}^{\infty} \sum_{\gamma=1}^{N_r} f_0(E_n)\, S_{F,\alpha\gamma}^*(E, E_n)\, S_{F,\beta\gamma}(E_p, E_n) \right|^2 \leq 0.$$

Here we took into account that the terms with squared Fermi functions vanish for $\alpha \neq \beta$ in Eq. (6.16). For instance, in the terms with $f_0^2(E_n)$ we can sum over m and δ. Then using Eq. (3.28b) we find ($\alpha \neq \beta$):

$$\sum_{m=-\infty}^{\infty} \sum_{\delta=1}^{N_r} S_{F,\alpha\gamma}(E, E_m) S_{\beta\delta}^*(E_p, E_m) = \delta_{\alpha\beta} \delta_{p0} = 0.$$

In the same way one can prove that the term with $f_0^2(E_m)$ is also zero.

The auto-correlator $\mathcal{P}_{\alpha\alpha}$ is a mean square of current fluctuations in the lead α, hence it is non-negative. From Eqs. (6.17) and (6.18)

$$\mathcal{P}_{\alpha\alpha}^{(th)} \geq 0, \qquad \mathcal{P}_{\alpha\alpha}^{(sh)} \geq 0. \tag{6.19}$$

The fact that thermal noise and shot noise separately satisfy the sum rule, Eq. (6.17), and the sign rule, Eqs. (6.18) and (6.19), justifies the splitting of noise into these two parts.

Additionally, thermal noise and shot noise depend differently on both the temperature T and the pump frequency Ω_0. Let us show this in a regime where the parameters of a scatterer vary slowly, $\Omega_0 \to 0$.

6.3 Noise in the adiabatic regime

The Floquet scattering matrix elements up to terms linear in Ω_0 are given in Eq. (3.50). Recall that the adiabatic expansion implies that the frozen scattering matrix $\hat{S}$ changes only a little on an energy scale of order $\hbar\Omega_0$, see Eq. (3.49).

6.3.1 Thermal noise

Substitute Eq. (3.50) into Eq. (6.15) and calculate the thermal noise up to terms linear in Ω_0 [178]

$$\mathcal{P}_{\alpha\beta}^{(th)} = \mathcal{P}_{\alpha\beta}^{(th,0)} + \mathcal{P}_{\alpha\beta}^{(th,\Omega_0)}, \tag{6.20a}$$

where

$$\mathcal{P}_{\alpha\beta}^{(th,0)} = k_B T \int_0^{\infty} dE \left(-\frac{\partial f_0}{\partial E}\right) \int_0^{\mathcal{T}} \frac{dt}{\mathcal{T}}$$

$$\times \frac{e^2}{h} \left(2\delta_{\alpha\beta} - |S_{\alpha\beta}(t,E)|^2 - |S_{\beta\alpha}(t,E)|^2\right), \tag{6.20b}$$

$$\mathcal{P}_{\alpha\beta}^{(th,\Omega_0)} = k_B T \int_0^\infty dE \left(-\frac{\partial f_0}{\partial E}\right) \int_0^{\mathcal{T}} \frac{dt}{\mathcal{T}}$$
$$\times e \left(\delta_{\alpha\beta} \frac{dI_\alpha(t,E)}{dE} - \frac{dI_{\alpha\beta}(t,E)}{dE} - \frac{dI_{\beta\alpha}(t,E)}{dE}\right).$$
(6.20c)

As expected, thermal noise is proportional to the temperature. Let us introduce a frozen conductance matrix that is averaged over time, see Eq. (5.41),

$$\hat{\bar{G}} = \int_0^{\mathcal{T}} \frac{dt}{\mathcal{T}} \hat{G}(t). \qquad (6.21)$$

Then the quantity $\mathcal{P}_{\alpha\beta}^{(th,0)}$ depends on its elements

$$\mathcal{P}_{\alpha\beta}^{(th,0)} = k_B T \left(2\delta_{\alpha\beta} G_0 - \bar{G}_{\alpha\beta} - \bar{G}_{\beta\alpha}\right), \qquad (6.22)$$

in the same way as equilibrium noise, Nyquist–Johnson noise, Eq. (2.61), depends on elements of the conductance matrix, $\hat{G}$, Eq. (1.54), in the stationary case. Therefore, $\mathcal{P}_{\alpha\beta}^{(th,0)}$ can be called *quasi-equilibrium noise*. To compare Eqs. (6.20b) and (2.61) we use the identity Eq. (2.66) and the fact that Eq. (6.20b) was derived under the conditions given in Eq. (6.14).

The second part of the thermal noise, Eq. (6.20c), indicates that the system is in fact not in equilibrium. The part $\mathcal{P}_{\alpha\beta}^{(th,\Omega_0)}$ can be called *non-equilibrium thermal noise*, since, on one hand, it is proportional to the temperature (hence thermal), and, on the other hand, it depends on currents generated by the dynamic scatterer (hence non-equilibrium). Since the spectral current powers $dI_{\alpha\beta}(t,E)/dE$, Eq. (5.9), and $dI_\alpha(t,E)/dE$, Eq. (4.20), are both linear in the pump frequency Ω_0, then $\mathcal{P}_{\alpha\beta}^{(th,\Omega_0)} \sim \Omega_0$.

6.3.2 Low-temperature shot noise

If the temperature is low enough,

$$k_B T \ll \hbar\Omega_0, \qquad (6.23)$$

then thermal noise can be ignored. In this regime the main source of noise is the dynamic scatterer generating *photon-assisted shot noise*. Another source of shot noise, the bias, is absent because of Eq. (6.13). Photon-assisted shot noise is non-equilibrium noise. This follows (in the same way as in the stationary case with bias) from the fact that the noise is due to scattered electrons that have a non-equilibrium distribution function $f_\alpha^{(out)}(E)$. It follows from Eq. (4.4), see also Eq. (4.5), the distribution function $f_\alpha^{(out)}(E)$ is non-equilibrium for energies different from the Fermi energy by an amount of the order of $\hbar\Omega_0$.

Let us calculate $\mathcal{P}_{\alpha\beta}^{(sh)}$, Eq. (6.16), in the lowest order in Ω_0. It is sufficient to use the Floquet scattering matrix elements in the zeroth order in Ω_0. For instance, to the required accuracy we find from Eq. (3.50):

$$\hat{S}_F(E_m, E_p) = \hat{S}_{m-p}(E) + \mathcal{O}(\Omega_0). \quad (6.24)$$

Recall, in the lowest adiabatic approximation the frozen scattering matrix $\hat{S}$ should be treated as energy independent over the scale of order $\hbar\Omega_0$. Therefore, under the condition given in Eq. (6.23) we can integrate over energy in Eq. (6.16) keeping the scattering matrix elements constant (for definiteness we will calculate them at $E = \mu$). Then the remaining integral over energy becomes trivial:

$$\int_0^\infty dE \left\{ f_0(E_n) - f_0(E_m) \right\}^2 = \begin{cases} \hbar\Omega_0(m-n), & m > n, \\ \hbar\Omega_0(n-m), & m < n. \end{cases} \quad (6.25)$$

Using Eqs. (6.24) and (6.25) in Eq. (6.16) we find,

$$\mathcal{P}_{\alpha\beta}^{(sh)} = \frac{e^2 \Omega_0}{4\pi} \sum_{\gamma,\delta=1}^{N_r} \sum_{n,m,p=-\infty}^{\infty}$$

$$\times |m-n| S_{\alpha\gamma,-n}^*(\mu) S_{\alpha\delta,-m}(\mu) S_{\beta\delta,p-m}^*(\mu) S_{\beta\gamma,p-n}(\mu). \quad (6.26)$$

So photon-induced shot noise is linear in the pump frequency Ω_0 (see also Ref. [95]). To simplify the above equation we proceed as follows. For each fixed n we consider the sum over m and split it into two parts, the sum over $m < n$ and the sum over $m > n$. Then we introduce a new index $q = m - n$ instead of m. After that we find for any quantity $X_{n,m}$ dependent on indices

n and m the following:

$$\sum_{m=-\infty}^{\infty} |m-n|\, X_{m,n} = \sum_{m=-\infty}^{n-1} (n-m)\, X_{m,n} + \sum_{m=n+1}^{\infty} (n-m)\, X_{m,n}$$

$$= \sum_{q=-\infty}^{-1} (-q)\, X_{q+n,n} + \sum_{q=1}^{\infty} q X_{q+n,n} = \sum_{q=1}^{\infty} q \left(X_{-q+n,n} + X_{q+n,n} \right).$$

The term with $m = n$ is zero due to the factor $m - n = n - n \equiv 0$. Then equation (6.26) is transformed into the following form,

$$\mathcal{P}^{(sh)}_{\alpha\beta} = \frac{e^2 \Omega_0}{4\pi} \sum_{q=1}^{\infty} \sum_{\gamma=1}^{N_r} \sum_{\delta=1}^{N_r} q \Big[\left\{ S^*_{\alpha\gamma}(\mu)\, S_{\alpha\delta}(\mu) \right\}_{-q} \left\{ S_{\beta\gamma}(\mu)\, S^*_{\beta\delta}(\mu) \right\}_{q}$$

$$+ \left\{ S^*_{\alpha\gamma}(\mu)\, S_{\alpha\delta}(\mu) \right\}_{q} \left\{ S_{\beta\gamma}(\mu)\, S^*_{\beta\delta}(\mu) \right\}_{-q} \Big]. \tag{6.27}$$

Going from Eq. (6.26) to Eq. (6.27) we have summed over n and p using the following identity valid for the Fourier coefficients of any periodic functions $A(t)$ and $B(t)$:

$$\sum_{n=-\infty}^{\infty} A_n \left(B_{n+q} \right)^* = (AB^*)_{-q}\,, \qquad \sum_{n=-\infty}^{\infty} A_{n+q} \left(B_n \right)^* = (AB^*)_{q}\,. \tag{6.28}$$

It is easy to check that Eq. (6.27) satisfies the symmetry given in Eq. (6.12), $\mathcal{P}^{(sh)}_{\alpha\beta} = \mathcal{P}^{(sh)}_{\beta\alpha}$. To show it we need to use $\gamma \leftrightarrow \delta$ in an expression for $\mathcal{P}^{(sh)}_{\beta\alpha}$.

6.3.3 High-temperature shot noise

At higher temperatures,

$$k_B T \gg \hbar \Omega_0\,, \tag{6.29}$$

thermal noise dominates. In this regime shot noise, Eq. (6.16), is only a small fraction of the total noise. However, thermal noise and shot noise depend differently on both the pump frequency Ω_0 and the temperature, which allows us in principle to distinguish them.

With Eq. (6.29) we can expand the difference of Fermi functions in Eq. (6.16) in powers of Ω_0. Up to the first non-vanishing term we get:

$$f_0(E_n) - f_0(E_m) = \hbar\Omega_0 \frac{\partial f_0(E)}{\partial E}(n-m).$$

Substituting this expansion and the adiabatic approximation for the Floquet scattering matrix, Eq. (6.24), into Eq. (6.16) we find the high-temperature shot noise ($k_B T \gg \hbar\Omega_0$):

$$\begin{aligned}
\mathcal{P}_{\alpha\beta}^{(sh)} &= \frac{e^2}{4\pi}\hbar\Omega_0^2 \int_0^\infty dE \left(\frac{\partial f_0}{\partial E}\right)^2 \sum_{q=-\infty}^{\infty} q^2 \\
&\times \sum_{\gamma=1}^{N_r}\sum_{\delta=1}^{N_r} \left\{S_{\alpha\gamma}^*(E) S_{\alpha\delta}(E)\right\}_q \left\{S_{\beta\gamma}(E) S_{\beta\delta}^*(E)\right\}_{-q}.
\end{aligned} \quad (6.30)$$

In this equation we keep the integration over energy since it is over the interval of order $k_B T \gg \hbar\Omega_0$ near the Fermi energy μ. The use of an adiabatic approximation, Eq. (6.24), does not put any restrictions on the energy dependence of the frozen scattering matrix, $\hat{S}$, over such an energy interval. The quadratic dependence of high-temperature shot noise on the pump frequency Ω_0 was shown in Ref. [81].

6.3.4 Shot noise within a wide temperature range

One can relax the restrictions of Eqs. (6.23) and (6.29) and calculate analytically the shot noise at an arbitrary ratio of the temperature and the energy quantum $\hbar\Omega_0$ dictated by pumping. This is possible if the scattering matrix can be treated as energy independent over the relevant energy interval:

$$\hbar\Omega_0, k_B T \ll \delta E. \quad (6.31)$$

Recall that δE is an energy interval over which the scattering matrix changes significantly.

So, with Eq. (6.31) when calculating the shot noise, Eq. (6.16), in the adiabatic regime [when we use Eq. (6.24)] we can calculate the scattering matrix elements at $E = \mu$ only. Then the corresponding integral over energy is calculated analytically,

$$\int_0^\infty dE \left\{ f_0(E_n) - f_0(E_m) \right\}^2 = (m-n)\hbar\Omega_0 \coth\left(\frac{(m-n)\hbar\Omega_0}{2k_BT}\right) - 2k_BT,$$

and finally we arrive at the following,

$$\mathcal{P}_{\alpha\beta}^{(sh)} = \frac{e^2}{h} \sum_{q=-\infty}^{\infty} F(q\hbar\Omega_0, k_BT) \tag{6.32a}$$

$$\times \sum_{\gamma=1}^{N_r} \sum_{\delta=1}^{N_r} \left\{ S_{\alpha\gamma}^*(\mu) S_{\alpha\delta}(\mu) \right\}_q \left\{ S_{\beta\gamma}(\mu) S_{\beta\delta}^*(\mu) \right\}_{-q},$$

where

$$F(q\hbar\Omega_0, k_BT) = \frac{q\hbar\Omega_0}{2} \coth\left(\frac{q\hbar\Omega_0}{2k_BT}\right) - k_BT = \begin{cases} \frac{|q|\hbar\Omega_0}{2}, & k_BT \ll \hbar\Omega_0, \\ \frac{(q\hbar\Omega_0)^2}{12k_BT} & k_BT \gg \hbar\Omega_0. \end{cases} \tag{6.32b}$$

This equation (6.32) reproduces both equation (6.27) for low-temperature shot noise, which is linear in Ω_0 and temperature independent, and equation (6.30) for high-temperature shot noise, which is quadratic in the pump frequency and, under the conditions of Eq. (6.31), inversely proportional to the temperature.

6.3.5 Noise as a function of the pump frequency

At zero temperature the dynamic scatterer generates only shot noise, which is linear in Ω_0. With increasing temperature thermal noise arises. It also depends on Ω_0. Therefore, the part $\delta\mathcal{P}_{\alpha\beta}^{(\Omega_0)}$ of the total high-temperature noise dependent on Ω_0 can be written as the sum of two contributions,

$$\delta\mathcal{P}_{\alpha\beta}^{(\Omega_0)} = \mathcal{P}_{\alpha\beta}^{(sh)} + \mathcal{P}_{\alpha\beta}^{(th,\Omega_0)}. \tag{6.33}$$

Let us compare these two terms. The non-equilibrium thermal noise, Eq. (6.20c), generated by the dynamic scatterer in the adiabatic regime ($\hbar\Omega_0 \ll \delta E$), is of the following order:

$$\mathcal{P}_{\alpha\beta}^{(th,\Omega_0)} \sim k_B T \frac{\hbar\Omega_0}{\delta E},$$

while the high-temperature shot noise, Eq. (6.32), can be estimated as follows:

$$\mathcal{P}_{\alpha\beta}^{(sh)} \sim \frac{(\hbar\Omega_0)^2}{k_B T}.$$

Their ratio is equal to

$$\frac{\mathcal{P}_{\alpha\beta}^{(sh)}}{\mathcal{P}_{\alpha\beta}^{(th,\Omega_0)}} \sim \frac{\hbar\Omega_0 \delta E}{(k_B T)^2}.$$

From this it is seen that at $k_B T \ll \sqrt{\hbar\Omega_0 \delta E}$ shot noise dominates. However, at higher temperatures non-equilibrium thermal noise determines the dependence of the total noise on the pump frequency Ω_0. Therefore, with increasing temperature one can expect the following [178]:

$$\delta\mathcal{P}_{\alpha\beta}^{(\Omega_0)} \sim \frac{e^2}{2h} \begin{cases} \hbar\Omega_0, & k_B T \ll \hbar\Omega_0, \\ \dfrac{(\hbar\Omega_0)^2}{6 k_B T}, & \hbar\Omega_0 \ll k_B T \ll \sqrt{\hbar\Omega_0 \delta E}, \\ \hbar\Omega_0 \dfrac{k_B T}{\delta E}, & \sqrt{\hbar\Omega_0 \delta E} \ll k_B T. \end{cases} \quad (6.34)$$

We stress that the linear dependence on Ω_0 at low and high temperatures is due to different physical reasons. At low temperatures it is due to shot noise, at high temperatures it is due to thermal noise.

Here we have presented the Floquet scattering theory for noise of quantum pumps. The same problem has also been investigated within different approaches: the random matrix theory [179–181], the full counting statistics[1] [122, 185–188, 190, 191], and the Green function formalism [189, 192–195]. Note the prediction [81, 122, 178, 185, 194] that in the quantized emission regime the shot noise will vanish. It seems that experiment confirms this [196]. It is also worth mentioning a deep connection between noise and the entanglement generated by the pump [197–200], see also a review in Ref. [201].

---

[1] The counting statistics of charge transfer in phase coherent conductors was originally introduced in Ref. [182] for the stationary case and in Ref. [183] for the periodically driven case. For a review, see Ref. [184].

Chapter 7

Energetics of a dynamic scatterer

7.1 DC heat current

By analogy with the DC charge current, Eq. (4.3), we define a DC energy current I_α^E in the lead α as the difference between the energy flow $I_\alpha^{E(out)}$ carried by non-equilibrium electrons from the scatterer to the reservoir and an equilibrium energy flow $I_\alpha^{E(in)}$ from the reservoir to the scatterer:

$$I_\alpha^E = I_\alpha^{E(out)} - I_\alpha^{E(in)}. \quad (7.1)$$

Here the corresponding energy currents are defined as [88]

$$I_\alpha^{E(in/out)} = \frac{1}{h} \int_0^\infty dE\, E\, f_\alpha^{(in/out)}(E), \quad (7.2)$$

where $f_\alpha^{(out)}(E)$ is a distribution function for scattered electrons and $f_\alpha^{(in)}(E) \equiv f_\alpha(E)$ is an equilibrium distribution function for incident electrons. We are interested in the DC heat current I_α^Q that is the total energy current I_α^E reduced by the convective energy flow of electrons carrying a DC charge current I_α:

$$I_\alpha^Q = I_\alpha^E - \mu_\alpha \frac{I_\alpha}{e}. \quad (7.3)$$

The division of I_α^E into heat I_α^Q and convective $\mu_\alpha I_\alpha/e$ flows can be explained on the basis of the particle and energy balance for the reservoir α with both the chemical potential μ_α and the temperature T_α fixed (for macroscopic samples, see, e.g., [202]). If the DC charge I_α and energy I_α^E currents enter the reservoir α, then its charge (the particle number) and its energy should change. At the same time the chemical potential and the temperature of the reservoir should change also. Let us analyze what should be done to keep μ_α and T_α fixed. To keep μ_α fixed one needs to remove an excess number of electrons with the rate I_α/e they enter the reservoir. For this reason the metallic contact, acting as a reservoir for the mesoscopic sample, is connected to another much bigger conductor. This connection is located far away from where the mesoscopic sample is connected, so that all the injected non-equilibrium electrons are now in equilibrium. Therefore, all electrons that are removed (to keep μ_α fixed) are in equilibrium and hence have an energy μ_α, see, e.g., [15]. It is clear that removing electrons is accompanied by removing energy with a rate $\mu_\alpha I_\alpha/e$. Note this convective energy flow, $\mu_\alpha I_\alpha/e$, is removed at equilibrium, hence it can be reversibly given back to the reservoir.

Now let us analyze how to maintain the reservoir's temperature fixed. In general the removed convective energy flow is not equal to the total energy flow I_α^E incoming to the reservoir α. To prevent heating of the reservoir it is also necessary to remove energy with a rate I_α^Q, Eq. (7.3). Since, as a rule, the reservoir cannot produce work then the only way to take energy out of it, I_α^Q (keeping the number of particles fixed), is to bring it into contact with a thermal bath. The energy exchange between the reservoir and the bath is essentially irreversible. This is the reason why we named as *a heat* that part of the energy flow denoted as I_α^Q. We emphasize that this part of the energy flow becomes a proper heat (i.e., it can change the temperature) only deep inside the reservoir, after thermalizing of non-equilibrium electrons. In the absence of a bath the reservoir's temperature would be changed under the action of a heat current I_α^Q, which, as we show, can be directed into the reservoir as well as out of it: The dynamic scatterer can either heat a reservoir α or cool it, even when the temperatures of all the reservoirs were originally the same.

Expressing the distribution function $f_\alpha^{(out)}(E)$ for scattered electrons in terms of the Floquet scattering matrix elements and the distribution functions $f_\beta(E)$ for incident electrons, Eq. (4.2), and using Eq. (3.40) for the DC charge current, $I_{\alpha,0}$, we finally get the following expression for the DC heat current I_α^Q, Eq. (7.3) [40]

$$I_\alpha^Q = \frac{1}{h} \int_0^\infty dE \, (E - \mu_\alpha) \sum_{n=-\infty}^{\infty} \sum_{\beta=1}^{N_r} |S_{F,\alpha\beta}(E, E_n)|^2 \{f_\beta(E_n) - f_\alpha(E)\}.$$

(7.4)

Here for the coefficient of $f_\alpha(E)$ we can use the identity,

$$\sum_{n=-\infty}^{\infty} \sum_{\beta=1}^{N_r} |S_{F,\alpha\beta}(E, E_n)|^2 = 1, \qquad (7.5)$$

which follows from Eq. (3.28b) at $m = 0$ and $\gamma = \alpha$. This identity is a consequence of the unitarity of the Floquet scattering matrix.

In addition we give also the following two equations for the DC heat current. The first one,

$$I_\alpha^Q = \frac{1}{h} \int_0^\infty dE \, (E_n - \mu_\alpha) \sum_{n=-\infty}^{\infty} \sum_{\beta=1}^{N_r} |S_{F,\alpha\beta}(E_n, E)|^2 \{f_\beta(E) - f_\alpha(E_n)\},$$

(7.6)

follows from Eq. (7.4) via the substitutions $E \to E_n$ and $n \to -n$. And the second one,

$$I_\alpha^Q = \frac{1}{h} \int_0^\infty dE \left\{ \sum_{n=-\infty}^{\infty} \sum_{\beta=1}^{N_r} (E_n - \mu_\alpha) |S_{F,\alpha\beta}(E_n, E)|^2 f_\beta(E) \right.$$

(7.7)

$$\left. - (E - \mu_\alpha) f_\alpha(E) \right\},$$

is obtained via the same substitution but made only to the term for f_β, while for the term with f_α we used the identity (7.5).

Now we use the last equation to show the existence of two quite general effects due to the dynamic scatterer. For better clarity we assume that all the reservoirs have the same chemical potential and the same temperature:

$$\mu_\alpha = \mu_0, \quad T_\alpha = T_0, \quad f_\alpha(E) = f_0(E), \quad \alpha = 1, \ldots, N_r. \qquad (7.8)$$

Therefore, all the possible energy/heat flows in the system are generated by the dynamic scatterer only. In presenting this problem we follow Ref. [203].

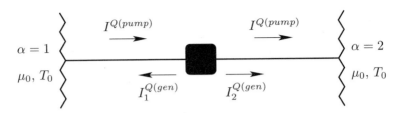

Fig. 7.1 The heat flows caused by the dynamic scatterer with two contacts. $I_\alpha^{Q(gen)}$ is the generated heat flowing into the reservoir α and $I^{Q(pump)}$ is the pumped heat. The heat production rate is $I_{tot}^Q = I_1^{Q(gen)} + I_2^{Q(gen)}$.

7.1.1 Heat generation by the dynamic scatterer

The first of the effects mentioned above is:
The periodic variation in time of the parameters of a mesoscopic scatterer is accompanied by the pumping of energy into an electron system, which leads to heating of the electron reservoirs [40, 81, 95, 204, 205].[1]

In other words, when a quantum pump is operating heat is produced as expected. To calculate the total heat production rate I_{tot}^Q we sum the heat currents I_α^Q flowing into all leads. Using Eq. (7.7) under the conditions given in Eq. (7.8) we find

$$I_{tot}^Q \equiv \sum_{\alpha=1}^{N_r} I_\alpha^Q = \frac{\Omega_0}{2\pi} \int_0^\infty dE\, f_0(E) \sum_{n=-\infty}^{\infty} n \sum_{\alpha=1}^{N_r} \sum_{\beta=1}^{N_r} |S_{F,\alpha\beta}(E_n, E)|^2 \,. \quad (7.9)$$

Since the sum of DC heat currents is not zero, in contrast to the sum of DC charge currents, see Eq. (4.11), we conclude that indeed the dynamic scatterer is a source of heat, Fig. 7.1. Taking into account the physical meaning of the quantities in Eq. (7.9) we can say that the quantity I_{tot}^Q is due to the energy absorbed by electrons scattered off a dynamic sample. This additional energy comes form the driving external forces/fields causing a change of parameters of the scatterer.

7.1.2 Heat transfer between the reservoirs

The second effect is:
The dynamic scatterer acts as a heat pump transferring heat between the

---

[1]The heating of pump is considered in Ref. [206].

electron reservoirs [207–210].[2]

This effect is quite analogous to the effect considered earlier of DC charge current generation. The dynamic scatterer can cause heat flows, $I_\alpha^{Q(pump)}$, which are directed from the scatterer in some leads and to the scatterer in other leads, Fig. 7.1. At the same time the sum of such flows in all the leads is zero,

$$\sum_{\alpha=1}^{N_r} I_\alpha^{Q(pump)} = 0, \tag{7.10}$$

as in the case with a DC charge current, Eq. (4.11).

Equation (7.10) means that the heat currents $I_\alpha^{Q(pump)}$ flow through the scatterer neither being accumulated nor disappearing, i.e., the dynamic scatterer is not a source (or a sink) for these heat flows. Its role consists only in providing the conditions under which the heat currents flowing out of the reservoirs can be redistributed in such a way that the heat can be taken out of some reservoirs, $I_{\alpha_1}^{Q(pump)} < 0$, and can be pushed into other reservoirs, $I_{\alpha_2}^{Q(pump)} > 0$. Note if some reservoir α_0 is at the zero temperature then in the lead connecting the scatterer to this reservoir the heat current is not negative, $I_{\alpha_0}^{Q(pump)} \geq 0$, since it is impossible to take heat out of such a reservoir.

To show the existence of a quantum pump heat effect we proceed as follows. Let us formally split the total heat production rate I_{tot}^Q into parts $I_\alpha^{Q(gen)}$ such that

$$I_{tot}^Q = \sum_{\alpha=1}^{N_r} I_\alpha^{Q(gen)}. \tag{7.11}$$

Comparing this equation with Eq. (7.9) we can write,

$$I_\alpha^{Q(gen)} = \frac{\Omega_0}{2\pi} \int_0^\infty dE \, f_0(E) \sum_{n=-\infty}^{\infty} n \sum_{\beta=1}^{N_r} |S_{F,\alpha\beta}(E_n, E)|^2. \tag{7.12}$$

One can interpret the quantity $I_\alpha^{Q(gen)}$ as that part of the generated heat which flows into the reservoir α. Then comparing Eqs. (7.12) and Eq. (7.7)

---

[2] A similar effect but for phonons is investigated in Refs. [211, 212]. Pumping of heat due to a moving potential (a moving domain wall) is addressed in Ref. [213].

(at $f_\alpha = f_0, \forall \alpha$) we can see that $I_\alpha^{Q(gen)}$ is different from the heat current I_α^Q flowing into the lead α. The difference,

$$I_\alpha^{Q(pump)} = I_\alpha^Q - I_\alpha^{Q(gen)}, \tag{7.13}$$

is exactly what we are looking for. This is the part of the heat current that is transferred between the reservoirs. This part is not related to the heat generated by the dynamic scatterer. Using Eqs. (4.7) and (4.10) in Eq. (4.11) we find,

$$I_\alpha^{Q(pump)} = \frac{1}{h} \int_0^\infty dE (E - \mu_0) f_0(E) \left\{ \sum_{n=-\infty}^\infty \sum_{\beta=1}^{N_r} |S_{F,\alpha\beta}(E_n, E)|^2 - 1 \right\}. \tag{7.14}$$

With the unitarity condition (3.28a) one can easily check that Eq. (7.14) satisfies the conservation law, Eq. (7.10).

According to Eq. (7.13) the heat flow I_α^Q in the lead α consists of two parts, Fig. 7.1. The first one, $I_\alpha^{Q(gen)}$, is a positive heat flow generated by the dynamic scatterer. The second one, $I_\alpha^{Q(pump)}$, is a transferred heat flow, which can be either positive (the heat flow is directed to the reservoir α) or negative (the heat flow is directed from the reservoir α). Note if $I_\alpha^{Q(pump)} < 0$ and the transferred heat flow is larger than the generated heat flow in the same lead, $\left|I_\alpha^{Q(pump)}\right| > I_\alpha^{Q(gen)}$, then the reservoir α is cooled, since $I_\alpha^Q = I_\alpha^{Q(pump)} + I_\alpha^{Q(gen)} < 0$.

We have split the heat flow I_α^Q into $I_\alpha^{Q(gen)}$ and $I_\alpha^{Q(pump)}$ to show that I_α^Q can be negative. Therefore, the electron reservoirs may not only be heated (which is intuitively expected since the functioning of a device, in our case a quantum pump, is accompanied by energy dissipation), but also can be cooled (which is a non-trivial effect). Strictly speaking we can rigorously calculate only the heat flow I_α^Q, Eq. (7.7), and the total generated heat rate I_{tot}^Q, Eq. (7.9). The split presented in Eqs. (7.12) and (7.14) is not unique, since equation (7.11) is not sufficient for an unambiguous definition of the quantities $I_\alpha^{Q(gen)}$. In the next section we consider an adiabatic regime and give additional physical arguments supporting the splitting of the total heat flow I_α^Q into the sum of generated and transferred heat flows.

7.2 Heat flows in the adiabatic regime

At $\varpi \ll 1$, Eq. (3.49), up to the terms linear in frequency, Ω_0, of an external drive the Floquet scattering matrix elements are [see Eqs. (3.44), (3.46a), and (3.48a)]

$$\left|S_{F,\alpha\beta}(E_n, E)\right|^2 = \left|S_{\alpha\beta,n}(E)\right|^2 + \frac{n\hbar\Omega_0}{2} \frac{\partial \left|S_{\alpha\beta,n}(E)\right|^2}{\partial E}$$

$$+ 2\hbar\Omega_0 \mathrm{Re}\left[S^*_{\alpha\beta,n}(E) A_{\alpha\beta,n}(E)\right] + \mathcal{O}\left(\Omega_0^2\right). \quad (7.15)$$

Substituting this expression into Eq. (7.6) we calculate the heat flow I_α^Q under the conditions given in Eq. (7.8) up to terms of order Ω_0^2. We consider separately the finite temperature and zero temperature cases.

7.2.1 High temperatures

If

$$k_B T_0 \gg \hbar\Omega_0 \quad (7.16)$$

then we can expand the difference of Fermi functions in Eq. (7.6) in powers of Ω_0:

$$f_0(E) - f_0(E_n) = \left(-\frac{\partial f_0}{\partial E}\right) n\hbar\Omega_0 - \frac{(n\hbar\Omega_0)^2}{2} \frac{\partial^2 f_0}{\partial E^2} + \mathcal{O}\left(\Omega_0^3\right). \quad (7.17)$$

Substituting this expansion into Eq. (7.6) and summing over n, using the properties of the Fourier coefficients, we calculate ($k_B T_0 \gg \hbar\Omega_0$)

$$I_\alpha^Q = I_\alpha^{Q(gen)} + I_\alpha^{Q(pump)} + \mathcal{O}\left(\Omega_0^3\right), \quad (7.18a)$$

where

$$I_\alpha^{Q(gen)} = \frac{\hbar}{4\pi} \int_0^\infty dE \left(-\frac{\partial f_0}{\partial E}\right) \int_0^\mathcal{T} \frac{dt}{\mathcal{T}} \left(\frac{\partial \hat{S}}{\partial t} \frac{\partial \hat{S}^\dagger}{\partial t}\right)_{\alpha\alpha}, \quad (7.18b)$$

and

$$I_\alpha^{Q(pump)} = \frac{1}{2\pi} \int_0^\infty dE \, (E - \mu_0) \left(-\frac{\partial f_0}{\partial E}\right) \int_0^{\mathcal{T}} \frac{dt}{\mathcal{T}} \text{Im} \left\{ \left(\hat{S} + 2\hbar\Omega_0\hat{A}\right) \frac{\partial \hat{S}^\dagger}{\partial t} \right\}_{\alpha\alpha}.$$
(7.18c)

The quantity $I_\alpha^{Q(pump)}$ satisfies the conservation law, Eq. (7.10). This follows from Eq. (4.11) for the DC charge current given in Eq. (4.22). Therefore, we get

$$\int_0^{\mathcal{T}} \frac{dt}{\mathcal{T}} \text{Im tr} \left\{ \left(\hat{S}(t,E) + 2\hbar\Omega_0\hat{A}(t,E)\right) \frac{\partial \hat{S}^\dagger(t,E)}{\partial t} \right\} = 0, \qquad (7.19)$$

which should hold for any E.

The separation of the contributions given in Eq. (7.18a) can be justified by the following arguments.

1. The generated heat flow $I_\alpha^{Q(gen)}$ is definitely positive in all leads, $\alpha = 1, \ldots, N_r$. This is exactly what is expected if heat is generated by the dynamic scatterer and dissipated in the reservoirs. To show its positiveness we rewrite Eq. (7.18b) in terms of the Fourier coefficients for the frozen scattering matrix elements:

$$I_\alpha^{Q(gen)} = \frac{\hbar\Omega_0^2}{4\pi} \int_0^\infty dE \left(-\frac{\partial f_0}{\partial E}\right) \sum_{n=-\infty}^\infty n^2 \sum_{\beta=1}^{N_r} |S_{\alpha\beta,n}(E)|^2. \qquad (7.20)$$

It is obvious that $I_\alpha^{Q(gen)} > 0$. From the above equation it follows that in the adiabatic regime the heat generated by the quantum pump is quadratic in the pump frequency Ω_0 [81].

2. The transferred heat flow $I_\alpha^{Q(pump)}$ vanishes at zero temperature, since it is impossible to take heat out of a reservoir with zero temperature in order to push it into another reservoir. This property follows from Eq. (7.18c) where at zero temperature $(E - \mu_0)\partial f_0/\partial E = 0$. From Eq. (7.18c) it follows that the transferred heat is linear in the pump frequency, $I_\alpha^{Q(pump)} \sim k_B T_0 \Omega_0$.

Note under the condition given in Eq. (7.16) it is possible to realize a regime where the energy ($\sim k_B T_0 \Omega_0$) taken out of a reservoir is larger than its heating ($\sim \Omega_0^2$). Then such a reservoir will be cooled. To characterize the cooling efficiency let us introduce a coefficient K_α equal to the ratio of the DC heat current in the lead α and the total work done by the driving forces. Since the volume of a system remains constant, the work mentioned is equal to the total heat generated by the scatterer. Therefore, the coefficient K_α is

$$K_\alpha = (-1)\frac{I_\alpha^Q}{I_{tot}^Q}, \tag{7.21}$$

where

$$I_{tot}^Q = \frac{\hbar}{4\pi}\int_0^\infty dE\left(-\frac{\partial f_0}{\partial E}\right)\int_0^{\mathcal{T}}\frac{dt}{\mathcal{T}}\,\mathrm{tr}\left(\frac{\partial \hat{S}}{\partial t}\frac{\partial \hat{S}^\dagger}{\partial t}\right). \tag{7.22}$$

The positive/negative sign corresponds to cooling/heating of the reservoir α.

7.2.2 Low temperatures

In the case of ultra-low temperatures,

$$k_B T_0 \ll \hbar\Omega_0, \tag{7.23}$$

during integration over the energy in Eq. (7.6) we can relax the energy dependence of the Floquet scattering matrix elements and calculate them at $E = \mu_0$. Such a simplification is possible because the integration over energy in Eq. (7.6) is over a window of the order of $k_B T_0$ near the Fermi energy. The scattering matrix changes significantly if the energy changes by the value of order δE. Taking into account Eq. (7.23) we find that in the adiabatic regime, Eq. (3.49), $k_B T_0 \ll \delta E$, which justifies the simplification used.

So, now Eq. (7.6) reads

$$I_\alpha^Q = \frac{1}{h}\sum_{n=-\infty}^\infty \sum_{\beta=1}^{N_r}\left|S_{F,\alpha\beta}(\mu_0 + n\hbar\Omega_0, \mu_0)\right|^2 \int_{\mu_0 - n\hbar\Omega_0}^{\mu_0} dE\,(E - \mu_0 + n\hbar\Omega_0)$$

$$= \frac{\hbar\Omega_0^2}{4\pi}\sum_{n=-\infty}^\infty n^2 \sum_{\beta=1}^{N_r}\left|S_{F,\alpha\beta}(\mu_0 + n\hbar\Omega_0, \mu_0)\right|^2.$$

Using Eq. (7.15) and making the inverse Fourier transformation we finally calculate ($k_B T_0 \ll \hbar\Omega_0$)

$$I_\alpha^Q = \frac{\hbar}{4\pi} \int_0^{\mathcal{T}} \frac{dt}{\mathcal{T}} \left(\frac{\partial \hat{S}(t,\mu_0)}{\partial t} \frac{\partial \hat{S}^\dagger(t,\mu_0)}{\partial t} \right)_{\alpha\alpha} + \mathcal{O}\left(\Omega_0^3\right). \quad (7.24)$$

Comparing the equation above with Eq. (7.18b) we conclude that at low temperatures, Eq. (7.23), the dynamic scatterer only heats the reservoirs. While the heat pump effect is absent, this is consistent with the conclusion made on the basis of Eq. (7.18c) calculated at zero temperature.

Chapter 8

Dynamic mesoscopic capacitor

A capacitor does not support a direct current. To model one we can consider a mesoscopic sample attached to only a single reservoir. We will call it *a mesoscopic capacitor* [214], since its capacitance depends not only on the geometry (as for a macroscopic capacitor) but also on the density of states (DOS) of electrons. By changing periodically in time either the potential of a sample via a near gate or the potential of a reservoir (or by changing both potentials simultaneously) one can generate an alternating current flowing between the sample and the reservoir. Because of gauge invariance the current depends on the potential difference rather than on each potential separately. Therefore, in what follows we consider a periodic potential applied to the sample and a stationary reservoir. The reservoir is in equilibrium and it is characterized by the Fermi distribution function $f_0(E)$ with chemical potential μ_0 and temperature $k_B T_0$.

8.1 General theory for a single-channel scatterer

For the sake of simplicity we consider the lead connecting the sample to the reservoir to be one-dimensional. We ignore spin-flip processes, therefore, electrons can be treated as spinless. Then the sample can be viewed as a single-channel scatterer, which has only one incoming and one outgoing orbital channel. In the stationary case the capacitor is characterized by a single scattering amplitude. If such a sample is driven by a periodic perturbation then the scattering amplitude becomes a matrix in the energy space with elements $S_F(E_n, E)$, where $E_n = E + n\hbar\Omega_0$ with n integer. We call this matrix *the Floquet scattering matrix*.

8.1.1 Scattering amplitudes

The scattering amplitudes $S_{in}(t, E)$ and $S_{out}(E, t)$ define elements of the Floquet scattering matrix as follows,

$$S_F(E + n\hbar\Omega_0, E) = S_{in,n}(E) \equiv \int_0^{\mathcal{T}} \frac{dt}{\mathcal{T}} e^{in\Omega_0 t} S_{in}(t, E), \qquad (8.1)$$

$$S_F(E, E - n\hbar\Omega_0) = S_{out,n}(E) \equiv \int_0^{\mathcal{T}} \frac{dt}{\mathcal{T}} e^{in\Omega_0 t} S_{out}(E, t), \qquad (8.2)$$

where $\mathcal{T} = 2\pi/\Omega_0$ is the period of the drive.

From the definition we get the following relation between in- and out-scattering amplitudes,

$$S_{in,n}(E) = S_{out,n}(E_n). \qquad (8.3)$$

In a time representation one can get

$$S_{in}(t, E) = \sum_{n=-\infty}^{\infty} \int_0^{\mathcal{T}} \frac{dt'}{\mathcal{T}} e^{in\Omega_0(t'-t)} S_{out}(E_n, t'),$$

$$\qquad (8.4)$$

$$S_{out}(E, t) = \sum_{n=-\infty}^{\infty} \int_0^{\mathcal{T}} \frac{dt'}{\mathcal{T}} e^{in\Omega_0(t'-t)} S_{in}(t', E_{-n}).$$

8.1.2 Unitarity conditions

The unitarity conditions read

$$\sum_{n=-\infty}^{\infty} S_F^*(E_n, E_m) S_F(E_n, E) = \sum_{n=-\infty}^{\infty} S_F(E_m, E_n) S_F^*(E, E_n) = \delta_{m,0}. \qquad (8.5)$$

Using Eqs. (8.1) and (8.2) we obtain from Eq. (8.5) the following relations between the amplitudes S_{in} and S_{out} [see Eq. (3.60)],

$$\int_0^{\mathcal{T}} \frac{dt}{\mathcal{T}} e^{im\Omega_0 t} S_{in}^*(t, E_m) S_{in}(t, E) = \int_0^{\mathcal{T}} \frac{dt}{\mathcal{T}} e^{im\Omega_0 t} S_{out}(E_m, t) S_{out}^*(E, t)$$
$$= \delta_{m,0}. \qquad (8.6)$$

Using the second equations in Eqs. (8.5) and (8.6) one can find

$$\sum_{m=-\infty}^{\infty} e^{-im\Omega_0 t} \sum_{n=-\infty}^{\infty} S_F(E_m, E_n) S_F^*(E, E_n)$$
$$= \sum_{m=-\infty}^{\infty} e^{-im\Omega_0 t} \left(S_{out}(E_m, t) S_{out}^*(E, t) \right)_m = \sum_{m=-\infty}^{\infty} e^{-im\Omega_0 t} \delta_{m,0} = 1. \qquad (8.7)$$

In fact we have proven the following useful identity as a direct consequence of the unitarity of scattering,

$$\sum_{n=-\infty}^{\infty} \int_0^{\mathcal{T}} \frac{dt'}{\mathcal{T}} e^{-in\Omega_0(t-t')} S_{in}(t, E_n) S_{in}^*(t', E_n) = 1, \qquad (8.8)$$

or equivalently,

$$\sum_{n=-\infty}^{\infty} e^{-in\Omega_0 t} S_{in}(t, E_n) S_{out,-n}^*(E) = 1, \qquad (8.9a)$$

$$\sum_{m=-\infty}^{\infty} \sum_{n=-\infty}^{\infty} e^{-im\Omega_0 t} S_{out,\,n+m}(E_m) S_{out,\,n}^*(E) = 1, \qquad (8.9b)$$

$$\sum_{m=-\infty}^{\infty} \sum_{n=-\infty}^{\infty} e^{-im\Omega_0 t} S_{in,\,n+m}(E_{-n}) S_{in,\,n}^*(E_{-n}) = 1. \qquad (8.9c)$$

8.1.3 Time-dependent current

The general expression for a time-dependent current [44] in the case of a periodically driven capacitor is

$$I(t) = \frac{e}{h} \int dE \sum_{n=-\infty}^{\infty} \{f_0(E) - f_0(E_n)\} \sum_{l=-\infty}^{\infty} e^{-il\Omega_0 t} S_F^*(E_n, E) S_F(E_{n+l}, E). \quad (8.10)$$

To simplify the expression above we shift $E \to E_n$ in the part dependent on $f_0(E_n)$. Then from Eq. (8.7) we conclude that this part is reduced to $f_0(E)$. Using Eq. (8.1) in the remaining part of Eq. (8.10) we arrive at the following [215]

$$I(t) = \frac{e}{h} \int dE f_0(E) \left\{ |S_{in}(t, E)|^2 - 1 \right\}. \quad (8.11)$$

Let us show that in the adiabatic regime this equation can be easily transformed into

$$I(t) = -\frac{ie}{2\pi} \int dE \left(-\frac{\partial f_0}{\partial E} \right) S(t, E) \frac{\partial S^*(t, E)}{\partial t}, \quad (8.12)$$

in accordance with a general theory developed in Ref. [43] (see, e.g., Ref. [147]). So in the adiabatic regime we use

$$S_{in}(t, E) = S(t, E) + \frac{i\hbar}{2} \frac{\partial^2 S(t, E)}{\partial t \partial E}, \quad (8.13)$$

in the first order in Ω_0 approximation with S being the frozen scattering amplitude.[1] To calculate $|S_{in}(t, E)|^2$ we use $|S|^2 = 1$ and, correspondingly, $\partial^2 |S|^2/(\partial t \partial E) = 0$. Also we use,

$$\frac{\partial S^*}{\partial t} \frac{\partial S}{\partial E} = \frac{\partial S^*}{\partial t} S S^* \frac{\partial S}{\partial E} = S^* \frac{\partial S}{\partial t} \frac{\partial S^*}{\partial E} S = \frac{\partial S}{\partial t} \frac{\partial S^*}{\partial E},$$

---

[1] See the first two terms on the RHS of Eq. (8.61) for a single-cavity capacitor but also Eq. (9.39) for a double-cavity capacitor. The appearance of an anomalous scattering amplitude A in the latter case does not affect Eq. (8.12).

and find (up to $\sim \Omega_0$ terms):

$$|S_{in}(t,E)|^2 \approx 1 - i\hbar \frac{\partial}{\partial E}\left(S\frac{\partial S^*}{\partial t}\right).$$

Substituting this equation into Eq. (8.11) and integrating over the energy E by parts we arrive at Eq. (8.12).

8.1.4 *Dissipation*

The (DC) heat flowing out of the driven capacitor is

$$I_E = \frac{1}{h}\int_0^\infty dE \sum_{n=-\infty}^\infty (E_n - \mu_0)\left[f_0(E) - f_0(E_n)\right]|S_F(E_n, E)|^2. \quad (8.14)$$

Using the unitarity of the Floquet scattering matrix, Eq. (8.5) with $m = 0$, and shifting the energy, $E_n \to E$, in the part with $f_0(E_n)$, we simplify

$$I_E = \frac{\Omega_0}{2\pi}\int_0^\infty dE\, f_0(E) \sum_{n=-\infty}^\infty n\,|S_F(E_n, E)|^2$$

$$+ \frac{1}{h}\int_0^\infty dE\,(E - \mu_0)\,f_0(E) \sum_{n=-\infty}^\infty |S_F(E_n, E)|^2$$

$$- \frac{1}{h}\int_0^\infty dE\,(E - \mu_0)\,f_0(E) \sum_{n=-\infty}^\infty |S_F(E, E_{-n})|^2$$

$$= \frac{\Omega_0}{2\pi}\int_0^\infty dE\, f_0(E) \sum_{n=-\infty}^\infty n\,|S_F(E_n, E)|^2.$$

Then we introduce the scattering amplitude $S_{in,n}(E) = S_F(E_n, E)$, use the property of the Fourier coefficients $\Omega_0 n S_{in,n}^*(E) = -i\left\{\partial S_{in}^*(t, E)/\partial t\right\}_{-n}$, and finally obtain [215]:

$$I_E = -\frac{i}{2\pi} \int_0^\infty dE\, f_0(E) \int_0^{\mathcal{T}} \frac{dt}{\mathcal{T}} S_{in}(t,E) \frac{\partial S_{in}^*(t,E)}{\partial t}. \quad (8.15)$$

Note Eqs. (8.11) and (8.15) are valid for arbitrary frequency and for arbitrary amplitude of the drive. The only disadvantage is that they involve an integration over all energies.

8.1.5 *Dissipation versus squared current*

It is interesting to note that in the adiabatic regime ($\Omega_0 \to 0$) at zero temperature the heat production I_E for the capacitor can be related to the average square current $\langle I^2 \rangle$. From Eq. (8.12) we find for zero temperature

$$I(t) = -\frac{ie}{2\pi} S(t,\mu_0) \frac{\partial S^*(t,\mu_0)}{\partial t}. \quad (8.16)$$

Using $S dS^* = -dS\, S^*$ following from the unitarity of the frozen scattering amplitude, $|S(t,E)|^2 = 1$, we represent the square current as

$$I^2(t) = -\frac{e^2}{4\pi^2} S \frac{\partial S^*}{\partial t} S \frac{\partial S^*}{\partial t} = \frac{e^2}{4\pi^2} S S^* \frac{\partial S}{\partial t} \frac{\partial S^*}{\partial t} = \frac{e^2}{4\pi^2} \left|\frac{\partial S}{\partial t}\right|^2,$$

and find for its average:

$$\langle I^2 \rangle = \int_0^{\mathcal{T}} \frac{dt}{\mathcal{T}} I^2(t) = \frac{e^2 \Omega_0^2}{4\pi^2} \sum_{n=1}^\infty n^2 \left\{ |S_n|^2 + |S_{-n}|^2 \right\}. \quad (8.17)$$

To calculate the heat current I_E in the adiabatic regime we use Eq. (8.14) with $f_0(E_n) \approx f_0(E) + n\hbar\Omega_0 \partial f_0(E)/\partial E + (n^2 \hbar^2 \Omega_0^2/2) \partial^2 f_0(E)/\partial E^2$. For $S_F(E_n, E) = S_{in,n}(E)$ we use Eq. (8.13) and find with accuracy up to Ω_0^2,

$$I_E = \frac{\Omega_0}{2\pi} \int_0^\infty dE\, (E - \mu_0) \left(-\frac{\partial f_0(E)}{\partial E}\right) \sum_{n=-\infty}^\infty n\, |S_n(E)|^2$$

$$+ \frac{\hbar \Omega_0^2}{4\pi} \int_0^\infty dE \left(-\frac{\partial f_0(E)}{\partial E}\right) \sum_{n=-\infty}^\infty n^2\, |S_n(E)|^2.$$

The first term is identically zero for a capacitor. To show this we take into account that for a single-channel capacitor the frozen scattering amplitude is of the following form

$$S(t, E) = e^{i\phi(t,E)}. \quad (8.18)$$

Then we find

$$\sum_{n=-\infty}^{\infty} in\Omega_0 |S_n(E)|^2 = \int_0^{\mathcal{T}} \frac{dt}{\mathcal{T}} S(t,E) \frac{\partial S^*(t,E)}{\partial t} = -i \int_0^{\mathcal{T}} \frac{dt}{\mathcal{T}} \frac{\partial \phi(t,E)}{\partial t} = 0.$$

Therefore, the heat flow generated by the dynamic capacitor is

$$I_E = \frac{\hbar \Omega_0^2}{4\pi} \int_0^{\infty} dE \left(-\frac{\partial f_0(E)}{\partial E}\right) \sum_{n=1}^{\infty} n^2 \left\{|S_n(E)|^2 + |S_{-n}(E)|^2\right\}. \quad (8.19)$$

Comparing Eqs. (8.17) and (8.19) we find at zero temperature [215]

$$I_E = R_q \langle I^2 \rangle, \quad (8.20)$$

where $R_q = h/(2e^2)$ is the relaxation resistance for a single-channel scatterer and spinless electrons, called the relaxation resistance quantum [214].

8.2 Chiral single-channel capacitor

Experiments demonstrate [129, 216] that a quantum capacitor in a two-dimensional electron gas in the integer quantum Hall effect regime is a promising device to realize sub-nanosecond, single- and few-electron, coherent quantum electronics. The quantum capacitor can be used as a single-particle emitter [129]. With such an emitter as an elementary block, several effects were predicted including shot-noise plateaus [217], two-particle emission and particle reabsorption [46], and a tunable two-particle Aharonov–Bohm effect [218]. The frequency-dependent noise of a chiral quantum capacitor was addressed experimentally [219] in a non-adiabatic regime and theoretically [215] in an adiabatic regime. Within the conceptually simple model of a quantum capacitor suggested in Ref. [219] the noise spectrum as well as the full counting statistics were calculated in Ref. [220].

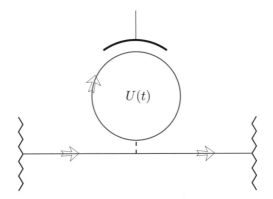

Fig. 8.1 The model for a single-cavity chiral quantum capacitor driven by a uniform potential $U(t)$. The dotted line denotes a QPC. The arrows indicate the direction of movement of electrons.

8.2.1 *Model and scattering amplitude*

We consider a model [129, 170, 216, 221] consisting of a single circular edge state of circumference L (a cavity) coupled via a quantum point contact (QPC) to a linear edge state which in turn flows out of a reservoir of electrons with temperature T_0 and Fermi energy μ_0, see Fig. 8.1. A potential that is periodic in time $U(t) = U(t + \mathcal{T})$ induced at a nearby gate is applied uniformly over the cavity. For the sake of simplicity we ignore the geometrical capacitance between the gate and the cavity.

Using the method presented in Section 3.5.3 one can calculate the elements of the Floquet scattering matrix as

$$S_F(E_n, E) = \int_0^{\mathcal{T}} \frac{dt}{\mathcal{T}} e^{in\Omega_0 t} S_{in}(t, E) , \qquad (8.21a)$$

$$S_{in}(t, E) = \sum_{q=0}^{\infty} e^{iqkL} S^{(q)}(t) , \qquad (8.21b)$$

$$S^{(0)} = r , \quad S^{(q>0)}(t) = \bar{t}^2 \, r^{q-1} \, e^{-i\Phi_q(t)} , \quad \Phi_q(t) = \frac{e}{\hbar} \int_{t-q\tau}^{t} dt' U(t') .$$

Here $r(E)/\bar{t}(E)$ is a reflection/transmission amplitude of a QPC connecting the cavity to the linear edge state and $\tau = m_e L/(\hbar k)$ is the time needed by an electron with energy E to make one turn around the cavity of length L.

The index q counts the number of turns which an electron completes until escaping the cavity. In the equation above we assume that $\hbar\Omega_0 \ll E$ and the reflection/transmission amplitude of a QPC changes in energy over the scale $\delta E \sim E$, which is much larger than $\hbar\Omega_0$. Correspondingly in what follows we neglect the terms of order $\hbar\Omega_0/\delta E$ and smaller.

To calculate a Floquet scattering matrix element, $S_F(E_n, E)$, we consider scattering of a plane wave, $e^{-iEt/\hbar + ikx}$, with unit amplitude and with energy E off the oscillating scatterer. We direct the x axis along the linear edge state and the y axis along the circular edge state of the cavity. We assume that the QPC connects points $x = 0$ and $y = 0$. Then the wave function is

$$\Psi(t,x) = \begin{cases} e^{-iEt/\hbar + ikx}, & x < 0, \\ e^{-iEt/\hbar} \sum_{n=-\infty}^{\infty} \sqrt{\frac{k}{k_n}} S_F(E_n, E) e^{-in\Omega_0 t + ik_n x}, & x > 0, \end{cases} \quad (8.22a)$$

$$\Psi(t,y) = e^{-iEt/\hbar} \sum_{n=-\infty}^{\infty} e^{-in\Omega_0 t} \sum_{l=-\infty}^{\infty} a_l \Upsilon_{n-l} e^{ik_l y}, \quad 0 < y < L, \quad (8.22b)$$

where Υ_p is a Fourier coefficient for $\Upsilon(t)$ dependent on the uniform periodic potential $U(t)$ of the cavity:

$$\Upsilon(t) = \exp\left(-\frac{ie}{\hbar} \int_{-\infty}^{t} dt' U(t')\right). \quad (8.23)$$

In what follows we suppose,

$$\epsilon = \frac{\hbar\Omega_0}{E} \ll 1. \quad (8.24)$$

Then up to zeroth order in ϵ we have [for spatial coordinates $x, y \ll L/(\epsilon\Omega_0 \tau)$]

$$\frac{k_n}{k} \approx 1, \quad e^{ik_n x} \approx e^{ikx} e^{in\Omega_0 x/v}, \quad (8.25)$$

where $v = \hbar k/m_e$ is an electron velocity.[2]

---

[2] In the case of edge states of the integer quantum Hall effect regime the velocity v can be understood as a velocity of an electron orbit (a drift orbit velocity or a skipping orbit velocity) [222, 223] in the crossed electrical and magnetic fields, which ranges from 10^4 m/s [224] to 10^5 m/s [225].

The following functions are periodic in time:

$$S_{in}(t, E) = \sum_{n=-\infty}^{\infty} e^{-in\Omega_0 t} S_F(E_n, E), \quad a(t) = \sum_{l=-\infty}^{\infty} e^{-il\Omega_0 t} a_l. \quad (8.26)$$

With these functions one can perform an inverse Fourier transformation in Eq. (8.22) and get

$$\Psi(t, x) = \begin{cases} e^{-iEt/\hbar + ikx}, & x < 0, \\ S_{in}\left(t - \dfrac{x}{v}, E\right) e^{-iEt/\hbar + ikx}, & x > 0, \end{cases} \quad (8.27a)$$

$$\Psi(t, y) = a\left(t - \dfrac{y}{v}\right) \Upsilon(t) e^{-iEt/\hbar + iky}, \quad 0 < y < L. \quad (8.27b)$$

The amplitudes of the wave function at $x = 0$ and $y = 0$ are related to each other through the scattering matrix of the QPC. If its elements r and $\bar{t}$ can be kept as energy independent over the scale of order $\hbar\Omega_0$ then different terms in Eq. (8.22) have the same boundary conditions at $x = 0$ and $y = 0$. Therefore, one can use a wave function directly in the form of Eq. (8.27):

$$\begin{pmatrix} S_{in}(t, E) \\ a(t)\Upsilon(t) \end{pmatrix} = \begin{pmatrix} r(E) & \bar{t}(E) \\ \bar{t}(E) & r(E) \end{pmatrix} \begin{pmatrix} 1 \\ a(t - \tau)\Upsilon(t) e^{ikL} \end{pmatrix}. \quad (8.28)$$

The time for a single turn, $\tau = L/v$, was introduced after Eq. (8.21).

We solve the system of equations (8.28) by recursion. The equation for $a(t)$,

$$a(t)\Upsilon(t) = \bar{t} + r\, a(t - \tau)\, \Upsilon(t)\, e^{ikL},$$

has the following solution:

$$a(t) = \bar{t}\,\Upsilon^*(t) + \bar{t} \sum_{q=1}^{\infty} r^q\, e^{iqkL}\, \Upsilon^*(t - q\tau). \quad (8.29)$$

Substituting Eq. (8.29) into Eq. (8.28) we find

$$S_{in}(t, E) = r + \bar{t}^2\, \Upsilon(t) \sum_{q=1}^{\infty} r^{q-1} e^{iqkL}\, \Upsilon^*(t - q\tau). \quad (8.30)$$

Then using Eq. (8.23) we arrive at Eq. (8.21b).

8.2.2 Unitarity

The Floquet scattering matrix is unitary. This puts the following constraint on the scattering amplitude S_{in} [42]:

$$\int_0^{\mathcal{T}} \frac{dt}{\mathcal{T}} |S_{in}(t,E)|^2 = 1. \tag{8.31}$$

Let us show that Eq. (8.21b) satisfies this condition:

$$\int_0^{\mathcal{T}} \frac{dt}{\mathcal{T}} |S_{in}(t,E)|^2 = A + B,$$

$$A = R + \frac{T^2}{R} \sum_{q=1}^{\infty} R^q = R + T = 1,$$

$$B = \int_0^{\mathcal{T}} \frac{dt}{\mathcal{T}} 2\Re \left\{ -T \sum_{q=1}^{\infty} r^q e^{i\{qkL - \Phi_q(t)\}} \right.$$

$$\left. + \frac{T^2}{R} \sum_{m=1}^{\infty} R^m \sum_{q=1}^{\infty} r^q e^{iqkL} e^{i\{\Phi_m(t) - \Phi_{m+q}(t)\}} \right\}$$

$$= 2T \Re \sum_{q=1}^{\infty} r^q e^{iqkL} \int_0^{\mathcal{T}} \frac{dt}{\mathcal{T}} \left\{ \frac{T}{R} \sum_{m=1}^{\infty} R^m e^{-i\Phi_q(t - m\tau)} - e^{-i\Phi_q(t)} \right\} = 0.$$

Here $T = |\bar{t}|^2$ and $R = |r|^2$ are the transmission and reflection probabilities, respectively. In the last line of the equation above we use, first, $\Phi_{m+q}(t) - \Phi_m(t) = \Phi_q(t - m\tau)$. Then using the periodicity of $\Phi_q(t)$ in time we make a shift $t - m\tau \to t$ in this term under the integration for a time period $\mathcal{T}$. After that one can sum over m and get zero.

Note in the stationary case, $\Upsilon(t) = 1$, the elements of the Floquet scattering matrix become $S_F(E_n, E) = \delta_{n,0} S(E)$, where the stationary scattering amplitude is

$$S(E) = r + \frac{\bar{t}^2 e^{ikL}}{1 - r e^{ikL}}. \tag{8.32a}$$

This quantity can be presented in the following form:

$$S(E) = -e^{ikL} \frac{r - Re^{-ikL}}{(r - Re^{-ikL})^*}, \tag{8.32b}$$

which is manifestly unitary.

8.2.3 *Gauge invariance*

Now we show that the model we use is gauge-invariant, i.e., we get the same current either applying a periodic potential $U(t)$ at the reservoir or applying a potential $-U(t)$ at the cavity.

We consider a stationary cavity but suppose that the periodic potential $U(t) = U(t + 2\pi/\Omega_0)$ is applied at the reservoir. In this case the state of an electron in the reservoir is the Floquet state, see Eqs. (3.27) and (5.18). Let the operator $\hat{a}'^{\dagger}(E)$ create an electron in the reservoir in the Floquet state,

$$\Psi_E(t, \vec{r}) = e^{i\vec{k}\vec{r}} e^{-i\frac{E}{\hbar}t} \sum_{n=-\infty}^{\infty} \Upsilon_n e^{-in\Omega t}, \tag{8.33}$$

where Υ_n is the Fourier coefficient for $\Upsilon(t)$ defined in Eq. (8.23). If $U(t) = U\cos(\Omega t)$ then $\Upsilon_n = J_n(eU/\hbar\Omega)$, where J_n is the Bessel function of the first kind of the nth order. The operators $\hat{a}'^{\dagger}(E)$ and $\hat{a}'(E)$ describe equilibrium fermions,

$$\langle \hat{a}'^{\dagger}(E), \hat{a}'(E') \rangle = \delta(E - E') f_0(E). \tag{8.34}$$

We assume also that there is no potential within the lead connecting the cavity and the reservoir. Therefore, the wave function for electrons in the lead is a plane wave, $\psi_E(t, x) = e^{ikx - i\frac{E}{\hbar}t}$. Note that the wave number k for $\psi_E(t, x)$ and the wave vector $\vec{k}$ for $\Psi_E(t, \vec{r})$ in the reservoir depend on energy differently. While in the lead, $k = \sqrt{2m_e E/\hbar^2}$ depends on the total energy E of an electron, in the reservoir k depends on the Floquet energy, E, and it is independent of an additional sideband energy $E_n - E = n\hbar\Omega$.

Let the operator $\hat{a}^{\dagger}(E)$ create an electron within the lead. Then matching wave functions with the same total energy, see Eq. (5.24), one can write

$$\hat{a}(E) = \sum_{n=-\infty}^{\infty} \Upsilon_n \hat{a}'(E_{-n}). \tag{8.35}$$

Note that we ignore the reflection due to the wave number changing. The corresponding reflection coefficient is as small as $(\hbar\Omega/\mu_0)^2 \ll 1$. We usually ignore such small quantities.

After scattering by the stationary cavity an electron acquires the scattering amplitude $S(E)$. Therefore, the operator $\hat{b}(E)$ annihilating the scattered electron with energy E is

$$\hat{b}(E) = S(E)\hat{a}(E) = \sum_{n=-\infty}^{\infty} S(E)\Upsilon_n \hat{a}'(E_{-n}). \tag{8.36}$$

Now we calculate the current, $I(t)$, flowing in the lead,

$$I(t) = \frac{e}{h} \iint_0^{\infty} dEdE'\, e^{i\frac{E-E'}{\hbar}t} \left\{ \langle \hat{b}^\dagger(E)\hat{b}(E') \rangle - \langle \hat{a}^\dagger(E)\hat{a}(E') \rangle \right\}. \tag{8.37}$$

The lth harmonic of this current is

$$I_l = \frac{e}{h} \int_0^{\mathcal{T}} \frac{dt}{\mathcal{T}} e^{il\Omega_0 t} \iint_0^{\infty} dEdE'\, e^{i\frac{E-E'}{\hbar}t} \left\{ \langle \hat{b}^\dagger(E)\hat{b}(E') \rangle - \langle \hat{a}^\dagger(E)\hat{a}(E') \rangle \right\}. \tag{8.38}$$

Using Eqs. (8.34)–(8.36) and making the shift $E \to E + n\hbar\Omega$ we finally calculate

$$I_l = \frac{e}{h} \int_0^{\infty} dE f_0(E) \sum_{n=-\infty}^{\infty} \Upsilon_n^\star \Upsilon_{n+l} \left\{ S^\star(E_n) S(E_{n+l}) - 1 \right\}. \tag{8.39}$$

To simplify the above equation we introduce a time-dependent function,

$$S(t, E) = \sum_{n=-\infty}^{\infty} S(E_n) \Upsilon_n e^{-in\Omega t}, \tag{8.40}$$

and take into account that $\sum_{n=-\infty}^{\infty} \Upsilon_n^* \Upsilon_{n+l} = \delta_{l,0}$. Then after the inverse Fourier transformation we get from Eq.(8.39):

$$I(t) = \frac{e}{h} \int dE f_0(E) \left\{ |S(t,E)|^2 - 1 \right\}. \quad (8.41)$$

This equation defines the same current as Eq. (8.11) in the case when the potential $-U(t)$ is applied to the cavity. To check we need to show that the function $S(t,E)$ differs from the function $S_{in}(t,E)$, Eq.(8.21b), first, by the phase factor (in fact by the factor $\Upsilon(t)$), which is irrelevant for the current, and, second, by the replacement $U \to -U$. So we substitute

$$S(E) = \sum_{q=0}^{\infty} e^{iqkL} S^{(q)}, \quad S^{(0)} = r, \quad S^{(q>0)}(t) = \bar{t}^2 r^{q-1},$$

into Eq. (8.40) and calculate

$$S(t,E) = \sum_{q=0}^{\infty} e^{iqkL} S'^{(q)}(t), \quad (8.42)$$

$$S'^{(0)}(t) = \Upsilon(t) r,$$

$$S'^{(q>0)}(t) = \bar{t}^2 r^{q-1} \sum_n e^{iqn\Omega\tau} \Upsilon_n e^{-in\Omega t}$$

$$= \bar{t}^2 r^{q-1} \Upsilon(t - q\tau) = \Upsilon(t) \bar{t}^2 r^{q-1} e^{-i\tilde{\Phi}_q(t)},$$

$$\tilde{\Phi}_q(t) = \frac{e}{\hbar} \int_{t-q\tau}^{t} dt' \left(-U(t') \right).$$

Here we used $k(E_n) \approx k(E) + n\Omega/v$ and $L/v = \tau \equiv h/\Delta$. Comparing Eqs. (8.21b) and (8.42) we see that

$$S(U(t), E) = \Upsilon(t) S_{in}(-U(t), E). \quad (8.43)$$

One can understand this equation as follows. The particle leaving a reservoir at time t has a phase $\Upsilon(t)$ induced by the oscillating potential. However, to calculate a current we need to count the particles leaving the cavity at time t. If the particle leaving the cavity at time t has spent q turns in the cavity then it left the reservoir at time $t - q\tau$. Such a particle has a time-dependent phase $\Upsilon(t - q\tau)$. The common phase for all the amplitudes is irrelevant for a measurable quantity. Therefore, one can ignore the largest time-dependent phase $\Upsilon(t)$. After such an artificial transformation the time-dependent phase becomes $\Upsilon^*(t)\Upsilon(t - q\tau)$. This is exactly the phase that a particle spending q turns in the cavity would feel if the potential $-U(t)$ were applied at the cavity instead of at the reservoir.

8.2.4 Time-dependent current

Replacing in Eq. (8.10) the Floquet scattering matrix elements by the Fourier coefficients for $S_{in}(t, E)$ we obtain the following [compare with Eq. (3.65)]:

$$I(t) = \frac{e}{h} \int_0^\infty dE \sum_{n=-\infty}^\infty \{f_0(E) - f_0(E_n)\} \int_0^{\mathcal{T}} \frac{dt'}{\mathcal{T}} e^{in\Omega_0(t-t')} S_{in}^*(t', E) S_{in}(t, E). \tag{8.44}$$

Then substituting Eqs. (8.21b) into Eq. (8.44) and integrating over the energy E we find the current as a sum of two terms $I^{(d)}$ and $I^{(nd)}$ [221]:

$$I(t) = I^{(d)}(t) + I^{(nd)}(t), \tag{8.45a}$$

$$I^{(d)}(t) = \frac{e^2}{h} T^2 \sum_{q=1}^\infty R^{q-1} \{U(t) - U(t - q\tau)\}, \tag{8.45b}$$

$$I^{(nd)} = \frac{eT^2}{\pi\tau} \Im \left\{ \sum_{s=1}^\infty \frac{\eta\left(s\frac{T_0}{T^*}\right) \{r e^{ik_F L}\}^s}{s} \sum_{q=1}^\infty R^{q-1} \left(e^{-i\Phi_s(t-q\tau)} - e^{-i\Phi_s(t)} \right) \right\}. \tag{8.45c}$$

Here $R = 1 - T$ is the reflection probability for a QPC, $k_F = \sqrt{2m_e\mu_0}/\hbar$, and $k_B T^* = \hbar/(\pi\tau) = \Delta/(2\pi^2)$. In Eq. (8.45) the time τ is calculated for electrons with Fermi energy, $E = \mu_0$. Such an approximation is valid in the zeroth order in $k_B T_0/\mu_0 \to 0$. The function $\eta(x) = x/\sinh(x)$ appeared after an integration over the energy:

$$\eta\left(2\pi^2 s \frac{k_B T_0}{\Delta}\right) = i\frac{2\pi s}{\Delta} \int_0^\infty dE f_0(E) e^{i2\pi \frac{E-\mu_0}{\Delta} s}. \quad (8.46)$$

Note that when we integrate over energy the term with $f_0(E_n)$ in Eq. (8.44) we make a shift $E_n \to E$ and expand the exponential factors in accordance with Eq. (8.25):

$$e^{iqk_n L} \approx e^{iqkL} e^{in\Omega_0 \tau},$$

where $k_n = k(E_n)$. We represent the double sum appearing after substituting Eq. (8.21) into Eq. (8.44) as

$$\sum_{q=1}^\infty \sum_{p=1}^\infty A_q B_p = \sum_{q=1}^\infty A_q B_q + \sum_{q=1}^\infty \sum_{s=1}^\infty A_q B_{q+s} + \sum_{p=1}^\infty \sum_{s=1}^\infty A_{p+s} B_p.$$

In Eq. (8.45) we assume that the energy scale δE, over which the reflection/transmission amplitude of a QPC changes significantly, is much larger than the temperature, $\delta E \gg k_B T_0$. Therefore, we calculate r and $\bar{t}$ at $E = \mu_0$. This is correct if $k_B T_0 \ll \mu_0$ since for a QPC $\delta E \sim E$ and only electrons with energies $E \sim \mu_0$ are relevant for transport.

The diagonal contribution $I^{(d)}$ arises due to the interference of photon-assisted amplitudes corresponding to the spatial paths with the same length [having the same index q in Eq. (8.21b)], which an electron follows in propagating through the system. This contribution is temperature independent. Since we neglect inelastic processes the temperature cannot be too high. The non-diagonal part, $I^{(nd)}$, is due to the interference of photon-assisted amplitudes corresponding to different numbers of turns, $q_1 \neq q_2$. This part is suppressed by the temperature (at $T_0 \gtrsim T^*$) since it is the sum of contributions that oscillate strongly with an electron energy. Therefore, at high temperatures, $T_0 \gg T^*$, only that part of the current that is linear in the cavity's potential is present, $I(t) \approx I^{(d)}(t)$. In contrast at $T_0 \ll T^*$ both parts, $I^{(d)}(t)$ and $I^{(nd)}(t)$, do contribute and the current is a non-linear function of $U(t)$.

The current $I(t)$ depends on the pump frequency Ω_0 with period $\delta \Omega_0 = 2\pi/\tau$. The corresponding periodicity is governed by the time τ for a single turn around the cavity. If $\tau = n\mathcal{T}$, hence $\Omega_0 = n\delta\Omega_0$, then the oscillating potential $U(t) = U(t+\mathcal{T})$ does not change the phase of the electrons contributing to the current. Such electrons enter the cavity, make several (q) turns, and escape. Therefore, they visit the cavity for a finite time period $\delta t = q\tau = qn\mathcal{T}$. These electrons see an effectively stationary cavity since the time-dependent phase they acquire is zero, $\Phi_q(t) = 0$. In such a case a current does not arise, $I(t) = 0$. At frequencies different from these particular values the phase accumulated by an electron within the cavity is dependent on time. Consequently, in accordance with the Friedel sum rule [16], the charge accumulated within the cavity is dependent on time, which in turn causes the appearance of a time-dependent current, $I(t) \neq 0$.

8.2.5 High-temperature current

Since at $T_0 \gg T^*$ the current, $I(t) \approx I^{(d)}(t)$, is linear in potential, we can introduce a frequency-dependent linear response function (the conductance),

$$G_l^{(d)} = \frac{I_l^{(d)}}{U_l}, \qquad (8.47)$$

where U_l and I_l are Fourier coefficients for the potential and current, respectively:

$$U(t) = \sum_{l=-\infty}^{\infty} U_l e^{-il\Omega_0 t}, \quad I^{(d)}(t) = \sum_{l=-\infty}^{\infty} I_l^{(d)} e^{-il\Omega_0 t}. \qquad (8.48)$$

Taking into account that

$$U(t - q\tau) = \sum_{l=-\infty}^{\infty} U_l e^{il\Omega_0 q\tau} e^{-il\Omega_0 t},$$

we calculate from Eq. (8.45b):

$$G_l^{(d)} = \frac{e^2}{h} T \frac{1 - e^{il\Omega_0 \tau}}{1 - R e^{il\Omega_0 \tau}}. \qquad (8.49)$$

The AC conductance $G_l^{(d)}$ shows a strong non-linear dependence on the frequency Ω_0 of the drive. The frequency affects both the magnitude and the phase of the response function. In particular, at $0 < l\Omega_0\tau \bmod 2\pi < \pi$

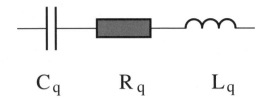

Fig. 8.2 An equivalent electrical circuit to model the low-frequency response of a quantum capacitor.

the response is capacitive-like. While at $\pi < l\Omega_0\tau \bmod 2\pi < 2\pi$ it is inductive-like. It is interesting that at $l\Omega_0\tau \bmod 2\pi = \pi$ the response is purely ohmic, $G_l^{(d)} = 1/R_q$ (for R_q see below).

In general to model a mesoscopic system via an equivalent electric circuit one needs to use some frequency-dependent element. However, at small frequencies, $\Omega_0\tau \ll 1$, one can model it as a capacitance $C_q^{(d)}$, a resistance $R_q^{(d)}$, and an inductance $L_q^{(d)}$ connected in series, Fig. 8.2. The resistance of such a circuit is equal to

$$\frac{1}{G(\omega)} = R_q + i\left\{\frac{1}{\omega C_q} - \omega L_q\right\}. \tag{8.50}$$

Comparing it to Eq. (8.49) at

$$\omega = l\Omega_0 \ll \frac{1}{\tau}, \tag{8.51}$$

we find

$$C_q^{(d)} = \frac{e^2}{\Delta}, \quad R_q^{(d)} = \frac{h}{e^2}\left(\frac{1}{T} - \frac{1}{2}\right), \quad L_q^{(d)} = \frac{h^2}{12e^2\Delta}, \tag{8.52}$$

where $\Delta = h/\tau \ll \mu_0$ is a level spacing in the isolated cavity. The upper index (d) indicates a high-temperature regime.

At lower temperatures, $T_0 \lesssim T^*$, both parts, $I^{(d)}$ and $I^{(nd)}$, contribute to the current. The current $I^{(nd)}$, Eq. (8.45c), is a non-linear function of both the magnitude and the frequency of the driving potential $U(t)$.

8.2.6 Linear response regime

For an oscillating potential with a small amplitude,

$$eU_l \ll l\hbar\Omega_0, \qquad (8.53)$$

one can simplify the expression for $I^{(nd)}$. At zero temperature we use Eq. (8.45c) and find a compact expression valid at an arbitrary frequency. On the other hand, at small driving frequencies, $\Omega_0 \tau \ll 1$, it is more convenient to expand directly the expressions for the scattering matrix, Eqs. (8.21), and then to calculate the current $I(t)$, Eq. (8.44). In this way we can obtain a simple expression allowing us to analyze the temperature dependence of a current.

8.2.6.1 Zero-temperature linear response current

We expand exponents depending on Φ_s in Eq. (8.45c) up to linear in U_l terms. Then, taking into account that at zero temperature $\eta(0) = 1$, one can sum over s and q. After that we can calculate $G_l^{(nd)} = I_l^{(nd)}/U_l$.

We use the following transformation taking into account that $U_l = U_{-l}^*$, since $U(t)$ is real:

$$I(t) = \Im \sum_{l=-\infty}^{\infty} U_l G'_l e^{-il\Omega_0 t}$$

$$= \sum_{l=-\infty}^{\infty} \frac{U_l G'_l e^{-il\Omega_0 t} - U_l^* G_l'^* e^{il\Omega_0 t}}{2i} = \sum_{l=-\infty}^{\infty} I_l e^{-il\Omega_0 t},$$

$$I_l = U_l G_l, \quad G_l = \frac{G'_l - G'^*_{-l}}{2i}.$$

Then the total zero-temperature AC conductance $G_l = G_l^{(d)} + G_l^{(nd)}$ is found to be

$$G_l = G_l^{(d)} \left\{ 1 + \frac{i}{l\Omega_0 \tau} \ln\left(\frac{1 + R e^{2il\Omega_0 \tau} - 2\sqrt{R} e^{il\Omega_0 \tau} \cos(\chi_F)}{1 + R - 2\sqrt{R}\cos(\chi_F)}\right) \right\}. \quad (8.54)$$

Here we use the following notation: $r = \sqrt{R} e^{i\chi_r}$ and $\chi_F = k_F L + \chi_r - 2\pi e U_0/\Delta$, where $|eU_0| \ll \mu_0$ is the average value of the oscillating potential. We stress that Eq. (8.54) is valid for small amplitude but arbitrary frequency of an oscillating potential.

To get the parameters of a low-frequency equivalent circuit (at low temperatures we denote them as C_q, R_q, and L_q) we evaluate Eq. (8.54) in the limit of $l\Omega_0 \tau \to 0$ and obtain after comparison with Eq. (8.50):

$$C_q = \frac{e^2}{\Delta} \frac{T}{1 + R - 2\sqrt{R}\cos(\chi_F)} \equiv e^2 \nu(\mu_0), \quad R_q = \frac{h}{2e^2},$$
(8.55)
$$L_q = \frac{h^2 \nu(\mu_0)}{12 e^2} \left\{ 1 + \frac{8R - 2(1+R)\sqrt{R}\cos(\chi_F) - 4R\cos^2(\chi_F)}{T^2} \right\}.$$

Here $\nu(E) = i/(2\pi) S(E) \partial S^*/\partial E$ is the DOS of a stationary cavity coupled to a linear edge state [for $S(E)$ see Eq. (8.32)].

Even neglecting the geometrical capacitance we arrived at the intrinsic, *quantum capacitance* C_q proportional to the density of states ν [214].[3]

The low-temperature charge relaxation resistance R_q in the case of a single-channel transport[4] is universal, in the sense that it is independent of the parameters of the capacitor. This universality was predicted in Ref. [214] and confirmed experimentally in Ref. [216]. It remains so even in the presence of the Coulomb blockade effect [234–237]. In contrast, at high temperatures the charge relaxation resistance R_q, Eq. (8.52), depends on the transmission T of a QPC connecting a cavity and a linear edge state (see also the discussion in Ref. [238]).

To describe the linear response of a mesoscopic system up to cubic in frequency terms we need to introduce *the quantum inductance* [239–242] L_q having a purely dynamic origin and, therefore, being proportional to the time spent by an electron in the cavity, $\sim h/(T\Delta)$ at $T \to 0$.[5]

8.2.6.2 Low-frequency linear response current

At a small frequency,

$$\Omega_0 \tau \ll 1, \qquad (8.56)$$

the Floquet scattering matrix, Eq. (8.21), can be expressed in terms of a stationary scattering matrix $S(E)$, Eq. (8.32), calculated at $k(U_0) = \sqrt{2m_e E}/\hbar - 2\pi e U_0/(L\Delta)$. To do so we expand $U'(t') = U(t') - U_0$ in the equation for $\Phi_q(t)$, Eq. (8.21b), in powers of $t' - t$,

---

[3] This constitutes a basis for the capacitance spectroscopy of mesoscopic systems [226–232].
[4] For the multi-channel case see Refs. [9, 170, 214, 233].
[5] See also Refs. [233, 243] where the low-frequency admittance of a sample coupled to more than one reservoir was investigated. In this case the inductance-like contribution to the emittance is linear in frequency.

$$U'(t') \approx U'(t) + (t' - t)\frac{dU'(t)}{dt} + \frac{(t'-t)^2}{2}\frac{d^2U'(t)}{dt^2}, \quad (8.57)$$

and integrate over t'. Then expanding corresponding exponents we calculate up to linear in $U'(t)$ and quadratic in Ω_0 terms:

$$S_{in}(t, E) \approx S(U_0, E) - eU'(t)\frac{\partial S(U_0, E)}{\partial E} - \frac{i\hbar}{2}\frac{edU'(t)}{dt}\frac{\partial^2 S(U_0, E)}{\partial E^2}$$
$$+ \frac{\hbar^2}{6}\frac{ed^2U'(t)}{dt^2}\frac{\partial^3 S(U_0, E)}{\partial E^3}. \quad (8.58)$$

Note that the first three terms in the right hand side of the equation above can be found from the adiabatic expansion Eq. (8.13) if one expands the frozen scattering matrix up to linear in $U'(t)$ terms,

$$S(t, E) \equiv S(U(t), E) \approx S(U_0, E) + U'(t)\frac{\partial S(U_0, E)}{\partial U_0},$$

discard $\sim \Omega_0^2$ terms, and take into account that in our model $\partial S/\partial U_0 = -e\,\partial S/\partial E$.

Substituting Eq. (8.58) into Eq. (8.44), where in addition we expand

$$f_0(E) - f_0(E_n) \approx -\frac{\partial f_0}{\partial E}n\hbar\Omega_0 - \frac{\partial^2 f_0}{\partial E^2}\frac{(n\hbar\Omega_0)^2}{2}, \quad (8.59)$$

we find the low-frequency conductance:

$$G_l = \int_0^\infty dE\left(-\frac{\partial f_0(E)}{\partial E}\right)G_l(E),$$
$$\quad (8.60)$$
$$G_l(E) = -ie^2 l\Omega_0\,\nu(E) + e^2\hbar\frac{(l\Omega_0)^2}{2}\nu^2(E)$$
$$- ie^2\hbar^2\frac{(l\Omega_0)^3}{6}\left\{\frac{1}{8\pi^2}\frac{\partial^2\nu(E)}{\partial E^2} - \nu^3(E)\right\}.$$

At zero temperature Eq. (8.60) gives the parameters of the equivalent electric circuit of Eq. (8.55). It is less evident but still true that at high temperatures ($T_0 \gg T^*$) one can find the parameters given in Eq. (8.52) from Eq. (8.60).

8.2.7 Non-linear low-frequency regime

In the limit of low frequencies, Eq. (8.56), one can go beyond the linear response regime, Eq. (8.53). Substituting Eq. (8.57) into Eq. (8.21b) and expanding up to terms of order Ω_0^2 we calculate the scattering matrix as

$$S_{in}(t, E) = S(t, E) + \frac{i\hbar}{2} \frac{\partial^2 S(t, E)}{\partial t \partial E} + \frac{\hbar^2}{6} \frac{ed^2 U(t)}{dt^2} \frac{\partial^3 S(t, E)}{\partial E^3}$$

$$- \frac{\hbar^2}{8} \left(\frac{edU(t)}{dt} \right)^2 \frac{\partial^4 S(t, E)}{\partial E^4} + \mathcal{O}\left\{ (\Omega_0 \tau)^3 \right\}. \quad (8.61)$$

Recall the frozen scattering matrix $S(t, E) = S(U(t), E)$. To calculate it one can use Eq. (8.32) and replace $kL \to kL - 2\pi eU(t)/\Delta$. Note Eq. (8.58) is non-linear in U_0 but linear in $U'(t) = U(t) - U_0$. In contrast Eq. (8.61) is non-linear in a full time-dependent potential $U(t)$.

Substituting Eqs. (8.61) and (8.59) into Eq. (8.44) we calculate the low-frequency current as [221]

$$I(t) = \int_0^\infty dE \left(-\frac{\partial f_0(E)}{\partial E} \right) \left\{ \mathcal{J}^{(1)}(t, E) + \mathcal{J}^{(2)}(t, E) + \mathcal{J}^{(3)}(t, E) \right\}, \quad (8.62\text{a})$$

$$\mathcal{J}^{(1)}(t, E) = e^2 \, \nu(t, E) \, \frac{dU(t)}{dt}, \quad (8.62\text{b})$$

$$\mathcal{J}^{(2)}(t, E) = -\frac{e^2 h}{2} \frac{\partial}{\partial t} \left\{ \nu^2(t, E) \frac{dU(t)}{dt} \right\}, \quad (8.62\text{c})$$

$$\mathcal{J}^{(3)}(t, E) = -\frac{e^2 h^2}{6} \frac{\partial^2}{\partial t^2} \left\{ \left(\frac{1}{8\pi^2} \frac{\partial^2 \nu(t, E)}{\partial E^2} - \nu^3(t, E) \right) \frac{dU(t)}{dt} \right\}$$

$$- \frac{e^3 h^2}{96\pi^2} \frac{\partial}{\partial t} \left\{ \frac{\partial^3 \nu(t, E)}{\partial E^3} \left(\frac{dU(t)}{dt} \right)^2 \right\}, \quad (8.62\text{d})$$

where the frozen DOS is

$$\nu(t,E) = \frac{i}{2\pi} S(t,E) \frac{\partial S^*(t,E)}{\partial E} = \frac{1}{\Delta}\left\{1 + 2\Re \sum_{q=1}^{\infty} r^q\, e^{iq[kL - 2\pi eU(t)/\Delta]}\right\}. \tag{8.63}$$

Note in Eq. (8.62) we use $\partial \nu/\partial t = -e(dU/dt)(\partial \nu/\partial E)$, since the DOS depends on time via an oscillating uniform potential only.

To illustrate the physical meaning of Eq. (8.62) it is instructive to rewrite this equation. Integrating over energy by parts one can represent it in the form of the continuity equation for a charge current,

$$I(t) + \frac{\partial Q(t)}{\partial t} = 0, \tag{8.64a}$$

$$Q(t) = e \int_0^\infty dE\, f_0(E)\, \nu_{dyn}(t,E), \tag{8.64b}$$

$$\nu_{dyn}(t,E) = \nu(t,E) - \frac{h}{2}\frac{\partial \nu^2(t,E)}{\partial t} + \frac{h^2}{6}\frac{\partial^2 \nu^3(t,E)}{\partial t^2}$$
$$- \frac{h^2}{96\pi^2}\frac{\partial^2}{\partial E^2}\left\{2\frac{\partial^2 \nu(t,E)}{\partial t^2} - \frac{\partial^2 \nu(t,E)}{\partial E^2}\left(\frac{dU}{dt}\right)^2\right\}, \tag{8.64c}$$

where $Q(t)$ is the charge accumulated on a mesoscopic capacitor and $\nu_{dyn}(t,E)$ is *the dynamic density of states*.

The dynamic DOS takes into account a retardation effect, i.e., the finite amount of time spent by an electron inside capacitor. As a result the charge $Q(t)$ accumulated on the capacitor depends on the frequency of the drive. At small driving frequencies, $\Omega_0 \to 0$, such a dependence (up to terms of order Ω_0^2) can be accounted by introducing an effective resistance R_q connected in series with a capacitance C_q. In a linear response regime these quantities are constant parameters, see Eq.(8.55) for low- and Eq. (8.52) for high-temperature regimes. In a non-linear regime these parameters become dependent on the driving potential, i.e., the capacitor is characterized by the non-linear dependence of a charge Q on a potential drop U_C and the resistor has a non-linear current-voltage (I-V) characteristic. In such a case it is more convenient to introduce the differential parameters, the differential capacitance, $C_\partial(U_C) = \partial Q(U_C)/\partial U_C$, and the differential resistance, $R_\partial(V) = \partial V/\partial I(V)$. In terms of these quantities the current $I(t)$ flowing into an equivalent electrical circuit subject to the potential $U(t) = U_C + V$ is (at $\Omega_0 \to 0$) [221]

$$I(t) = C_\partial \frac{dU}{dt} - R_\partial C_\partial \frac{\partial}{\partial t}\left(C_\partial \frac{dU}{dt}\right). \qquad (8.65)$$

Comparing Eqs. (8.62) and (8.65) we find

$$C_\partial(t) = e^2 \int_0^\infty dE \left(-\frac{\partial f_0(E)}{\partial E}\right) \nu(t,E), \qquad (8.66a)$$

$$R_\partial(t) = \frac{h}{2e^2} \frac{\int_0^\infty dE \left(-\frac{\partial f_0}{\partial E}\right) \frac{\partial}{\partial t}\left(\nu^2(t,E)\frac{dU}{dt}\right)}{\int_0^\infty dE \left(-\frac{\partial f_0}{\partial E}\right) \nu(t,E) \int_0^\infty dE \left(-\frac{\partial f_0}{\partial E}\right) \frac{\partial}{\partial t}\left(\nu(t,E)\frac{dU}{dt}\right)}.$$
$$(8.66b)$$

We conclude, in a non-linear low-frequency regime the DOS defines the intrinsic capacitance of a mesoscopic sample (which is coupled in series with a geometrical one if any). That is in accordance with Ref. [214] where the linear response regime was considered. The difference is that in a non-linear regime the DOS is related to a differential capacitance while in a linear response regime the DOS is related to an ordinary capacitance. Another difference we found concerns the effective resistance. In a linear regime it has a universal value at zero temperature, $R_q = h/(2e^2)$, see Ref. [214] and Eq. (8.55). In contrast in a non-linear regime R_∂ becomes as dependent on the sample's properties (on the DOS) as on the potential $U(t)$.

Note the third contribution in Eq. (8.62), $\mathcal{J}^{(3)}$, defines a differential inductance $L_\partial(t) = \partial \Phi/\partial I$ (where Φ is a magnetic flux). The corresponding equation can be calculated straightforwardly. There is insufficient space to show it here.

8.2.8 Transient current caused by a step potential

Let the potential change abruptly at some time moment t_0:[6]

---

[6]For different approaches to this problem see, e.g., Refs. [220, 244–246].

$$U(t) = \begin{cases} 0, & t < t_0, \\ U_0, & t > t_0. \end{cases} \quad (8.67)$$

Strictly speaking we suppose that the potential U jumps from zero to U_0 for some time interval $\delta t \gg \hbar\mu_0^{-1}$. The last inequality allows us to use the scattering matrix, Eq. (8.21), valid if all the relevant energy scales are much smaller than the Fermi energy μ_0. On the other hand δt should be sufficiently small compared with the intrinsic time scales (τ, RC-time, etc.) that we can speak about an abrupt change.

Using Eq. (8.21) in Eq. (8.44) we can represent the current as a sum of two contributions [see Eq. (3.138)]:

$$I(t) = I^{(d)}(t) + I^{(nd)}(t), \quad (8.68a)$$

$$I^{(d)}(t) = -i\frac{e}{2\pi} \sum_{q=0}^{\infty} S^{(q)}(t) \frac{\partial S^{(q)*}(t)}{\partial t}, \quad (8.68b)$$

$$I^{(nd)}(t) = \frac{e}{\pi\tau} \Im \sum_{s=1}^{\infty} \frac{\eta\left(s\frac{T}{T^*}\right) e^{iskFL}}{s} C_s(t), \quad C_s(t) = \sum_{q=0}^{\infty} S^{(q+s)}(t) S^{(q)*}(t). \quad (8.68c)$$

These equations are equivalent to equations (8.45).

Notice, Eq. (8.44) for a time-dependent current was originally derived using the Floquet scattering theory. However, it can be cast into a form that does not depend on the periodicity of the drive, see in general Eq. (3.67) and in particular Eqs. (8.45) and (8.68) as examples. Then one can use this equation to calculate an aperiodic current also. Therefore, we use Eq. (8.68) to analyze a transient current caused by the potential Eq. (8.67).

8.2.8.1 High-temperature current

At high temperatures, $T_0 \gg T^*$, only the diagonal current $I^{(d)}(t)$ survives. For the potential $U(t)$, Eq. (8.67), this current is (for $t \geq t_0$) [221]

$$I^{(d)}(t) = \frac{e^2 U_0}{\Delta} \frac{T}{\tau} R^{N(t)}, \quad (8.69)$$

where $N(t) = [t/\tau]$ is the integer part of the ratio t/τ.

As we see at high temperatures the current, $I(t) = I^{(d)}(t)$, decays in time in a step-like manner: It is constant over the time interval τ and it exponentially decreases with increasing time. Over a time scale larger than τ one can write $I(t) \sim I_0 e^{-(t-t_0)/\tau_D}$, where $I_0 = e^2 U_0 T/h$ and

$$\tau_D = \frac{\tau}{\ln\left(\frac{1}{R}\right)}, \qquad (8.70)$$

is a decay time. For a small transparency of a QPC, $T \to 0$, the decay time is $\tau_D \approx \tau/T$.

8.2.8.2 *Low-temperature current*

At lower temperatures, $T_0 \lesssim T^*$, the current $I(t) = I^{(d)}(t) + I^{(nd)}(t)$ still decays in time but in addition it shows fast oscillations with a period $h/(eU_0)$. To calculate $I^{(nd)}(t)$ we need to know $C_s(t)$, Eq. (8.68). We substitute Eq. (8.67) into Eq. (8.21b) and find

$$S^{(q)} = \begin{cases} r, & q = 0, \\ t^2 r^{q-1} \times \begin{cases} e^{-i\frac{eU_0}{\hbar}\tau q}, & 1 \le q \le N, \\ e^{-i\frac{eU_0}{\hbar}t}, & N+1 \le q. \end{cases} \end{cases} \qquad (8.71)$$

Then we calculate

$$S^{(s)} S^{(0)*} = -Tr^s \times \begin{cases} e^{-i\frac{eU_0}{\hbar}\tau s}, & 1 \le s \le N,\ 1 \le N, \\ e^{-i\frac{eU_0}{\hbar}t}, & N+1 \le s,\ \forall N, \end{cases} \qquad (8.72)$$

and

$$S^{(q+s)}S^{(q)*} = T^2 R^{q-1} r^s$$

$$\times \begin{cases} e^{-i\frac{eU_0}{\hbar}\tau s}, & 1 \le q \le N-s,\ 1 \le s \le N-1,\ 2 \le N, \\ e^{i\frac{eU_0}{\hbar}(\tau q-t)}, & \begin{cases} N-s+1 \le q \le N,\ s \le N,\ 1 \le N, \\ 1 \le q \le N,\ N+1 \le s,\ 1 \le N, \end{cases} \\ 1, & N+1 \le q,\ \forall s,\ \forall N. \end{cases} \quad (8.73)$$

Finally we find

$$C_s = \gamma_N(t)\, T\, r^s\, R^N \left\{ 1 - \frac{\theta(N-s)}{\left(R e^{i\frac{eU_0}{\hbar}\tau}\right)^s} \right\} - \chi(t)\theta(s-N-1), \quad (8.74)$$

$$\gamma_N(t) = 1 - T\, e^{-i\frac{eU_0}{\hbar}t}\, \frac{e^{i\frac{eU_0}{\hbar}\tau(N+1)}}{1 - R e^{i\frac{eU_0}{\hbar}\tau}}, \quad \chi(t) = e^{-i\frac{eU_0}{\hbar}t}\, \frac{1 - e^{i\frac{eU_0}{\hbar}\tau}}{1 - R e^{i\frac{eU_0}{\hbar}\tau}}.$$

Then we calculate from Eq. (8.68c):

$$I^{(nd)}(t) = \frac{eT}{\pi\tau} \int_0^\infty dE \left(-\frac{\partial f_0}{\partial E}\right) \Im\left\{ J^{(1)}(t,E) + J^{(2)}(t,E) + J^{(3)}(t,E) \right\}, \quad (8.75a)$$

$$J^{(1)}(t,E) = -R^N \ln\left(1 - re^{ikL}\right) \left\{ 1 - Te^{-i2\pi\frac{eU_0}{\Delta}\frac{t}{\tau}}\, \frac{e^{i2\pi\frac{eU_0}{\Delta}(N+1)}}{1 - Re^{i2\pi\frac{eU_0}{\Delta}}} \right\}, \quad (8.75b)$$

$$J^{(2)}(t,E) = -R^N \sum_{s=1}^{N\ge 1} \frac{e^{is\left(kL - 2\pi\frac{eU_0}{\Delta}\right)}}{s}\, \frac{r^s}{R^s} \left\{ 1 - Te^{-i2\pi\frac{eU_0}{\Delta}\frac{t}{\tau}}\, \frac{e^{i2\pi\frac{eU_0}{\Delta}(N+1)}}{1 - Re^{i2\pi\frac{eU_0}{\Delta}}} \right\}, \quad (8.75c)$$

$$J^{(3)}(t,E) = (-1) \sum_{s=N+1}^\infty \frac{e^{iskL}\, r^s}{s}\, e^{-i2\pi\frac{eU_0}{\Delta}\frac{t}{\tau}}\, \frac{1 - e^{i2\pi\frac{eU_0}{\Delta}}}{1 - Re^{-i2\pi\frac{eU_0}{\Delta}}}. \quad (8.75d)$$

Note, the equations above differ from Eq. (8.68c). Using Eq. (8.46) we reintroduced an integration over the energy E and then used the following identity:

$$\int_0^\infty dE f_0(E) e^{\frac{i2\pi s}{\Delta}(E-\mu_0)} = \frac{\Delta}{i2\pi s} \int_0^\infty dE \left(-\frac{\partial f_0}{\partial E}\right) e^{\frac{i2\pi s}{\Delta}(E-\mu_0)}. \quad (8.76)$$

The current $I^{(nd)}(t)$, Eq. (8.75), can be greatly simplified if $eU_0 = n\Delta$:

$$J^{(1)}(t, E) = -R^N \ln\left(1 - re^{ikL}\right)\left\{1 - e^{-i2\pi \frac{eU_0}{\Delta} \frac{t}{\tau}}\right\}, \quad (8.77a)$$

$$J^{(2)}(t, E) = (-1)R^N \sum_{s=1}^{N\geq 1} \frac{e^{iskL}}{s} \frac{r^s}{R^s}\left\{1 - e^{-i2\pi \frac{eU_0}{\Delta} \frac{t}{\tau}}\right\}, \quad (8.77b)$$

$$J^{(3)}(t, E) = 0. \quad (8.77c)$$

If in addition the temperature is low [such that only $E = \mu_0$ is relevant in Eq. (8.75a)] and the Fermi level lies exactly in the middle between the cavity's levels, $r^s e^{isk_F L} = (-1)^s R^{s/2}$, then the total current, $I(t) = I^{(d)}(t) + I^{(nd)}(t)$, is

$$I(t) = \frac{Q}{\tau} T R^{N(t)}\left\{1 + \frac{\sin\left(2\pi n \frac{t}{\tau}\right)}{\pi n} \zeta(t)\right\}, \quad (8.78a)$$

$$\zeta(t) = -\ln\left(1 + \sqrt{R}\right) - \theta(N-1) \sum_{s=1}^{N} \frac{(-1)^s}{sR^{\frac{s}{2}}},$$

$t < \tau$: $\quad (8.78b)$

$$I(t) = \frac{Q}{\tau} T \left\{1 - \frac{\sin\left(2\pi n \frac{t}{\tau}\right)}{\pi n} \ln\left(1 + \sqrt{R}\right)\right\},$$

$t \gg \tau$: $\quad (8.78c)$

$$I(t) \approx \frac{Q}{\tau} T R^N \left\{1 - \frac{\sin\left(2\pi n \frac{t}{\tau}\right)}{\pi n} \ln \sqrt{R}\right\}.$$

8.2.8.3 Emitted charge

Let us calculate the charge

$$Q = \int_0^\infty dt \{I^{(d)}(t) + I^{(nd)}(t)\}, \qquad (8.79)$$

emitted from the cavity under the action of the potential $U(t)$, Eq. (8.67). So, we integrate the current $I(t)$ over a time interval of duration τ (over which N is constant) and then sum over N from zero to infinity,

$$Q = \sum_{N=0}^{\infty} \int_{N\tau}^{(N+1)\tau} dt \{I^{(d)}(t) + I^{(nd)}(t)\}. \qquad (8.80)$$

After some simple algebra we calculate

$$Q = \frac{e^2 U_0}{\Delta} + \frac{e}{\pi} \int_0^\infty dE \left(-\frac{\partial f_0}{\partial E}\right) \Im \ln\left(\frac{1 - re^{i(kL - 2\pi \frac{eU_0}{\Delta})}}{1 - re^{ikL}}\right). \qquad (8.81)$$

At $T_0 \gg T^*$ the above equation gives $Q = e^2 U_0/\Delta$.

At lower temperatures we consider the limit $T \to 0$ when the density of states can be approximated as a sum of delta-function peaks centered at the eigen energies E_n of an isolated cavity. At $\mu_0 \gg \Delta$ the spectrum near the Fermi energy is equidistant, $E_n = E_0 + n\Delta$. Then we use in Eq. (8.81),

$$\frac{1}{\pi} \Im \ln\left(1 - e^{i 2\pi \frac{E - E_0}{\Delta}}\right) = -\frac{1}{2} + \left\{\left\{\frac{E - E_0}{\Delta}\right\}\right\},$$

where $\{\{X\}\}$ is the fractional part of X, and find

$$Q = \frac{e^2 U_0}{\Delta} + e \int_0^\infty dE \left(-\frac{\partial f_0}{\partial E}\right) \left(\left\{\left\{\frac{E - E_0 - eU_0}{\Delta}\right\}\right\} - \left\{\left\{\frac{E - E_0}{\Delta}\right\}\right\}\right). \qquad (8.82)$$

From this equation it follows that at zero temperature the emitted (absorbed) charge is quantized, Fig. 8.3. For instance, if μ_0 is centered exactly in the middle of two levels, then we get

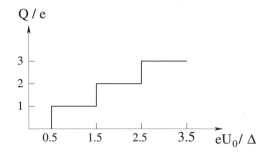

Fig. 8.3 The dependence of an emitted charge Q on the height U_0 of a potential step at zero temperature. The Fermi level is centered in the middle between the levels of the cavity.

$$Q = e\left[\frac{1}{2} + \frac{eU_0}{\Delta}\right], \quad k_B T_0 = 0, \quad T \to 0, \quad (8.83)$$

where $[X]$ is the integer part of X.

At finite but small temperatures, $k_B T_0 \ll \Delta$, the deviation δQ from this quantized value is

$$\delta Q = \mathrm{sgn}(1 - 2v_0) \frac{1 - e^{(|1-2v_0|-1)\frac{\Delta}{k_B T_0}}}{e^{|1-2v_0|\frac{\Delta}{2k_B T_0}} + 1}, \quad (8.84)$$

where $v_0 = \{\{eU_0/\Delta\}\}$ lies within the interval $0 \le v_0 < 1$, $\mathrm{sgn}(X) = +1$ for $X > 0$ and -1 for $X < 0$. The function $\delta Q(v_0)$ has the following asymptotes:

$$\delta Q(v_0) = \begin{cases} v_0 \frac{2\Delta}{k_B T_0} e^{-\frac{\Delta}{2k_B T_0}}, & v_0 \to 0, \\ \pm \frac{1}{2}, & v_0 = \frac{1}{2} \mp 0, \\ -(1-v_0) \frac{2\Delta}{k_B T_0} e^{-\frac{\Delta}{2k_B T_0}}, & v_0 \to 1. \end{cases} \quad (8.85)$$

The violation of charge quantization is exponentially small at low temperatures unless we are at the transition point from one plateau to another.

Next we consider how the quantization of an emitted charge is affected by the finiteness of a QPC's transmission coefficient. For the scattering amplitude $S(E)$, Eq. (8.32), at $T \ll 1$ the density of states can be approximated by the sum of Breit–Wigner resonances [247] with width $\Gamma = T\Delta/(4\pi) \ll \Delta$,

$$\nu(E) = \frac{1}{\pi} \sum_n \frac{\Gamma}{(E - E_n)^2 + \Gamma^2}. \qquad (8.86)$$

Then at zero temperature, $k_B T_0 = 0$, the deviation δQ from Eq. (8.83) is

$$\delta Q = -\theta(2v_0 - 1) + \frac{1}{2}\left\{1 - \frac{\arctan\frac{(1-2v_0)\,2\pi}{T}}{\arctan\frac{2\pi}{T}}\right\}, \qquad (8.87)$$

where $\theta(X)$ is the Heaviside theta-function equal to zero for $X < 0$ and unity for $X > 0$. The asymptotes for $\delta Q(v_0)$ are

$$\delta Q(v_0) = \begin{cases} v_0 \frac{T}{\pi^2}, & v_0 \to 0, \\ \pm\frac{1}{2}, & v_0 = \frac{1}{2} \mp 0, \\ -(1 - v_0)\frac{T}{\pi^2}, & v_0 \to 1. \end{cases} \qquad (8.88)$$

In contrast to the temperature, the effect of a finite QPC transmission is more crucial, since δQ is linear in transmission T.

Chapter 9

Quantum circuits with mesoscopic capacitor as a particle emitter

Below we consider the model introduced in Section 8.2.1 in the large amplitude ($\sim \Delta$) adiabatic regime at low temperatures. We are particularly interested in the limit of a small transparency, $T \to 0$. In this case electrons and holes emitted by the cavity are well separated in time and we can treat them as separate particles in a quite intuitive manner. On the other hand, as we show, they are subject to the Pauli exclusion principle and they are able to interfere. Therefore, they are quantum particles.

9.1 Quantized emission regime

First we show that at zero temperature the current generated by a capacitor slowly driven by a large-amplitude periodic potential, $U(t) = U(t + \mathcal{T})$, consists of a series of positive and negative pulses corresponding to the emission of electrons and holes. When we speak about an electron emitted by the cavity we mean the following. With increasing potential energy, $eU(t)$, the position of quantum levels in a cavity changes. One of the occupied levels can rise above the Fermi level. Then an electron occupying this level leaves the cavity. Therefore, the stream of electrons in a linear edge state (which the capacitor is connected to) is increased by one: An electron is emitted. In contrast, when $eU(t)$ decreases, an empty level can sink below the Fermi level. Then one additional electron enters the cavity leaving a hole in the stream of electrons in a linear edge state: A hole is emitted. Therefore, the quantum capacitor can serve as a single-particle source (SPS). Since after a period $\mathcal{T}$ the charge on the capacitor returns to its initial value, such an SPS emits the same number of electrons and holes, i.e., it is a source of quantized alternating currents [129]. The advantage of this source, compared to pumps generating quantized direct currents

[125–127], is the possibility of working in the adiabatic regime where, as we show below, first, the emitted particles have a smaller energy and, second, one can achieve pronounced effects.

Note, if the same cavity is coupled to two counter-propagating edge states having different potentials, then the electron flow and the hole flow can be separated [248].

Let the capacitor, see Fig. 8.1, be driven by the potential

$$U(t) = U_0 + U_1 \cos\left(\Omega_0 t + \varphi\right). \tag{9.1}$$

To describe a capacitor in an adiabatic regime it is enough to know its frozen scattering amplitude. In our model $S(t, E)$ is given in Eq. (8.32b) where we need to replace $kL \to kL - 2\pi eU(t)/\Delta$. If in addition the temperature is zero,

$$k_B T_0 = 0, \tag{9.2}$$

then we need the scattering amplitude at $E = \mu_0$ only. We rewrite $S(t) = S(t, \mu_0)$ as

$$S(t) = e^{i\theta_r} \frac{\sqrt{1-T} - e^{i\phi(t)}}{1 - \sqrt{1-T}e^{i\phi(t)}}, \tag{9.3}$$

where θ_r is the phase of the reflection amplitude of the QPC connecting the SPS to the linear edge state, $r = \sqrt{R}\,e^{i\theta_r}$, $\phi(t) = \phi(\mu_0) - 2\pi eU(t)/\Delta$ is a phase accumulated by an electron with energy $E = \mu_0$ during one trip along the cavity, and $\phi(\mu_0) = \theta_r + k_F L$.

To proceed analytically we assume that the amplitude U_1 of an oscillating potential is chosen in such a way that during a period only one level of the SPS crosses the Fermi level. The time of crossing t_0 is defined by the following condition: $\phi(t_0) = 0 \mod 2\pi$. Introducing the deviation of a phase from its resonance value, $\delta\phi(t) = \phi(t) - \phi(t_0)$, we obtain the scattering amplitude in the limit

$$T \to 0, \tag{9.4}$$

as

$$S(t) = -e^{i\theta_r} \frac{T + 2i\delta\phi(t)}{T - 2i\delta\phi(t)} + \mathcal{O}(T^2). \tag{9.5}$$

We keep only terms in the leading order in T.

There are two times when the resonance condition occurs (two times of crossing). The first time is when the level rises above the Fermi level and the second one is when the level sinks below the Fermi level. We denote these times as $t_0^{(-)}$ and $t_0^{(+)}$, respectively. At time $t_0^{(-)}$ one electron is emitted by the cavity, while at time $t_0^{(+)}$ one electron enters the cavity, a hole is emitted.

We suppose that the constant part of the potential U_0 accounts for detuning of the nearest electron level E_n in the SPS from the Fermi level. Then the resonance times can be found from the following equation:

$$E_n + eU\left(t_0^{(\mp)}\right) = \mu_0 \quad \Rightarrow \quad U_0 + U_1 \cos\left(\Omega_0 t_0^{(\mp)} + \varphi\right) = 0. \qquad (9.6)$$

For $|eU_0| < \Delta/2$ and $|eU_0| < |eU_1| < \Delta - |eU_0|$ we find

$$t_0^{(\mp)} = \mp t_0^{(0)} - \frac{\varphi}{\Omega_0}, \quad t_0^{(0)} = \frac{1}{\Omega_0} \arccos\left(-\frac{U_0}{U_1}\right). \qquad (9.7)$$

The deviation from the resonance time, $\delta t^{(\mp)} = t - t_0^{(\mp)}$, can be related to the deviation from the resonance phase, $\delta\phi^{(\mp)} = \mp M \Omega_0 \delta t^{(\mp)}$, where $\mp M = d\phi/dt|_{t=t_0^{(\mp)}}/\Omega_0 = \mp 2\pi |e|\Delta^{-1}\sqrt{U_1^2 - U_0^2}$. With these definitions we can rewrite Eq. (9.5) as

$$S(t) = e^{i\theta_r} \begin{cases} \dfrac{t - t_0^{(+)} - i\Gamma_\tau}{t - t_0^{(+)} + i\Gamma_\tau}, & \left|t - t_0^{(+)}\right| \lesssim \Gamma_\tau, \\[6pt] \dfrac{t - t_0^{(-)} + i\Gamma_\tau}{t - t_0^{(-)} - i\Gamma_\tau}, & \left|t - t_0^{(-)}\right| \lesssim \Gamma_\tau, \\[6pt] 1, & \left|t - t_0^{(\mp)}\right| \gg \Gamma_\tau. \end{cases} \qquad (9.8)$$

Here Γ_τ is (a half of) a time interval during which the level rises above or sinks below the Fermi level:

$$\Omega_0 \Gamma_\tau = \frac{T\Delta}{4\pi \left|eU_1 \sin\left(\Omega_0 t_0^{(0)} + \varphi\right)\right|} = \frac{T\Delta}{4\pi |e|\sqrt{U_1^2 - U_0^2}}. \qquad (9.9)$$

Equation (9.8) assumes that the overlap between the resonances is small,

$$\left|t_0^{(+)} - t_0^{(-)}\right| \gg \Gamma_\tau. \qquad (9.10)$$

Substituting Eq. (9.8) into Eq. (8.12) we find the adiabatic current at zero temperature (for $0 < t < \mathcal{T}$):

$$I(t) = \frac{e}{\pi} \left\{ \frac{\Gamma_\tau}{\left(t - t_0^{(-)}\right)^2 + \Gamma_\tau^2} - \frac{\Gamma_\tau}{\left(t - t_0^{(+)}\right)^2 + \Gamma_\tau^2} \right\}. \qquad (9.11)$$

This current consists of the two pulses of Lorentzian shape with half-width Γ_τ corresponding to the emission of an electron and a hole. Integrating over time it is easy to check that the first pulse carries a charge e while the second pulse carries a charge $-e$.

In this regime the frozen density of states, Eq. (8.63), is

$$\nu(t, \mu_0) = \frac{4}{\Delta \mathcal{T}} \left\{ \frac{\Gamma_\tau^2}{\left(t - t_0^{(-)}\right)^2 + \Gamma_\tau^2} + \frac{\Gamma_\tau^2}{\left(t - t_0^{(+)}\right)^2 + \Gamma_\tau^2} \right\}. \qquad (9.12)$$

With this equation one can estimate an adiabaticity condition, i.e., the condition under which the current $I^{(2)} \sim \Omega_0^2$ is small compared to a current that is linear in Ω_0, $I^{(1)}$, see Eq. (8.62). We use $\nu \sim 1/(\mathcal{T}\Delta)$. In a linear response regime we have $I^2 \sim e^2 h \nu^2 \, d^2 U/dt^2$, and correspondingly find

$$\varpi_{lin} \sim \frac{I^{(2)}}{I^{(1)}} \sim h \nu \Omega_0 \sim \frac{\tau \Omega_0}{\mathcal{T}} \ll 1. \qquad (9.13a)$$

While in a non-linear regime to leading order in $\Omega_0 \Gamma_\tau$ we can write $I^{(2)} \sim e^2 h \nu \left(\partial \nu / \partial t\right) (dU/dt)$. Then using $\partial \nu / \partial t \sim 1/(\Gamma_\tau \mathcal{T} \Delta)$ we calculate

$$\varpi_{n/lin} \sim \frac{I^{(2)}}{I^{(1)}} \sim \frac{h}{\Gamma_\tau \mathcal{T} \Delta} \sim \frac{\tau \Omega_0}{\mathcal{T}^2} \ll 1. \qquad (9.13b)$$

Comparing Eqs. (9.13a) and (9.13b) we conclude that in a quantized emission regime the adiabaticity condition is more restrictive compared to a linear response regime. For instance, if Eq. (9.13a) can be rewritten as $\tau_D \ll \mathcal{T}$, then Eq. (9.13b) can be rewritten as $\tau_D \ll \Gamma_\tau$.

We calculate also the heat production rate I_E in a quantized emission regime. For the current $I(t)$, Eq. (9.11), we find to leading order in $\Omega_0 \Gamma_\tau \ll 1$,

$$\mathcal{T}\langle I^2\rangle = \frac{e^2}{\pi}\frac{1}{\Gamma_\tau}, \tag{9.14}$$

and substituting into Eq. (8.20) we finally find [215]

$$I_E = \frac{\hbar}{\Gamma_\tau}\frac{1}{\mathcal{T}}. \tag{9.15}$$

This heat flow is due to additional (over the μ_0) energy $\hbar/(2\Gamma_\tau)$ carried by each particle (an electron or a hole) emitted during a period $\mathcal{T}$. Note that this energy is small compared to the amplitude of the driving potential, $\hbar/(2\Gamma_\tau) \ll eU_1$, which follows from the adiabaticity condition (9.13b) taking into account that $eU_1 \sim \Delta$.

9.1.1 Simple model for a single-particle state emitted by an adiabatic capacitor

At zero temperature the current emitted by an adiabatic dynamic capacitor, Eq. (8.16), depends only on the scattering amplitude calculated at the Fermi energy, $E = \mu$. This means that in a quantized emission regime the current is significant only when one of the cavity's levels crosses the Fermi level. Then one can say that the cavity injects particles (electrons and holes) at the surface of the Fermi sea. The Fermi sea in itself remains undisturbed [244, 249, 250]. These injected particles can be characterized with the following single-particle wave function a distance x away from the cavity,

$$\Psi(t,x) = \sqrt{n_0}\, S\left(t - \frac{x}{v_\mu}\right), \tag{9.16}$$

where $n_0 = k_\mu/(2\pi)$ is a density of 1D spinless chiral fermions, $v_\mu = \hbar k_\mu/m$ is a velocity of an electron with Fermi energy, and $S(t)$ is given in Eq. (9.8).[1] Then with Eq. (9.16) the current $I(t)$, see Eqs. (8.16) and (9.11), can be calculated using the standard quantum-mechanical equation for a single-particle current: $I(t) = -e\hbar/m\,\Im\{\Psi\partial\Psi^*/\partial x\}$. We stress that the wave function $\Psi(t,x)$, Eq. (9.16), is a model one. The only justification for it is that the current $I(t)$, Eq. (8.16), which is the result of calculations for a

---

[1] To understand how $t - x/v$ appears, see Eq. (8.27a) and the related derivation.

many-particle system of non-interacting fermions, can also be calculated as a current of a single particle. Therefore, we hope that the wave function $\Psi(t,x)$, Eq. (9.16), is sufficient to understand local (in time and space) properties of a single-particle state emitted by the dynamic quantum capacitor in an adiabatic regime at zero temperature.

9.2 Shot noise quantization

We will show that the quantized alternating current generated by an SPS results in a quantized shot noise [217, 244, 249] for the geometry of Fig. 9.1.

We calculate a zero-frequency symmetrized correlation function power $\mathcal{P}_{12}$ for currents $I_1(t)$ and $I_2(t)$ flowing into the contacts 1 and 2. At zero temperature it reads [see Eq. (6.27)]

$$\mathcal{P}_{12} = \frac{e^2 \Omega}{4\pi} \sum_{q=-\infty}^{\infty} |q| \sum_{\gamma,\delta=1}^{2} \left\{ \tilde{S}_{0,1\gamma} \tilde{S}_{0,1\delta}^* \right\}_q^* \left\{ \tilde{S}_{0,2\gamma} \tilde{S}_{0,2\delta}^* \right\}_q. \tag{9.17}$$

The frozen scattering matrix $\hat{\tilde{S}}_0(t)$ for the entire system is

$$\hat{\tilde{S}}_0(t) = \begin{pmatrix} e^{ik_F L_{11}} S(t) r_C & e^{ik_F L_{12}} t_C \\ e^{ik_F L_{21}} S(t) t_C & e^{ik_F L_{22}} r_C \end{pmatrix}, \tag{9.18}$$

where $L_{\gamma\delta}$ is the length of a path along the linear edge states from the contact δ to the contact γ, r_C/t_C is the reflection/transmission amplitude of the central QPC, and $S(t)$ is the scattering amplitude of the capacitor. Recall that at zero temperature we need all of the quantities only at $E = \mu_0$. After some simple algebra we find

$$\mathcal{P}_{12} = -\mathcal{P}_0 \sum_{q=1}^{\infty} q \left\{ |S_q|^2 + |S_{-q}|^2 \right\}, \tag{9.19}$$

where

$$\mathcal{P}_0 = e^2 R_C T_C \frac{\Omega_0}{2\pi}. \tag{9.20}$$

To calculate the shot noise we need the Fourier coefficients,

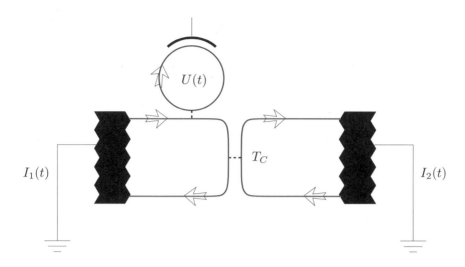

Fig. 9.1 The mesoscopic electron collider with a single-particle source in one of its branches. The cavity is connected to a linear edge state which in turn is connected to another linear edge state via the central QPC with transmission T_C. The arrows indicate the direction of movement of the electrons. The potential $U(t)$ induced by the back-gate generates an alternating current $I(t)$, which is split at the central QPC into the currents $I_1(t)$ and $I_2(t)$ flowing to the leads.

$$S_q = \int_0^{\mathcal{T}} \frac{dt}{\mathcal{T}} e^{iq\Omega_0 t} S(t), \tag{9.21}$$

which in the limit

$$\Gamma_\tau \ll \mathcal{T}, \tag{9.22}$$

can be calculated as follows. The function $S(t)$, Eq. (9.8), is almost constant and changes only in a tiny interval ($\sim \Gamma_\tau$) around times $t_0^{(\mp)}$. Since only integrating over those small intervals plays a role in Eq. (9.21), we can formally extend the integral, $\int_0^{\mathcal{T}} \to \int_{-\infty}^{\infty}$, and evaluate it closing the contour in the complex t-plane, $\int_{-\infty}^{\infty} \to \oint$, in the upper, $\operatorname{Im} t > 0$, for $q > 0$, or in the lower, $\operatorname{Im} t < 0$, for $q < 0$, semi-plane. The corresponding contour integral is evaluated using the Cauchy integral,

$$\frac{1}{2\pi i}\oint dt \sum_{j=1}^{N_p} \frac{f_j(t)}{(t-t_{pj})^{n_j+1}} = \sum_{j=1}^{N_p} \frac{1}{n_j!} \frac{d^{n_j} f_j}{dt^{n_j}}\bigg|_{t=t_{pj}}, \tag{9.23}$$

where t_{pj} is a pole of the n_jth order and N_p is the number of poles that lie inside the integration contour.

The function $S(t)$, Eq. (9.8), has poles $t_p^{(-)} = t_0^{(-)} + i\Gamma_\tau$ and $t_p^{(+)} = t_0^{(+)} - i\Gamma_\tau$ in the upper and lower semi-planes of the complex variable t, respectively. A simple evaluation gives

$$S_q = -2\Omega_0 \Gamma_\tau \, e^{-|q|\Omega_0 \Gamma_\tau} \, e^{i\theta_r} \begin{cases} e^{iq\Omega_0 t_0^{(-)}}, & q > 0, \\ e^{iq\Omega_0 t_0^{(+)}}, & q < 0. \end{cases} \quad (9.24)$$

Substituting the above equation into Eq. (9.19) and evaluating the following sum to leading order in a small parameter $\epsilon = \Omega_0 \Gamma_\tau$,

$$\sum_{q=1}^{\infty} q \, |S_q|^2 = 1 + \mathcal{O}\left(\epsilon^2\right), \quad (9.25)$$

we finally find the noise

$$\mathcal{P}_{12} = -2\mathcal{P}_0, \quad (9.26)$$

which is independent of both the parameters of the SPS and the parameters of the driving potential.

If the amplitude U_1 of the oscillating potential $U(t)$, Eq. (9.1), increases, for instance if n electrons and n holes are emitted during each period $\mathcal{T}$, then the noise is n times larger, [2] $\mathcal{P}_{12} = -2n\mathcal{P}_0$, see the upper solid (black) line in Fig. 9.5 in Section 9.4.

Remarkably the noise produced by the SPS is quantized. The increment $\mathcal{P}_0$, Eq. (9.26), depends on the frequency Ω_0 of the oscillating voltage and on the transparency T_C of the central QPC. Therefore the quantization is not universal.

9.2.1 Probability interpretation of the shot noise

The noise $\mathcal{P}_{12}$, Eq. (9.26), can be understood as the shot noise due to one electron and one hole emitted by the source during the period $\mathcal{T} = 2\pi/\Omega_0$.

---

[2] The authors of Ref. [249] considered the Lorentzian current pulses generated by carefully shaped external voltage pulses across two-terminal conductors and showed the shot noise is proportional to the number of excitations. The operator algebra describing these excitations is also derived.

The shot noise originates because in each particular event the indivisible particle has either to be reflected from or transmitted through the central QPC [31]. Since an electron and a hole are emitted at different times they are uncorrelated and contribute to the noise independently. Since the electron-hole symmetry is not violated in our system they contribute to the noise equally, leading to the factor 2 in Eq. (9.26). Further for definiteness we consider an electron contribution,

$$\mathcal{P}_{12}^{(e)} = -\mathcal{P}_0 = -e^2 R_C T_C \frac{\Omega_0}{2\pi}. \quad (9.27)$$

The hole contribution can be considered similarly.

To interpret $\mathcal{P}_{12}^{(e)}$ we introduce the following probabilities, which are evaluated by averaging over many periods. First, we introduce a single-particle probability $\mathcal{N}_\alpha$ that is the probability of detecting an electron at the reservoir $\alpha = 1, 2$ during a period. Taking into account that the SPS emits only one electron during a period, we calculate for the circuit under consideration, Fig. 9.1:

$$\mathcal{N}_1 = R_C, \quad \mathcal{N}_2 = T_C. \quad (9.28)$$

Second, we introduce a two-particle probability $\mathcal{N}_{\alpha\beta}$ that is the probability of detecting two particles at different contacts during a period. Since in our case there is only one electron emitted during a period, we have

$$\mathcal{N}_{12} = 0. \quad (9.29)$$

And, finally, we introduce the following correlation function,

$$\delta\mathcal{N}_{12} = \mathcal{N}_{12} - \mathcal{N}_1 \mathcal{N}_2. \quad (9.30)$$

From Eqs. (9.28)–(9.30) we find

$$\delta\mathcal{N}_{12} = -R_C T_C. \quad (9.31)$$

Comparing the above equation and Eq. (9.27) we find the following relation between the noise power and the particle number correlator

$$\mathcal{P}_{12} = \frac{e^2 \Omega_0}{2\pi} \delta\mathcal{N}_{12}. \quad (9.32)$$

The equations (9.30) and (9.32) show how the current cross-correlator $\mathcal{P}_{12}$ relates to the two-particle detection probability $\mathcal{N}_{12}$. We will show that this relation holds also for circuits with several SPSs when $\mathcal{N}_{12} \neq 0$.

9.3 Two-particle source

Two cavities placed in series, Fig. 9.2, and driven by the potentials $U_L(t)$ and $U_R(t)$ with the same period $\mathcal{T}$ can serve as a two-particle source. Depending on the phase difference between the potentials $U_L(t)$ and $U_R(t)$ such a double-cavity capacitor can emit electron and hole pairs, or electron-hole pairs, or emit single particles, electrons and holes [46].

9.3.1 Scattering amplitude

If the cavities are placed at a small distance, $L_{LR} \approx 0$, from each other, then the Floquet scattering amplitude of the capacitor is

$$S_F^{(2)}(E_n, E) = \sum_{m=-\infty}^{\infty} S_{R,F}(E_n, E_m) S_{L,F}(E_m, E). \tag{9.33}$$

Here $S_{j,F}(E_n, E)$ is the Floquet scattering amplitude for a single cavity, $j = L, R$. Then introducing the amplitude $S_{in}^{(2)}(t, E)$ whose Fourier coefficients define the elements of the Floquet scattering matrix of a double-cavity capacitor,

$$S_F^{(2)}(E_n, E) = \int_0^{\mathcal{T}} \frac{dt}{\mathcal{T}} e^{in\Omega_0 t} S_{in}^{(2)}(t, E), \tag{9.34}$$

and using Eq. (8.21) for a single-cavity scattering amplitude, we find

$$S_{in}^{(2)}(t, E) = \sum_{p=0}^{\infty} e^{ipkL_R} S_R^{(p)}(t) \sum_{r=0}^{\infty} e^{irkL_L} S_L^{(r)}(t - p\tau), \tag{9.35}$$

where L_j is the length of the cavity $j = L, R$.

9.3.2 Adiabatic approximation

In the limit of a slow excitation, $\Omega_0 \to 0$, we can approximate the single-cavity Floquet scattering matrix as

$$S_{j,F}(E_n, E) = S_{j,n}(E) + \frac{\hbar\Omega_0 n}{2} \frac{\partial S_{j,n}(E)}{\partial E} + \mathcal{O}\left(\Omega_0^2\right), \tag{9.36}$$

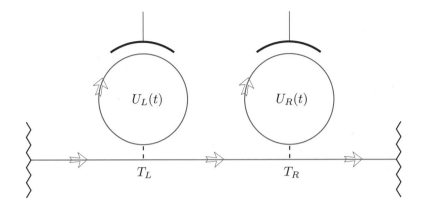

Fig. 9.2 The model for a double-cavity chiral quantum capacitor. Each cavity is driven by the periodic potential $U_j(t) = U_j(t + \mathcal{T})$, $j = L, R$. Both cavities are connected via the corresponding QPCs with transmission T_L and T_R, respectively, to the same linear edge state. The arrows indicate the direction of movement of electrons.

where $S_{j,n}(E)$ is the nth Fourier coefficient for the frozen scattering matrix, $S_j(t, E)$, of a single cavity. For the double-cavity capacitor we have to write

$$S_F^{(2)}(E_n, E) = S_n^{(2)}(E) + \frac{\hbar\Omega_0 n}{2}\frac{\partial S_n^{(2)}(E)}{\partial E} + \hbar\Omega_0 A_n(E) + \mathcal{O}\left(\Omega_0^2\right), \quad (9.37)$$

where $S_n^{(2)}$ is a Fourier coefficient for the frozen scattering matrix of a double-cavity system,

$$S^{(2)}(t, E) = S_R(t, E)S_L(t, E). \quad (9.38)$$

Correspondingly, the inverse Fourier transform gives

$$S_{in}^{(2)}(t, E) = S^{(2)}(t, E) + \frac{i\hbar}{2}\frac{\partial^2 S^{(2)}(t, E)}{\partial t \partial E} + \hbar\Omega_0 A(t, E). \quad (9.39)$$

To find the anomalous scattering amplitude $A(t, E)$ for a double-cavity system, we substitute Eq. (9.36) into Eq. (9.33). Then after the inverse Fourier transformation we find

$$S_{in}^{(2)}(t, E) = S_R(t, E)\, S_L(t, E) + i\hbar\frac{\partial S_L}{\partial t}\frac{\partial S_R}{\partial E}$$

$$+ \frac{i\hbar}{2}\left\{S_L\frac{\partial^2 S_R}{\partial t \partial E} + S_R\frac{\partial^2 S_L}{\partial t \partial E}\right\}. \quad (9.40)$$

Comparing Eqs. (9.39) and (9.40) we finally get

$$\hbar\Omega_0 \, A(t,E) = \frac{i\hbar}{2}\left\{\frac{\partial S_L}{\partial t}\frac{\partial S_R}{\partial E} - \frac{\partial S_L}{\partial E}\frac{\partial S_R}{\partial t}\right\}. \quad (9.41)$$

9.3.2.1 Time-dependent current

In Section 4.1.3 we calculate the time-dependent current generated by a double-cavity capacitor up to Ω_0^2 terms

$$I^{(2)}(t) = \frac{e}{2\pi}\int_0^\infty dE \left(-\frac{\partial f_0}{\partial E}\right)\left\{\Im\left(S^{(2)}\frac{\partial S^{(2)*}}{\partial t}\right) + 2\hbar\Omega_0 \Im\left(A\frac{\partial S^{(2)*}}{\partial t}\right)\right.$$
$$\left. + \frac{\partial}{\partial t}\left(\frac{\hbar}{2}\frac{\partial S^{(2)}}{\partial E}\frac{\partial S^{(2)*}}{\partial t} - i\hbar\Omega_0 \, S^{(2)} A^*\right)\right\}. \quad (9.42)$$

Using Eqs. (9.38) and (9.41) we find

$$I^{(2)}(t) = e^2 \int_0^\infty dE \left(-\frac{\partial f_0}{\partial E}\right)\left\{J^{(2,1)}(t,E) + J^{(2,2)}(t,E)\right\}. \quad (9.43a)$$

$$J^{(2,1)}(t,E) = \nu_L(t,E)\frac{dU_L(t)}{dt} + \nu_R(t,E)\frac{dU_R(t)}{dt}, \quad (9.43b)$$

$$J^{(2,2)}(t,E) = -\frac{\hbar}{2}\frac{\partial}{\partial t}\left\{\nu_L^2\frac{dU_L}{dt} + \nu_R^2\frac{dU_R}{dt} + 2\nu_L\nu_R\frac{dU_L}{dt}\right\}. \quad (9.43c)$$

Here $\nu_j(t,E)$ is the frozen DOS of the cavity $j = L, R$.

9.3.3 Mean square current

To recognize a regime when both cavities emit particles simultaneously we calculate the mean square current [46]

$$\langle I^2 \rangle = \int_0^\mathcal{T} \frac{dt}{\mathcal{T}} \left(I^{(2)}(t)\right)^2. \quad (9.44)$$

To leading order in Ω_0 we should keep only $J^{(2,1)}$ in Eq. (9.43). Alternatively one can express an adiabatic current directly in terms of the

Fourier coefficients $S_{0,q}^{(2)}$ for the double-cavity frozen scattering amplitude, Eq. (9.38). Then, by analogy with Eq. (8.17), we get at zero temperature

$$\langle I^2 \rangle = \frac{e^2 \Omega_0^2}{4\pi^2} \sum_{q=1}^{\infty} q^2 \left\{ \left| S_{0,q}^{(2)} \right|^2 + \left| S_{0,-q}^{(2)} \right|^2 \right\}. \qquad (9.45)$$

To calculate the Fourier coefficients,

$$S_q^{(2)} = \int_0^{\mathcal{T}} \frac{dt}{\mathcal{T}} e^{iq\Omega_0 t} S_L(t) S_R(t), \qquad (9.46)$$

we proceed as we calculated Eq. (9.24). For amplitudes $S_j(t)$ we use Eq. (9.8) with lower indices $_L$ and $_R$ indicating cavity-specific quantities θ_{rj}, $\Gamma_{\tau j}$, and $t_{0j}^{(\mp)}$, $j = L, R$. We assume that each cavity emits only one electron and one hole during a period. Then the functions $S_j(t)$ for $0 < t < \mathcal{T}$ have one pole, $t_{pj}^{(-)} = t_{0j}^{(-)} + i\Gamma_{\tau j}$, in the upper and one pole, $t_{pj}^{(+)} = t_{0j}^{(+)} - i\Gamma_{\tau j}$, in the lower semi-plane of a complex variable t. Therefore, we calculate

$$S_q^{(2)} = \begin{cases} S_R\left(t_{pL}^{(-)}\right) S_{L,q} + S_L\left(t_{pR}^{(-)}\right) S_{R,q}, & q > 0, \\ S_R\left(t_{pL}^{(+)}\right) S_{L,q} + S_L\left(t_{pR}^{(+)}\right) S_{R,q}, & q < 0, \end{cases} \qquad (9.47)$$

where, $S_{j,q}$ are given in Eq. (9.24) with θ_r, Γ_τ, and $t_0^{(\mp)}$ being replaced by θ_{rj}, $\Gamma_{\tau j}$, and $t_{0j}^{(\mp)}$, respectively. The squared Fourier coefficient is

$$\left| S_q^{(2)} \right|^2 = \left| S_R\left(t_{pL}^{(\chi)}\right) \right|^2 |S_{L,q}|^2 + \left| S_L\left(t_{pR}^{(\chi)}\right) \right|^2 |S_{R,q}|^2 + \xi_q^{(\chi)}, \qquad (9.48)$$

$$\xi_q^{(\chi)} = 2\Re \left\{ S_R\left(t_{pL}^{(\chi)}\right) S_{L,q} S_L^*\left(t_{pR}^{(\chi)}\right) S_{R,q}^* \right\},$$

where $\chi = -$ for $q > 0$ and $\chi = +$ for $q < 0$.

To proceed further we need to define more precisely whether two cavities emit particles at similar or at different times. Hence, we introduce the difference of times,

$$\Delta t_{L,R}^{(\chi,\chi')} = t_{0L}^{(\chi)} - t_{0R}^{(\chi')}, \qquad (9.49)$$

where $\chi = \mp$ and $\chi' = \mp$ depending on the particle (an electron or a hole) of interest, and compare $\Delta t_{L,R}^{(\chi,\chi')}$ to the duration of current pulses $\Gamma_{\tau L}$, $\Gamma_{\tau R}$.

9.3.3.1 Emission of separate particles

First, we assume that all the particles are emitted at different times,

$$\left| \Delta t_{L,R}^{(\chi,\chi')} \right| \gg \Gamma_{\tau L}, \Gamma_{\tau R}. \qquad (9.50)$$

In this case,

$$S_j\left(t_{0\bar{j}}^{(\chi)}\right) = e^{i\theta_{rj}}, \qquad (9.51)$$

where $j \neq \bar{j}$, and from Eq. (9.48) we find

$$\left|S_{0,q}^{(2)}\right|^2 = 4\Omega_0^2 \left\{\Gamma_{\tau L}^2 e^{-2|q|\Omega_0\Gamma_{\tau L}} + \Gamma_{\tau R}^2 e^{-2|q|\Omega_0\Gamma_{\tau R}}\right\} + \xi_q^{(\chi)}, \qquad (9.52)$$

$$\xi_q^{(\chi)} = 8\,\Omega_0^2\,\Gamma_{\tau L}\Gamma_{\tau R}\,e^{-|q|\Omega_0(\Gamma_{\tau L}+\Gamma_{\tau R})} \cos\left(q\Omega_0 \Delta t_{L,R}^{(\chi,\chi)}\right).$$

Next we need to sum over q in Eq. (9.45).

It is convenient to introduce the following quantities:

$$A_{1,j} = \sum_{q=1}^{\infty} e^{-2q\Omega_0\Gamma_{\tau j}} = \frac{e^{-2\Omega_0\Gamma_{\tau j}}}{1-e^{-2\Omega_0\Gamma_{\tau j}}} = \frac{1}{2\Omega_0\Gamma_{\tau j}} + \mathcal{O}(1), \qquad (9.53)$$

$$A_2 = \sum_{q=1}^{\infty} e^{-q\Omega_0\Gamma_{\tau\Sigma}} \cos(q\Omega_0\Delta t) = \frac{(-1)}{2} \frac{\cos(\Omega_0\Delta t) - e^{-\Omega_0\Gamma_{\tau\Sigma}}}{\cos(\Omega_0\Delta t) - \cosh(\Omega_0\Gamma_{\tau\Sigma})}$$

$$\stackrel{\Delta t \gg \Gamma_\tau}{=} -\frac{1}{2} + \mathcal{O}(\Omega_0\Gamma_{\tau\Sigma}), \qquad (9.54)$$

$$A_3 = \sum_{q=1}^{\infty} e^{-q\Omega_0\Gamma_{\tau\Sigma}} \sin(q\Omega_0\Delta t) = \frac{(-1)}{2} \frac{\sin(\Omega_0\Delta t)}{\cos(\Omega_0\Delta t) - \cosh(\Omega_0\Gamma_{\tau\Sigma})}$$

$$\stackrel{\Delta t \gg \Gamma_\tau}{=} \frac{1}{2}\cot\left(\frac{\Omega_0\Delta t}{2}\right) + \mathcal{O}(\Omega_0\Gamma_{\tau\Sigma}), \qquad (9.55)$$

where $\Gamma_{\tau\Sigma} = \Gamma_{\tau L} + \Gamma_{\tau R}$ and $\Delta t = \Delta t_{L,R}^{(\chi,\chi)}$. Then we see to leading order in $\Omega_0 \Gamma_{\tau j} \ll 1$, that the term with $\xi_q^{(\chi)}$ does not contribute to the sum over q unless $\Delta t_{L,R}^{(\chi,\chi)} \lesssim \Gamma_{\tau j}$.

Substituting Eq. (9.52) into Eq. (9.45) and using the following sum,

$$\sum_{q=1}^{\infty} q^2 e^{-2q\Omega_0 \Gamma_{\tau j}} = \frac{1}{4\Omega_0^2} \frac{\partial^2 A_{1,j}}{\partial \Gamma_{\tau j}^2} \approx \frac{1}{4\Omega_0^3 \Gamma_{\tau j}^3},$$

we find

$$\mathcal{T} \langle I^2 \rangle = \frac{e^2}{\pi} \left(\frac{1}{\Gamma_{\tau L}} + \frac{1}{\Gamma_{\tau R}} \right). \tag{9.56}$$

Comparing the above equation with the single-cavity result, Eq. (9.14), we conclude: If all particles are emitted at different times then both cavities contribute additively to $\langle I^2 \rangle$. Note, because of Eq. (8.20) the same is correct with respect to the generated heat flow.

9.3.3.2 Particle reabsorption regime

Let one cavity of the capacitor emit an electron (a hole) at the same time as the other cavity emits a hole (an electron). We expect that the source comprising both cavities does not generate a current, since the particle emitted by the first cavity is absorbed by the second cavity. The subsequent calculations of both the quantity $\langle I^2 \rangle$ and the shot noise (in Section 9.3.4.2) support such an expectation.

So, we suppose that

$$\left| \Delta t_{L,R}^{(+,-)} \right|, \left| \Delta t_{L,R}^{(-,+)} \right| \lesssim \Gamma_{\tau L}, \Gamma_{\tau R},$$

$$\left| \Delta t_{L,R}^{(-,-)} \right|, \left| \Delta t_{L,R}^{(+,+)} \right| \gg \Gamma_{\tau L}, \Gamma_{\tau R}. \tag{9.57}$$

In this case $\xi_q^{(\chi)}$ in Eq. (9.48) still does not contribute, since it depends on $\Delta t_{1,2}^{(\chi,\chi)}$, which is large. The other quantities that we need to calculate in Eq. (9.48) are the following:

$$S_L\left(t_{pR}^{(-)}\right) = e^{i\theta_{rL}} \frac{\Delta t_{L,R}^{(+,-)} + i(\Gamma_{\tau L} - \Gamma_{\tau R})}{\Delta t_{L,R}^{(+,-)} - i(\Gamma_{\tau L} + \Gamma_{\tau R})}, \quad S_L\left(t_{pR}^{(+)}\right) = e^{i\theta_{rL}} \frac{\Delta t_{L,R}^{(-,+)} - i(\Gamma_{\tau L} - \Gamma_{\tau R})}{\Delta t_{L,R}^{(-,+)} + i(\Gamma_{\tau L} + \Gamma_{\tau R})},$$

$$S_R\left(t_{pL}^{(-)}\right) = e^{i\theta_{rR}} \frac{\Delta t_{L,R}^{(-,+)} + i(\Gamma_{\tau L} - \Gamma_{\tau R})}{\Delta t_{L,R}^{(-,+)} + i(\Gamma_{\tau L} + \Gamma_{\tau R})}, \quad S_R\left(t_{pL}^{(+)}\right) = e^{i\theta_{rR}} \frac{\Delta t_{L,R}^{(+,-)} - i(\Gamma_{\tau L} - \Gamma_{\tau R})}{\Delta t_{L,R}^{(+,-)} - i(\Gamma_{\tau L} + \Gamma_{\tau R})}.$$

After squaring we find

$$\left|S_L\left(t_{pR}^{(-)}\right)\right|^2 = \left|S_R\left(t_{pL}^{(+)}\right)\right|^2 = \gamma\left(\Delta t_{L,R}^{(+,-)}\right),$$

$$\left|S_L\left(t_{pR}^{(+)}\right)\right|^2 = \left|S_R\left(t_{pL}^{(-)}\right)\right|^2 = \gamma\left(\Delta t_{L,R}^{(-,+)}\right),$$

where

$$\gamma(\Delta t) = \frac{(\Delta t)^2 + (\Gamma_{\tau L} - \Gamma_{\tau R})^2}{(\Delta t)^2 + (\Gamma_{\tau L} + \Gamma_{\tau R})^2}. \tag{9.58}$$

Remarkably $\gamma(\Delta t)$ is independent of q, i.e., when an electron and a hole are emitted close together then all the photon-assisted probabilities are reduced by the same factor. Therefore, we can immediately write instead of Eq. (9.56) the following equation:

$$\mathcal{T}\langle I^2\rangle = \frac{e^2}{2\pi}\left(\frac{1}{\Gamma_{\tau L}} + \frac{1}{\Gamma_{\tau R}}\right)\left\{\gamma\left(\Delta t_{L,R}^{(-,+)}\right) + \gamma\left(\Delta t_{L,R}^{(+,-)}\right)\right\}. \tag{9.59}$$

For identical cavities, $\Gamma_{\tau L} = \Gamma_{\tau R}$, emitting in synchronization, $\Delta t_{L,R}^{(-,+)} = \Delta t_{L,R}^{(+,-)} = 0$, the mean square current vanishes. Therefore, one can say that in this case the second (R) cavity reabsorbs all the particles emitted by the first (L) cavity.

9.3.3.3 Two-particle emission regime

Next we consider the cases when two particles of the same kind are emitted near simultaneously. Due to the Pauli exclusion principle it is impossible to have two (spinless) electrons (or two holes) in the same state. Therefore, the second emitted particle should have more energy than the first one. To be more precise, the electron pair (or the hole pair) has more energy than the sum of the energies of two separately emitted electrons (holes). Therefore, the heat flow I_E should be enhanced and, because of Eq. (8.20), the mean square current also should be enhanced [compared to Eq. (9.56)].

We assume

$$\left|\Delta t_{L,R}^{(-,-)}\right|, \left|\Delta t_{L,R}^{(+,+)}\right| \lesssim \Gamma_{\tau L}, \Gamma_{\tau R},$$

$$\left|\Delta t_{L,R}^{(-,+)}\right|, \left|\Delta t_{L,R}^{(+,-)}\right| \gg \Gamma_{\tau L}, \Gamma_{\tau R}. \quad (9.60)$$

In this case the two poles of $S^{(2)}(t) = S_L(t)S_R(t)$ as a function of the complex time t in the upper (or in the lower) semi-plane are close to each other, which affects the calculations significantly. Now the quantities A_2, Eq. (9.54), and A_3, Eq. (9.55), become of order $A_{1,j}$, Eq. (9.53). Therefore, we should keep $\xi_q^{(\chi)}$ in Eq. (9.48). From Eq. (9.60) it follows that $\Omega_0 \Delta t_{L,R}^{(\chi,\chi)} \ll 1$ and we find

$$A_2 = \frac{\Omega_0 (\Gamma_{\tau L} + \Gamma_{\tau R})}{\left(\Omega_0 \Delta t_{L,R}^{(\chi,\chi)}\right)^2 + \Omega_0^2 (\Gamma_{\tau L} + \Gamma_{\tau R})^2} + \mathcal{O}(1), \quad (9.61a)$$

$$A_3 = \frac{\Omega_0 \Delta t_{L,R}^{(\alpha,\alpha)}}{\left(\Omega_0 \Delta t_{L,R}^{(\chi,\chi)}\right)^2 + \Omega_0^2 (\Gamma_{\tau L} + \Gamma_{\tau R})^2} + \mathcal{O}(1). \quad (9.61b)$$

Also we will use the following quantities,

$$S_L\left(t_{pR}^{(-)}\right) = e^{i\theta_{rL}} \frac{\Delta t_{L,R}^{(-,-)} - i(\Gamma_{\tau L} + \Gamma_{\tau R})}{\Delta t_{L,R}^{(-,-)} + i(\Gamma_{\tau L} - \Gamma_{\tau R})}, \quad S_L\left(t_{pR}^{(+)}\right) = e^{i\theta_{rL}} \frac{\Delta t_{L,R}^{(+,+)} + i(\Gamma_{\tau L} + \Gamma_{\tau R})}{\Delta t_{L,R}^{(+,+)} - i(\Gamma_{\tau L} - \Gamma_{\tau R})},$$

$$S_R\left(t_{pL}^{(-)}\right) = e^{i\theta_{rR}} \frac{\Delta t_{L,R}^{(-,-)} + i(\Gamma_{\tau L} + \Gamma_{\tau R})}{\Delta t_{L,R}^{(-,-)} + i(\Gamma_{\tau L} - \Gamma_{\tau R})}, \quad S_R\left(t_{pL}^{(+)}\right) = e^{i\theta_{rR}} \frac{\Delta t_{L,R}^{(+,+)} - i(\Gamma_{\tau L} + \Gamma_{\tau R})}{\Delta t_{L,R}^{(+,+)} - i(\Gamma_{\tau L} - \Gamma_{\tau R})},$$

$$S_R\left(t_{pL}^{(-)}\right) S_L^*\left(t_{pR}^{(-)}\right) = e^{i(\theta_{rR} - \theta_{rL})} \frac{\left(\Delta t_{L,R}^{(-,-)}\right)^2 - (\Gamma_{\tau L} + \Gamma_{\tau R})^2 + 2i\Delta t_{L,R}^{(-,-)}(\Gamma_{\tau L} + \Gamma_{\tau R})}{\left(\Delta t_{L,R}^{(-,-)}\right)^2 + (\Gamma_{\tau L} - \Gamma_{\tau R})^2},$$

$$S_R\left(t_{pL}^{(+)}\right) S_L^*\left(t_{pR}^{(+)}\right) = e^{i(\theta_{rR} - \theta_{rL})} \frac{\left(\Delta t_{L,R}^{(+,+)}\right)^2 - (\Gamma_{\tau L} + \Gamma_{\tau R})^2 - 2i\Delta t_{L,R}^{(+,+)}(\Gamma_{\tau L} + \Gamma_{\tau R})}{\left(\Delta t_{L,R}^{(-,-)}\right)^2 + (\Gamma_{\tau L} - \Gamma_{\tau R})^2},$$

and some squares,

$$\left|S_L\left(t_{pR}^{(-)}\right)\right|^2 = \left|S_R\left(t_{pL}^{(-)}\right)\right|^2 = \frac{\left(\Delta t_{L,R}^{(-,-)}\right)^2 + (\Gamma_{\tau L} + \Gamma_{\tau R})^2}{\left(\Delta t_{L,R}^{(-,-)}\right)^2 + (\Gamma_{\tau L} - \Gamma_{\tau R})^2}, \quad (9.62)$$

$$\left|S_L\left(t_{pR}^{(+)}\right)\right|^2 = \left|S_R\left(t_{pL}^{(+)}\right)\right|^2 = \frac{\left(\Delta t_{L,R}^{(+,+)}\right)^2 + (\Gamma_{\tau L}+\Gamma_{\tau R})^2}{\left(\Delta t_{L,R}^{(+,+)}\right)^2 + (\Gamma_{\tau L}-\Gamma_{\tau R})^2}. \quad (9.63)$$

In Eq. (9.45) we need to calculate the following sum

$$\sum_{q=1}^{\infty} q^2 \left|S_q^{(2)}\right|^2 = \Phi_1 + \Phi_2, \quad (9.64a)$$

where

$$\Phi_1 = \left|S_L\left(t_{pR}^{(-)}\right)\right|^2 \sum_{q=1}^{\infty} q^2 \left\{|S_{L,q}|^2 + |S_{R,q}|^2\right\} = \left|S_L\left(t_{pR}^{(-)}\right)\right|^2$$

$$(9.64b)$$

$$\times \sum_{j=L,R} \Gamma_{\tau j}^2 \frac{\partial^2 A_{1,j}}{\partial \Gamma_{\tau j}^2} = \frac{\left(\Delta t_{L,R}^{(-,-)}\right)^2 + (\Gamma_{\tau L}+\Gamma_{\tau R})^2}{\left(\Delta t_{L,R}^{(-,-)}\right)^2 + (\Gamma_{\tau L}-\Gamma_{\tau R})^2} \frac{1}{\Omega_0}\left\{\frac{1}{\Gamma_{\tau L}}+\frac{1}{\Gamma_{\tau R}}\right\},$$

and

$$\Phi_2 = \sum_{q=1}^{\infty} q^2 \xi_q^{(-)} = 2\sum_{q=1}^{\infty} q^2 \Re\left\{S_R\left(t_{pL}^{(-)}\right) S_{L,q} S_L^*\left(t_{pR}^{(-)}\right) S_{R,q}^*\right\}$$

$$= \frac{8\Omega_0^2 \Gamma_{\tau L}\Gamma_{\tau R}}{\left(\Delta t_{L,R}^{(-,-)}\right)^2+(\Gamma_{\tau L}-\Gamma_{\tau R})^2} \quad (9.64c)$$

$$\times \left\{\frac{\left(\Delta t_{L,R}^{(-,-)}\right)^2-(\Gamma_{\tau L}+\Gamma_{\tau R})^2}{\Omega_0^2}\frac{\partial^2 A_2}{\partial \Gamma_{\tau L}^2}-\frac{2\Delta t_{L,R}^{(-,-)}(\Gamma_{\tau L}+\Gamma_{\tau R})}{\Omega_0^2}\frac{\partial^2 A_3}{\partial \Gamma_{\tau L}^2}\right\}.$$

From Eqs. (9.61) we calculate

$$\frac{\partial A_2}{\partial \Gamma_{\tau L}} = \frac{\left(\Delta t_{L,R}^{(\chi,\chi)}\right)^2-(\Gamma_{\tau L}+\Gamma_{\tau R})^2}{\Omega_0\left(\left(\Delta t_{L,R}^{(\chi,\chi)}\right)^2+(\Gamma_{\tau L}+\Gamma_{\tau R})^2\right)^2}, \quad \frac{\partial A_3}{\partial \Gamma_{\tau L}} = \frac{-2\Delta t_{L,R}^{(\chi,\chi)}(\Gamma_{\tau L}+\Gamma_{\tau R})}{\Omega_0\left(\left(\Delta t_{L,R}^{(\chi,\chi)}\right)^2+(\Gamma_{\tau L}+\Gamma_{\tau R})^2\right)^2},$$

and

$$\frac{\partial^2 A_2}{\partial \Gamma_{\tau L}^2} = \frac{-2(\Gamma_{\tau L}+\Gamma_{\tau R})\left(3\left(\Delta t_{L,R}^{(\chi,\chi)}\right)^2-\Gamma_{\tau\Sigma}^2\right)}{\Omega_0\left(\left(\Delta t_{L,R}^{(\chi,\chi)}\right)^2+\Gamma_{\tau\Sigma}^2\right)^3}, \quad \frac{\partial^2 A_3}{\partial \Gamma_{\tau L}^2} = \frac{-2\Delta t_{L,R}^{(\chi,\chi)}\left(\left(\Delta t_{L,R}^{(\chi,\chi)}\right)^2-3\Gamma_{\tau\Sigma}^2\right)}{\Omega_0\left(\left(\Delta t_{L,R}^{(\chi,\chi)}\right)^2+\Gamma_{\tau\Sigma}^2\right)^3}.$$

Using the above equations in Eq. (9.64c), we find

$$\Phi_2 = \frac{8\Gamma_{\tau L}\Gamma_{\tau R}\Pi}{\Omega_0 \left(\left(\Delta t_{L,R}^{(-,-)}\right)^2 + (\Gamma_{\tau L} - \Gamma_{\tau R})^2\right) \left(\left(\Delta t_{L,R}^{(-,-)}\right)^2 + (\Gamma_{\tau L} + \Gamma_{\tau R})^2\right)^3},$$

with

$$\Pi = -2\Gamma_{\tau\Sigma}\left\{\left(\left(\Delta t_{L,R}^{(-,-)}\right)^2 - \Gamma_{\tau\Sigma}^2\right)\left(3\left(\Delta t_{L,R}^{(-,-)}\right)^2 - \Gamma_{\tau\Sigma}^2\right)\right.$$
$$\left. -2\left(\Delta t_{L,R}^{(-,-)}\right)^2\left(\left(\Delta t_{L,R}^{(-,-)}\right)^2 - 3\Gamma_{\tau\Sigma}^2\right)\right\} = -2\Gamma_{\tau\Sigma}\left(\left(\Delta t_{L,R}^{(-,-)}\right)^2 + \Gamma_{\tau\Sigma}^2\right)^2.$$

After simplification it becomes

$$\Phi_2 = \frac{-16\Gamma_{\tau L}\Gamma_{\tau R}(\Gamma_{\tau L} + \Gamma_{\tau R})}{\Omega_0\left(\left(\Delta t_{L,R}^{(-,-)}\right)^2 + (\Gamma_{\tau L} - \Gamma_{\tau R})^2\right)\left(\left(\Delta t_{L,R}^{(-,-)}\right)^2 + (\Gamma_{\tau L} + \Gamma_{\tau R})^2\right)}. \tag{9.65}$$

Next, substituting Eqs. (9.64b) and (9.65) into Eq. (9.64a) we get

$$\sum_{q=1}^{\infty} q^2 \left|S_q^{(2)}\right|^2 = \frac{(\Gamma_{\tau L}+\Gamma_{\tau R})\left\{\left(\left(\Delta t_{L,R}^{(-,-)}\right)^2+(\Gamma_{\tau L}+\Gamma_{\tau R})^2\right)^2 - 16\Gamma_{\tau L}^2\Gamma_{\tau R}^2\right\}}{\Omega_0\Gamma_{\tau L}\Gamma_{\tau R}\left(\left(\Delta t_{L,R}^{(-,-)}\right)^2+(\Gamma_{\tau L}-\Gamma_{\tau R})^2\right)\left(\left(\Delta t_{L,R}^{(-,-)}\right)^2+(\Gamma_{\tau L}+\Gamma_{\tau R})^2\right)}$$

$$= \frac{\Gamma_{\tau\Sigma}\left(\left(\Delta t_{L,R}^{(-,-)}\right)^2+\Gamma_{\tau\Sigma}^2+4\Gamma_{\tau L}\Gamma_{\tau R}\right)}{\Omega_0\Gamma_{\tau L}\Gamma_{\tau R}\left(\left(\Delta t_{L,R}^{(-,-)}\right)^2+\Gamma_{\tau\Sigma}^2\right)} = \frac{1}{\Omega_0}\left(\frac{1}{\Gamma_{\tau L}} + \frac{1}{\Gamma_{\tau R}}\right)\left\{2 - \gamma\left(\Delta t_{L,R}^{(-,-)}\right)\right\},$$

where $\gamma(\Delta t)$ is defined in Eq. (9.58). The sum $\sum_{q=1}^{\infty} q^2 \left|S_{-q}^{(2)}\right|^2$ gives the same result but with $\Delta t_{1,2}^{(-,-)}$ being replaced by $\Delta t_{1,2}^{(+,+)}$. Finally, from Eq. (9.45) we have the mean square current,

$$\mathcal{T}\langle I^2 \rangle = \frac{e^2}{2\pi}\left(\frac{1}{\Gamma_{\tau L}} + \frac{1}{\Gamma_{\tau R}}\right)\left\{4 - \gamma\left(\Delta t_{L,R}^{(-,-)}\right) - \gamma\left(\Delta t_{L,R}^{(+,+)}\right)\right\}. \tag{9.66}$$

For identical cavities, $\Gamma_{\tau L} = \Gamma_{\tau R}$, emitting electrons and holes in synchronization, $\Delta t_{L,R}^{(-,-)} = \Delta t_{L,R}^{(+,+)} = 0$, the mean square current, hence the heat production rate, is twice as large compared to the one in the regime of separately emitted particles, Eq. (9.56).

Combining Eq. (9.59) with Eq. (9.66) we obtain an equation describing all the considered regimes [46]:

$$\mathcal{T}\langle I^2\rangle = \frac{e^2}{2\pi}\left(\frac{1}{\Gamma_{\tau L}} + \frac{1}{\Gamma_{\tau R}}\right)\left\{2 + \gamma\left(\Delta t_{L,R}^{(-,+)}\right) + \gamma\left(\Delta t_{L,R}^{(+,-)}\right)\right.$$
$$\left. - \gamma\left(\Delta t_{L,R}^{(-,-)}\right) - \gamma\left(\Delta t_{L,R}^{(+,+)}\right)\right\}. \tag{9.67}$$

Note that this equation is in the leading order in $\Omega_0\Gamma_{\tau j} \ll 1$. The higher order corrections arise from the current $J^{(2,2)}$ in Eq. (9.43) and from approximations we made evaluating the Fourier coefficients, Eq. (9.21).

9.3.4 *Shot noise of a two-particle source*

Let the double-cavity capacitor be connected to a linear edge state, which in turn is connected to another linear edge state via a central QPC with transmission T_C, Fig. 9.3. Our aim is to investigate how the shot noise, arising when emitted particles (electrons and holes) are scattered at the central QPC, depends on the regime of emission of the double-cavity capacitor.

By analogy with the single-cavity capacitor, Eq. (9.19), we can write

$$\mathcal{P}_{12} = -\mathcal{P}_0 \sum_{q=1}^{\infty} q\left\{\left|S_q^{(2)}\right|^2 + \left|S_{-q}^{(2)}\right|^2\right\}, \tag{9.68a}$$

and taking into account that $S^{(2)} = S_L S_R$:

$$\mathcal{P}_{12} = -\mathcal{P}_0 \sum_{q=1}^{\infty} q\left\{\left|(S_L S_R)_q\right|^2 + \left|(S_L S_R)_{-q}\right|^2\right\}. \tag{9.68b}$$

To evaluate this cross-correlator for the different emission regimes we proceed as in Section 9.3.3.

9.3.4.1 *Emission of separate particles*

If all the particles are emitted at different times, Eq. (9.50), then calculating the sum over q we can neglect the term $\xi_q^{(\chi)}$ in Eq. (9.52). Then using the following sum (to the leading order in $\Omega_0\Gamma_{\tau j} \ll 1$),

$$\sum_{q=1}^{\infty} q e^{-2q\Omega_0\Gamma_{\tau j}} = \frac{-1}{2\Omega_0}\frac{\partial A_{1,j}}{\partial \Gamma_{\tau j}} \approx \frac{1}{4\Omega_0^2\Gamma_{\tau j}^2},$$

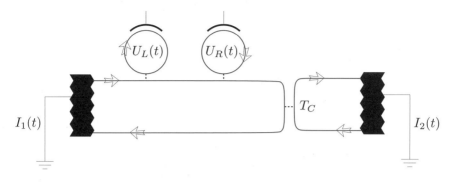

Fig. 9.3 The mesoscopic electron collider with a two-particle source in one of its branches. The double-cavity quantum capacitor is connected to the linear edge state, which in turn is connected via the central QPC with transmission T_C to another linear edge state. The arrows indicate the direction of movement of the electrons. The potentials $U_L(t)$ and $U_R(t)$ induced by the back-gates generate an alternating current $I(t)$, which is split at the central QPC into the currents $I_1(t)$ and $I_2(t)$ flowing to the leads.

(see Eq. (9.53) for $A_{1,j}$) we calculate

$$\mathcal{P}_{12} = -4\mathcal{P}_0. \tag{9.69}$$

This noise is due to four particles (two electrons and two holes) emitted by both cavities during the period $\mathcal{T} = 2\pi/\Omega_0$.

9.3.4.2 Particle reabsorption regime

Under the conditions given in Eq. (9.57) all the photon-assisted probabilities are reduced by the same factor, see Eq. (9.58). Therefore, we can immediately write instead of Eq. (9.69) the following equation:

$$\mathcal{P}_{12} = -2\mathcal{P}_0 \left\{ \gamma\left(\Delta t_{L,R}^{(-,+)}\right) + \gamma\left(\Delta t_{L,R}^{(+,-)}\right) \right\}. \tag{9.70}$$

If an electron and a hole are emitted at similar times, then both the noise $\mathcal{P}_{12}$, Eq. (9.70), and the mean square current $\langle I^2 \rangle$, Eq. (9.59), are suppressed. On the other hand their ratio remains the same as in the regime of emission of separate particles. This tells us that in the reabsorption regime the rarely emitted (not absorbed) electrons and holes remain uncorrelated.

9.3.4.3 Two-particle emission regime

The shot noise is not changed if two electrons (two holes) are emitted at similar times, Eq. (9.60). This means that the two particles are scattered at the central QPC independently as when they are emitted at different times. Therefore, despite the fact that the energy of two electrons (two holes) emitted simultaneously is enhanced compared to the sum of energies of two separately emitted electrons (holes), they remain uncorrelated and do not constitute a pair.

To calculate Eq. (9.68) we use Eqs. (9.48), (9.61), and (9.62) and evaluate the following sum

$$\sum_{q=1}^{\infty} q \left| S_q^{(2)} \right|^2 = F_1 + F_2, \tag{9.71a}$$

where

$$F_1 = \left| S_L \left(t_{pR}^{(-)} \right) \right|^2 \sum_{q=1}^{\infty} q \left\{ |S_{L,q}|^2 + |S_{R,q}|^2 \right\} = -2\Omega_0 \left| S_L \left(t_{pR}^{(-)} \right) \right|^2 \tag{9.71b}$$

$$\times \sum_{j=L,R} \Gamma_{\tau j}^2 \frac{\partial A_{1,j}}{\partial \Gamma_{\tau j}} = 2 \frac{\left(\Delta t_{L,R}^{(-,-)} \right)^2 + (\Gamma_{\tau L} + \Gamma_{\tau R})^2}{\left(\Delta t_{L,R}^{(-,-)} \right)^2 + (\Gamma_{\tau L} - \Gamma_{\tau R})^2},$$

and

$$F_2 = \sum_{q=1}^{\infty} q \xi_q^{(-)} = 8\Omega_0^2 \Gamma_{\tau L} \Gamma_{\tau R}$$

$$\times \mathrm{Re} \left\{ S_R \left(t_{pL}^{(-)} \right) S_L^* \left(t_{pR}^{(-)} \right) e^{i(\theta_{rL} - \theta_{rR})} \sum_{q=1}^{\infty} q e^{-q\Omega_0 \Gamma_{\tau \Sigma}} e^{iq\Omega_0 \Delta t_{L,R}^{(-,-)}} \right\}. \tag{9.71c}$$

Using the product $S_R \left(t_{pL}^{(-)} \right) S_L^* \left(t_{pR}^{(-)} \right)$ given just before Eq. (9.62) we write

$$F_2 = \frac{8\Omega_0^2 \Gamma_{\tau L} \Gamma_{\tau R}}{\left(\Delta t_{L,R}^{(-,-)} \right)^2 + (\Gamma_{\tau L} - \Gamma_{\tau R})^2} \left\{ -2\Delta t_{L,R}^{(-,-)} \Gamma_{\tau \Sigma} \sum_{q=1}^{\infty} q e^{-q\Omega_0 \Gamma_{\tau \Sigma}} \sin \left(q\Omega_0 \Delta t_{L,R}^{(-,-)} \right) \right.$$

$$\left. + \left(\left(\Delta t_{L,R}^{(-,-)} \right)^2 - \Gamma_{\tau \Sigma}^2 \right) \sum_{q=1}^{\infty} q e^{-q\Omega_0 \Gamma_{\tau \Sigma}} \cos \left(q\Omega_0 \Delta t_{L,R}^{(-,-)} \right) \right\},$$

and rewrite it, using Eqs. (9.54) and (9.55):

$$F_2 = \frac{8\Omega_0^2 \Gamma_{\tau L}\Gamma_{\tau R}}{\left(\Delta t_{L,R}^{(-,-)}\right)^2 + (\Gamma_{\tau L}-\Gamma_{\tau R})^2} \left\{ \frac{2\Delta t_{L,R}^{(-,-)}\Gamma_{\tau\Sigma}}{\Omega_0} \frac{\partial A_3}{\partial \Gamma_{\tau L}} - \frac{\left(\Delta t_{L,R}^{(-,-)}\right)^2 - \Gamma_{\tau\Sigma}^2}{\Omega_0} \frac{\partial A_2}{\partial \Gamma_{\tau L}} \right\}.$$

In the regime under consideration the derivatives $\partial A_2/\partial \Gamma_{\tau L}$ and $\partial A_3/\partial \Gamma_{\tau L}$ are given just below Eq. (9.64). Then we find

$$F_2 = \frac{(-8\Gamma_{\tau L}\Gamma_{\tau R})\left\{ \left(2\Delta t_{L,R}^{(-,-)}\Gamma_{\tau\Sigma}\right)^2 + \left(\left(\Delta t_{L,R}^{(-,-)}\right)^2 - \Gamma_{\tau\Sigma}^2\right)^2 \right\}}{\left\{\left(\Delta t_{L,R}^{(-,-)}\right)^2 + (\Gamma_{\tau L}-\Gamma_{\tau R})^2\right\} \left(\left(\Delta t_{L,R}^{(-,-)}\right)^2 + \Gamma_{\tau\Sigma}^2\right)^2}$$

$$= \frac{-8\Gamma_{\tau L}\Gamma_{\tau R}}{\left(\Delta t_{L,R}^{(-,-)}\right)^2 + (\Gamma_{\tau L}-\Gamma_{\tau R})^2}.$$

Substituting the above equation and Eq. (9.71b) into Eq. (9.71a) we get

$$\sum_{q=1}^{\infty} q\left|S_q^{(2)}\right|^2 = \frac{2\left(\Delta t_{L,R}^{(-,-)}\right)^2 + 2(\Gamma_{\tau L}+\Gamma_{\tau R})^2 - 8\Gamma_{\tau L}\Gamma_{\tau R}}{\left(\Delta t_{L,R}^{(-,-)}\right)^2 + (\Gamma_{\tau L}-\Gamma_{\tau R})^2}$$

$$= 2\frac{\left(\Delta t_{L,R}^{(-,-)}\right)^2 + (\Gamma_{\tau L}-\Gamma_{\tau R})^2}{\left(\Delta t_{L,R}^{(-,-)}\right)^2 + (\Gamma_{\tau L}-\Gamma_{\tau R})^2} = 2.$$

The same result applies for negative harmonics, $\sum_{q=1}^{\infty} q\left|S_{-q}^{(2)}\right|^2 = 2$.

Thus the noise power, Eq. (9.68), is $\mathcal{P}_{LR} = -4\mathcal{P}_0$. This is the same as in the regime when the particles are emitted at different times, Eq. (9.69). Therefore, the noise is not sensitive to whether two electrons (two holes) are emitted simultaneously or not. Note equation (9.70) is applicable for all the considered regimes.

9.4 Mesoscopic electron collider

Consider the circuit presented in Fig. 9.4 where two quantum capacitors (two SPSs) are placed in arms located at different sides of the central QPC. The particles emitted by the different SPSs are uncorrelated, hence they

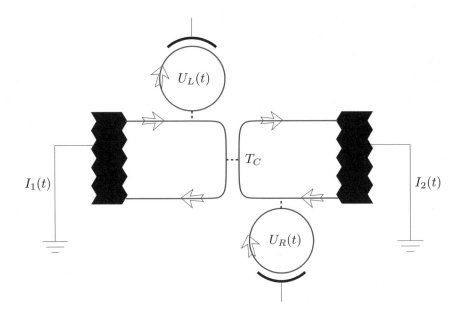

Fig. 9.4 The mesoscopic electron collider with single-particle sources in each of its branches. The two quantum capacitors are connected to the different linear edge states, which in turn are connected together via the central QPC with transmission T_C. The arrows indicate the direction of movement of electrons. The potentials $U_L(t)$ and $U_R(t)$ induced by the back-gates generate alternating currents $I_1(t)$ and $I_2(t)$ at the leads.

contribute to the noise independently. However, if the two SPSs emit electrons (holes) simultaneously, then these particles become correlated after scattering at the central QPC. The correlations arise due to the Pauli exclusion principle: Two electrons (holes) cannot be scattered to the same edge state, instead they are necessarily scattered to different edge states and they arrive at different contacts. Therefore, in this regime the system comprising two SPSs and the central QPC serves as a two-particle source emitting only particles in pairs whose constituents are directed to different contacts.

By merely changing the phase difference between the potentials driving the two cavities,[3] one can switch the statistics of particles emitted during a period from classical to quantum (fermionic).

---

[3]This phase difference controls the difference of emission times of particles exiting the cavities.

9.4.1 Shot noise suppression

The elements of the frozen scattering matrix $\hat{\tilde{S}}_0(t)$ for the circuit under study, Fig. 9.4, are

$$\hat{\tilde{S}}_0(t) = \begin{pmatrix} e^{ik_F L_{11}} S_L(t) r_C & e^{ik_F L_{12}} S_R(t) t_C \\ e^{ik_F L_{21}} S_L(t) t_C & e^{ik_F L_{22}} S_R(t) r_C \end{pmatrix}, \quad (9.72)$$

where $S_j(t)$ is the frozen scattering amplitude of the capacitor $j = L, R$. Other quantities are the same as in Eq. (9.18). With this scattering matrix we calculate from Eq. (9.17)

$$\mathcal{P}_{12} = -\mathcal{P}_0 \sum_{q=1}^{\infty} q \left\{ |(S_L^* S_R)_q|^2 + |(S_L^* S_R)_{-q}|^2 \right\}. \quad (9.73)$$

Comparing Eq. (9.73) with Eq. (9.68b) we see that the only difference is in the replacement $S_L \to S_L^*$. From Eq. (9.8) we conclude that the complex conjugate scattering amplitude corresponds to the emission of a hole (an electron) if the bare scattering amplitude corresponds to the emission of an electron (a hole). Therefore, we can use the results of Section 9.3.4 if we replace $\Delta t_{L,R}^{(-,-)} \to \Delta t_{L,R}^{(+,-)}$, etc.

If two capacitors emit particles at different times or they emit an electron and a hole at similar times, then the noise,

$$\mathcal{P}_{12} = -4\mathcal{P}_0, \quad (9.74)$$

is due to independent contributions of four uncorrelated particles emitted during a period by the both capacitors. Note the possible collision of an electron and a hole at the central QPC does not affect the shot noise, since an electron and a hole have different energies (above and below[4] the Fermi energy, respectively) and they are not subject to the Pauli exclusion principle, which could lead to the appearance of correlations crucial for noise.

In contrast, if two electrons (two holes) are emitted at similar times, $\Delta t_{L,R}^{(-,-)} = t_{0L}^{(-)} - t_{0R}^{(-)} \lesssim \Gamma_{\tau L}, \Gamma_{\tau R}$ ($\Delta t_{L,R}^{(+,+)} = t_{0L}^{(+)} - t_{0R}^{(+)} \lesssim \Gamma_{\tau L}, \Gamma_{\tau R}$) then the noise is suppressed [217]

$$\mathcal{P}_{12} = \mathcal{P}_{12}^{(e)} + \mathcal{P}_{12}^{(h)}, \quad (9.75a)$$

---

[4]There is no contradiction with the fact that a hole carries a positive heat, see Eq. (9.15), since heat is defined as the extra energy obtained by a reservoir with fixed chemical potential. To maintain it fixed we need to add one electron with energy μ_0 after a hole has entered the reservoir.

where an electron's contribution, $\mathcal{P}_{12}^{(e)}$, and a hole's, $\mathcal{P}_{12}^{(h)}$, are

$$\mathcal{P}_{12}^{(e)} = -2\mathcal{P}_0\gamma\left(\Delta t_{L,R}^{(-,-)}\right) = -2\mathcal{P}_0\left\{1 - \frac{4\Gamma_{\tau L}\Gamma_{\tau R}}{\left(t_{0L}^{(-)} - t_{0R}^{(-)}\right)^2 + \Gamma_{\tau\Sigma}^2}\right\}, \quad (9.75b)$$

$$\mathcal{P}_{12}^{(h)} = -2\mathcal{P}_0\gamma\left(\Delta t_{L,R}^{(+,+)}\right) = -2\mathcal{P}_0\left\{1 - \frac{4\Gamma_{\tau L}\Gamma_{\tau R}}{\left(t_{0L}^{(+)} - t_{0R}^{(+)}\right)^2 + \Gamma_{\tau\Sigma}^2}\right\}. \quad (9.75c)$$

We represent the noise as the sum of electron and hole parts since they contribute independently.

When each of the time differences $\Delta t_{L,R}^{(\chi,\chi)}$ is larger than the sum of half-widths of current pulses, then the two sources contribute to the shot noise independently, Fig. 9.5, lower solid (green) line. In this case Eq. (9.75) leads to Eq. (9.74). In contrast, if there is some overlap in time between the particle wave packets arriving at the central QPC, $\Delta t_{L,R}^{(-,-)} \sim \Gamma_{\tau L} + \Gamma_{\tau R}$ ($\Delta t_{L,R}^{(+,+)} \sim \Gamma_{\tau L} + \Gamma_{\tau R}$), then correlations between electrons (holes) arise and the noise decreases. In the case of full overlap, $\Delta t_{L,R}^{(\chi,\chi)} = 0$ and $\Gamma_{\tau L} = \Gamma_{\tau R}$, the noise is suppressed down to zero:

$$\mathcal{P}_{12}^{(e)} = 0, \quad \text{if} \quad t_{0L}^{(-)} = t_{0R}^{(-)}, \quad (9.76a)$$

$$\mathcal{P}_{12}^{(h)} = 0, \quad \text{if} \quad t_{0L}^{(+)} = t_{0R}^{(+)}. \quad (9.76b)$$

In Fig. 9.5 the dashed (red) line shows the noise generated by the two identical sources as a function of the amplitude $U_{L,1}$ of the potential acting on the capacitor L. If $U_{L,1} \neq U_{R,1}$ then the times when particles are emitted by the different sources are different. In this case both sources contribute to the noise independently. However, if $eU_{L,1}$ approaches $eU_{R,1} = 0.5\Delta_R$ then the time differences $\Delta t_{L,R}^{(\chi,\chi)} \to 0$ result in a suppression of the shot noise.

In contrast to the case considered in Section 9.3.4.2, where the noise decreases together with the current, here the noise vanishes while the current remains non-zero, $I_1(t) \neq 0$, $I_2(t) \neq 0$. Taking into account the conservation law for a zero-frequency noise power, $\sum_{\beta=1,2} \mathcal{P}_{\alpha\beta} = 0$, we derive from

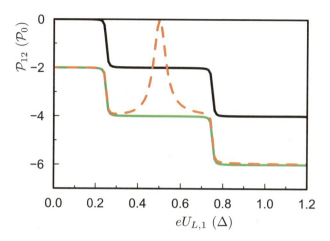

Fig. 9.5 The current cross-correlator $\mathcal{P}_{12}$, Eq. (9.73), as a function of the amplitude $U_{L,1}$ of a potential $U_L(t) = U_{L,0} + U_{L,1}\cos(\Omega_0 t + \varphi_L)$ acting upon the left capacitor, see Fig. 9.4. The three lines shown differ in the way that the right capacitor is driven: The right capacitor can be stationary (upper solid black line); driven by the potential $U_R(t) = U_{R,0} + U_{R,1}\cos(\Omega_0 t + \varphi_R)$, which is out of phase, $\varphi_R = \pi$, and has an amplitude $eU_{R,1} = 0.5\Delta_R$ (lower solid green line); or driven by the in-phase potential, $\varphi_R = 0$, with amplitude $eU_{R,1} = 0.5\Delta_R$ (dashed red line). Other parameters are $eU_{L,0} = eU_{R,0} = 0.25\Delta_R$ ($\Delta_L = \Delta_R$), $\varphi_L = 0$, and $T_L = T_R = 0.1$.

Eqs. (9.76) that $\mathcal{P}_{11}^{(x)} = \mathcal{P}_{22}^{(x)} = 0$, where $x = e, h$. In other words, there are noiseless (regular) electron (hole) flows entering the contacts $\alpha = 1, 2$. This regularity is due to the following. First, electrons (holes) are regularly emitted by the sources. And, second, due to the Pauli exclusion principle, each two electrons (holes) incident upon the QPC will be scattered into different contacts.

While electrons (holes) emitted by the different SPSs are statistically independent, after the collision at the central QPC the electrons (holes) become correlated (i.e., indistinguishable in a quantum-statistical sense) since they lose their origin: It is impossible to indicate which SPS has emitted the electron (hole) arriving at the given contact. Thus the disappearance of the shot noise [217] indicates an appearance of Fermi correlations between electrons (holes) after colliding at the QPC. This effect looks similar to the Hong, Ou, and Mandel [251] effect in optics. However, for electrons the probability of detecting particles at the different contacts peaks while for photons it shows a dip [251].

9.4.2 *Probability analysis*

For definiteness we will concentrate on electrons. Recall that during the period $\mathcal{T} = 2\pi/\Omega_0$ each SPS emits one electron at $t_{0L}^{(-)}$ and $t_{0R}^{(-)}$, respectively, for the L and R sources. The single-particle probability $\mathcal{N}_\alpha$, i.e., the probability of detecting an electron at the contact $\alpha = 1, 2$ during a period, is independent of the time difference $\Delta t_{L,R}^{(-,-)} = t_{0L}^{(-)} - t_{0R}^{(-)}$. In contrast, the two-particle probability $\mathcal{N}_{12}$, i.e., the probability of detecting electrons at both contacts during a period depends crucially on this time difference. Moreover at $\Delta t_{L,R}^{(-,-)} = 0$ the two-particle probability $\mathcal{N}_{12}$ becomes *the joint detection probability* introduced by Glauber [252], which is the probability of detecting two particles at two contacts simultaneously. Here simultaneous detection means detection during a time interval $\Gamma_{\tau j}$ which is much smaller than the distance (in time), $\sim \mathcal{T}$, between the particles arriving subsequently at the same contact.

9.4.2.1 *Single-particle probabilities*

At $\Delta t_{L,R}^{(-,-)} \gg \Gamma_\tau = \Gamma_{\tau L} = \Gamma_{\tau R}$ the particles emitted by the different sources remain distinguishable and we can write

$$\mathcal{N}_1 = \mathcal{N}_1^{(L)} + \mathcal{N}_1^{(R)}, \tag{9.77a}$$

$$\mathcal{N}_2 = \mathcal{N}_2^{(L)} + \mathcal{N}_2^{(R)}, \tag{9.77b}$$

where the upper indices (L) and (R) indicate the origin of the electron. The single-particle probability can be calculated as the square of the single-particle amplitude for the particle emitted by an SPS to arrive at the given contact, $\mathcal{N}_\alpha^{(j)} = |\mathcal{A}_{\alpha j}|^2$, with

$$\mathcal{A}_{1L} = e^{ik_F L_{1L}} r_C, \quad \mathcal{A}_{1R} = e^{ik_F L_{1R}} t_C,$$

$$\mathcal{A}_{2L} = e^{ik_F L_{2L}} t_C, \quad \mathcal{A}_{2R} = e^{ik_F L_{2R}} r_C, \tag{9.78}$$

where $L_{\alpha j} = L_{\alpha C} + L_{Cj}$ is the distance from the source $j = L, R$ through the quantum point contact C to the contact $\alpha = 1, 2$ along the linear edge state, see Fig. 9.4 and compare with Eq. (9.72). Then we find

$$\mathcal{N}_1^{(L)} = R_C, \quad \mathcal{N}_1^{(R)} = T_C, \tag{9.79a}$$

$$\mathcal{N}_2^{(L)} = T_C, \quad \mathcal{N}_2^{(R)} = R_C, \tag{9.79b}$$

and

$$\mathcal{N}_1 = \mathcal{N}_2 = 1. \tag{9.80}$$

For $\Delta t_{L,R}^{(-,-)} = 0$ we cannot distinguish which SPS an electron came from. However, apparently one electron should be detected at each contact. Therefore, Eq. (9.80) remains valid.

9.4.2.2 Two-particle probability for a classical regime

At $\Delta t_{L,R}^{(-,-)} \gg \Gamma_\tau = \Gamma_{\tau L} = \Gamma_{\tau R}$ electrons emitted by the different SPSs remain uncorrelated. Therefore, we can write

$$\mathcal{N}_{12}^{(LR)} = \mathcal{N}_1^{(L)} \mathcal{N}_2^{(R)}, \quad \mathcal{N}_{12}^{(RL)} = \mathcal{N}_1^{(R)} \mathcal{N}_2^{(L)}. \tag{9.81}$$

For $\mathcal{N}_\alpha^{(j)}$ see Eqs. (9.79). Taking into account that the two-electron probability can be represented as

$$\mathcal{N}_{12} = \mathcal{N}_{12}^{(L)} + \mathcal{N}_{12}^{(LR)} + \mathcal{N}_{12}^{(RL)} + \mathcal{N}_{12}^{(R)}, \tag{9.82}$$

and that a single electron cannot be detected at two distant places,

$$\mathcal{N}_{12}^{(L)} = \mathcal{N}_{12}^{(R)} = 0, \tag{9.83}$$

we find

$$\mathcal{N}_{12} = \mathcal{N}_{12}^{(LR)} + \mathcal{N}_{12}^{(RL)} = R_C^2 + T_C^2. \tag{9.84}$$

Note $\mathcal{N}_{12} < 1$ since with probability $R_C T_C$ the two electrons can reach the same (either 1 or 2) contact.

Using Eq. (9.80) we find $\delta \mathcal{N}_{12} = \mathcal{N}_{12} - \mathcal{N}_1 \mathcal{N}_2 = -2 R_C T_C$, which is, by virtue of Eq. (9.32), consistent with the shot noise due to electrons, $\mathcal{P}_{12}^{(e)} = -2\mathcal{P}_0$; see Eq. (9.74) for the total noise. Alternatively we can proceed as follows. Since in this regime electrons emitted by the different SPSs are statistically independent, the particle number correlation function, $\delta \mathcal{N}_{12} = \mathcal{N}_{12} - \mathcal{N}_1 \mathcal{N}_2$, can be represented as the sum

$$\delta\mathcal{N}_{12} = \delta\mathcal{N}_{12}^{(L)} + \delta\mathcal{N}_{12}^{(R)}, \qquad (9.85)$$

where the correlation functions due to single-particle contributions are

$$\delta\mathcal{N}_{12}^{(L)} = -\mathcal{N}_1^{(L)}\mathcal{N}_2^{(L)}, \quad \delta\mathcal{N}_{12}^{(R)} = -\mathcal{N}_1^{(R)}\mathcal{N}_2^{(R)}. \qquad (9.86)$$

Using Eqs. (9.79) we again find $\delta\mathcal{N}_{12} = -2R_C T_C$.

9.4.2.3 *Two-particle probability for a quantum regime*

If $\Delta t_{L,R}^{(-,-)} = 0$ then electrons collide at the central QPC and become correlated, i.e., they acquire fermionic statistics. Therefore, we cannot use Eq. (9.81). Strictly speaking, we cannot even introduce the upper indices, since we cannot indicate the origin of an electron arriving at a given contact. In this regime we can still use Eq. (9.80). Since there are no events with two electrons arriving at the same contact, we find

$$\mathcal{N}_{12} = 1. \qquad (9.87)$$

This quantum result is independent of the parameters of the central QPC in contrast to its classical counterpart, Eq. (9.84). Using Eqs. (9.80) and (9.87) we calculate the particle number correlation function: $\delta\mathcal{N}_{12} = 0$. This is consistent with the zero noise result, Eq. (9.76a), if one uses Eq. (9.32) relating the shot noise and the particle number correlation function.

The result given in Eq. (9.87) can also be calculated as a two-particle probability, $\mathcal{N}_{12} = |\mathcal{A}^{(2)}|^2$. Note due to collisions at the central QPC the electrons are indistinguishable. Then the two scattering processes for two electrons, described by the following two-particle amplitudes $\mathcal{A}_a^{(2)} = \mathcal{A}_{1L}\mathcal{A}_{2R}$ and $\mathcal{A}_b^{(2)} = -\mathcal{A}_{1R}\mathcal{A}_{2L}$,[5] result in the same final state. Therefore, these amplitudes should be added, $\mathcal{A}^{(2)} = \mathcal{A}_a^{(2)} + \mathcal{A}_b^{(2)}$, and the two-particle amplitude can be written as the Slater determinant,

$$\mathcal{A}^{(2)} = \det\begin{vmatrix} \mathcal{A}_{1L} & \mathcal{A}_{1R} \\ \mathcal{A}_{2L} & \mathcal{A}_{2R} \end{vmatrix}. \qquad (9.88)$$

---

[5] The minus sign is due to fermionic statistics: Two electrons are interchanged in the incoming scattering channels.

Using the single-particle amplitudes given in Eq. (9.78) and taking into account that $L_{1L} + L_{2R} = L_{1R} + L_{2L}$ (due to crossing of the trajectories at the central QPC) and $r_C t_C^* = -r_C^* t_C$ (due to unitarity) we arrive at Eq. (9.87).

Comparing Eqs. (9.87) and (9.80) one can see that $\mathcal{N}_{12} = \mathcal{N}_1 \mathcal{N}_2$. This equation seems to tell us that the arrival of electrons at one contact is not correlated with the arrival of electrons at the other contact. However, we found that electrons arrive at the contacts in pairs, i.e., the electrons are strongly correlated. This seeming inconsistency is due to a special value of single-particle probabilities, $\mathcal{N}_j = 1$. In the next section we consider a circuit with $\mathcal{N}_j < 1$ when the single particles as well as the pairs of correlated (due to collisions at the central QPC) particles do contribute to the noise. We will show that the colliding particles are positively correlated [253].

9.5 Noisy mesoscopic electron collider

In Fig. 9.6 we show a set-up where the regular flows, emitted by the two quantum capacitors S_L and S_R, start fluctuating (noisy) after passing the quantum point contacts L and R, respectively. There are events with two, one, or zero particles entering the central part of the circuit (CPC) and contributing to the cross-correlator $\mathcal{P}_{12}$ of the currents $I_1(t)$ and $I_2(t)$ flowing into the contacts 1 and 2, respectively. Under the conditions when electrons (holes) emitted by the different SPSs can collide at the quantum point contact C, there are different contributions to $\mathcal{P}_{12}$.[6] Namely, there are single- and two-particle contributions. In the case when two particles enter the CPC they become correlated (after colliding at the contact C) and cause a two-particle contribution to the noise. In contrast, if only one particle (either from the contact L or from the contact R) enters the CPC, then it causes a negative single-particle contribution, see Eq. (9.27). If $T_L = T_R$ the cross-correlator is zero, $\mathcal{P}_{12} = 0$. Therefore, the two-particle contribution is positive: After colliding at the quantum point contact C two electrons (holes) become positively correlated.

---

[6]If electrons (holes) do not collide at the contact C then there are only negative single-particle contributions to the noise.

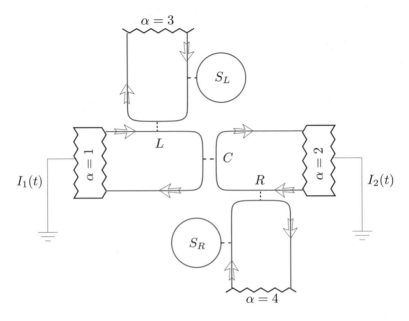

Fig. 9.6 The mesoscopic electron collider with noisy single-particle sources. The two noisy flows originating from the quantum point contacts L and R can collide at the quantum point contact C.

9.5.1 Current cross-correlator suppression

The elements of the frozen scattering matrix $\hat{\tilde{S}}_0(t)$, which we need to calculate $\mathcal{P}_{12}$, for the circuit shown in Fig. 9.6 are

$$\tilde{S}_{0,11}(t) = e^{ik_F L_{11}} r_L r_C\,, \qquad \tilde{S}_{0,12}(t) = e^{ik_F L_{12}} r_R t_C\,,$$
(9.89a)
$$\tilde{S}_{0,13}(t) = e^{ik_F L_{13}} S_L(t) t_L r_C\,, \quad \tilde{S}_{0,14}(t) = e^{ik_F L_{14}} S_R(t) t_R t_C\,,$$

$$\tilde{S}_{0,21}(t) = e^{ik_F L_{21}} r_L t_C\,, \qquad \tilde{S}_{0,22}(t) = e^{ik_F L_{22}} r_R r_C\,,$$
(9.89b)
$$\tilde{S}_{0,23}(t) = e^{ik_F L_{23}} S_L(t) t_L t_C\,, \quad \tilde{S}_{0,24}(t) = e^{ik_F L_{24}} S_R(t) t_R r_C\,,$$

where the lower indices L, R, and C denote the corresponding quantum point contacts. Using these elements in Eq. (9.17) we calculate by analogy with Eqs. (9.75):

$$\mathcal{P}_{12} = \mathcal{P}_{12}^{(e,1)} + \mathcal{P}_{12}^{(e,2)} + \mathcal{P}_{12}^{(h,1)} + \mathcal{P}_{12}^{(h,2)}\,, \qquad (9.90a)$$

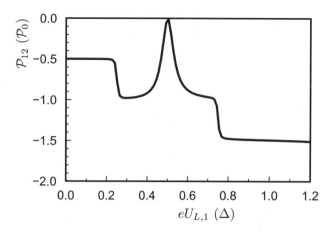

Fig. 9.7 The current cross-correlator $\mathcal{P}_{12}$ as a function of the amplitude $U_{L,1}$ of a potential $U_L(t) = U_{L,0} + U_{L,1}\cos(\Omega_0 t + \varphi_L)$ driving the single-particle source S_L, see Fig. 9.6. The parameters of the single-particle sources S_L and S_R are the same as in Fig. 9.5 but $\varphi_L = \varphi_R$. Other parameters are $T_L = T_R = 0.5$.

where the single-particle,

$$\mathcal{P}_{12}^{(e,1)} = \mathcal{P}_{12}^{(h,1)} = -\mathcal{P}_0\left(T_L^2 + T_R^2\right), \tag{9.90b}$$

and the two-particle,

$$\mathcal{P}_{12}^{(e,2)} = 2\mathcal{P}_0 T_L T_R \frac{4\Gamma_{\tau L}\Gamma_{\tau R}}{\left(t_{0L}^{(-)} - t_{0R}^{(-)}\right)^2 + (\Gamma_{\tau L} + \Gamma_{\tau R})^2}, \tag{9.90c}$$

$$\mathcal{P}_{12}^{(h,2)} = 2\mathcal{P}_0 T_L T_R \frac{4\Gamma_{\tau L}\Gamma_{\tau R}}{\left(t_{0L}^{(+)} - t_{0R}^{(+)}\right)^2 + (\Gamma_{\tau L} + \Gamma_{\tau R})^2}, \tag{9.90d}$$

contributions to the current cross-correlator are introduced. The cross-correlator $\mathcal{P}_{12}$ is shown in Fig. 9.7. Equation (9.90a) describes the shot noise at the second plateau ($\mathcal{P}_{12} \sim -\mathcal{P}_0$) of this plot.

In the classical regime, $\Delta t_{L,R}^{(\chi,\chi)} \gg \Gamma_{\tau L}, \Gamma_{\tau R}$, when the emitted particles remain statistically independent, the two-particle contribution is not present, $\mathcal{P}_{12}^{(x,2)} \sim \mathcal{O}\left(\Gamma_{\tau j}/\Delta t_{L,R}^{(\chi,\chi)}\right)^2 \approx 0$, where $x = e, h$, and the cross-correlator,

$$\mathcal{P}_{12} = -2\mathcal{P}_0\left\{T_L^2 + T_R^2\right\}, \tag{9.91}$$

is due to single particles only. In this classical regime the measurements at contacts 1 and 2 can give the following outcomes (during one period separately for electrons and holes): (i) Two particles can be detected at the same contact, (ii) two particles can be detected at different contacts, (iii) one particle can be detected at either of the contacts, and (iv) no particle can be detected at all.

While if electrons (holes) can collide at the central QPC, $\Delta t_{L,R}^{(\chi,\chi)} = 0$, the cross-correlator is suppressed, compare with to Eq. (9.91). Assuming $\Gamma_{\tau L} = \Gamma_{\tau R} \equiv \Gamma_\tau$ we find from Eq. (9.90a),

$$\mathcal{P}_{12} = -2\mathcal{P}_0 \left(T_R - T_L\right)^2. \tag{9.92}$$

The current cross-correlator becomes zero in the symmetric case, $T_L = T_R$. We should stress that the current flowing into either of the contacts is noisy, $\langle \delta I_\alpha^2 \rangle > 0$, $\alpha = 1, 2$, despite the fact that the current cross-correlator is suppressed. This suppression is due to a positive two-particle contribution compensating a negative single-particle one. This regime differs from the classical one considered above in two points: (i) There are no events with two electrons (holes) detected at the same contact, (ii) if two electrons (holes) are detected at different contacts they are detected simultaneously.

9.5.2 Probability analysis

As before, we concentrate on electrons. Holes can be considered in the same way.

9.5.2.1 Single-particle probabilities

The single-particle probabilities are insensitive to whether electrons emitted by different sources can collide at the central quantum point contact C or not. Therefore, we assume $\Delta t_{L,R}^{(-,-)} \gg \Gamma_{\tau L}, \Gamma_{\tau R}$ and use the following single-particle amplitudes in Eqs. (9.77):

$$\mathcal{A}_{1L} = e^{ik_F L_{1L}} t_L r_C, \quad \mathcal{A}_{1R} = e^{ik_F L_{1R}} t_R t_C,$$
$$\mathcal{A}_{2L} = e^{ik_F L_{2L}} t_L t_C, \quad \mathcal{A}_{2R} = e^{ik_F L_{2R}} t_R r_C. \tag{9.93}$$

Then we find

$$\mathcal{N}_1^{(L)} = T_L R_C, \quad \mathcal{N}_1^{(R)} = T_R T_C, \tag{9.94a}$$

$$\mathcal{N}_2^{(L)} = T_L T_C, \quad \mathcal{N}_2^{(R)} = T_R R_C, \tag{9.94b}$$

and

$$\mathcal{N}_1 = T_L + T_C (T_R - T_L),$$

$$\mathcal{N}_2 = T_R - T_C (T_R - T_L). \tag{9.95}$$

Apparently $\mathcal{N}_1 + \mathcal{N}_2 = T_L + T_R$: The number of electrons entering the CPC in the number reaching contacts 1 and 2.

9.5.2.2 Two-particle probability for a classical regime

We can use the results of Section 9.4.2.2. Substituting Eqs. (9.94) into Eq. (9.81) and then into Eq. (9.84) we arrive at the following:

$$\mathcal{N}_{12} = T_L T_R \left(R_C^2 + T_C^2 \right). \tag{9.96}$$

This equation differs from Eq. (9.84) by the factor $T_R T_L$, which is the probability for two particles to enter the CPC and to contribute to $\mathcal{N}_{12}$.

Calculating $\delta\mathcal{N}_{12} = \mathcal{N}_{12} - \mathcal{N}_1 \mathcal{N}_2$ with Eqs. (9.95) and (9.96) we find,

$$\delta\mathcal{N}_{12} = -R_C T_C \left(T_L^2 + T_R^2 \right), \tag{9.97}$$

which is consistent with a single-electron contribution to the cross-correlator, Eq. (9.90b) by virtue of Eq. (9.32).

Alternatively Eq. (9.97) can be written as Eq. (9.85) with

$$\delta\mathcal{N}_{12}^{(L)} = -R_C T_C T_L^2, \quad \delta\mathcal{N}_{12}^{(R)} = -R_C T_C T_R^2. \tag{9.98}$$

9.5.2.3 Two-particle probability for a quantum regime

If $\Delta t_{L,R}^{(-;-)} = 0$ then electrons that reach the contacts 1 and 2 are correlated. Therefore, instead of Eq. (9.85) we should write

$$\delta \mathcal{N}_{12} = \delta \mathcal{N}_{12}^{(L)} + \delta \mathcal{N}_{12}^{(R)} + \delta \mathcal{N}_{12}^{\widehat{(LR)}}, \qquad (9.99)$$

where the cross-correlation functions due to a single-particle contribution are given in Eq. (9.98) and the cross-correlation function due to a two-particle contribution is

$$\delta \mathcal{N}_{12}^{\widehat{(LR)}} = \mathcal{N}_{12} - \mathcal{N}_1^{(L)} \mathcal{N}_2^{(R)} - \mathcal{N}_1^{(R)} \mathcal{N}_2^{(L)}. \qquad (9.100)$$

The two-particle probability $\mathcal{N}_{12} = \left| \mathcal{A}^{(2)} \right|^2$, with the two-particle amplitude (in the case of indistinguishable electrons) being the Slater determinant, Eq. (9.88). Using the single-particle amplitudes given in Eq. (9.93) we calculate

$$\mathcal{N}_{12} = T_L T_R. \qquad (9.101)$$

Note this equation is independent of the parameters of the central QPC, which can be used as an indication of the quantum regime. We stress that in a quantum regime the two-particle probability becomes the Glauber joint detection probability [252].

Equation (9.101) can be understood as follows: If and only if two electrons enter the CPC (one electron from L and one electron from R) then they necessarily collide at the central QPC and reach different contacts. Therefore, the probability of detecting one electron at contact 1 and one electron at contact 2 is equal to the probability for two electrons to enter the CPC.

Using Eqs. (9.101) and (9.94) we calculate the two-particle cross-correlation function, Eq. (9.100),

$$\delta \mathcal{N}_{12}^{\widehat{(LR)}} = 2 T_L T_R R_C T_C, \qquad (9.102)$$

which, by virtue of Eq. (9.32), is consistent with the two-particle contribution to the cross-correlation function, Eq. (9.90c), at $\Gamma_{\tau L} = \Gamma_{\tau R}$ and $t_{0L}^{(-)} = t_{0R}^{(-)}$.

With Eqs. (9.98) and (9.102) we calculate the total particle cross-correlation function, Eq. (9.99):

$$\delta \mathcal{N}_{12} = - R_C T_C \left(T_R - T_L \right)^2, \qquad (9.103)$$

which is consistent with an electron's contribution to the current cross-correlation function $\mathcal{P}_{12}^{(e)} = \mathcal{P}_{12}/2 = -\mathcal{P}_0 \left(T_R - T_L \right)^2$; see Eq. (9.92) for the total current cross-correlation function.

9.6 Two-particle interference effect

Let us consider a circuit with two Mach–Zehnder interferometers (MZIs), Fig. 9.8, and show that the particles emitted by the SPSs can even show such a subtle effect as a two-particle interference effect [218].

In contrast to previous sections, we now consider a non-adiabatic regime: We take into account the time necessary for electrons (holes) to propagate along the circuit's branches, while the process of emission by the SPS is treated adiabatically. If the difference between the times for propagation along the different arms, U and D, of an interferometer is larger than $\Gamma_{\tau j}$, then single-particle interference is suppressed. Then the currents flowing into the contacts are insensitive to magnetic fluxes through the interferometers. However, if the parameters of the circuit are adjusted in such a way that the particles emitted by S_L and S_R can collide at the outputs $L1$ and $R2$, then the current cross-correlation function becomes sensitive to the magnetic fluxes Φ_L and Φ_R of both interferometers. This effect is a manifestation of two-particle interference taking place in the system.

9.6.1 *Model and definitions*

The circuit, Fig. 9.8, has four contacts and, therefore, it is described by a 4×4 scattering matrix $\hat{S}_{in}(t, E)$ defining the corresponding elements of the Floquet scattering matrix, $S_{F,\alpha\beta}(E_n, E) = S_{in,\alpha\beta,n}(E)$, $\alpha, \beta = 1, 2, 3, 4$. All the contacts are in equilibrium. They are described by the same Fermi distribution function, $f_i(E) = f_0(E), \forall i$, with chemical potential μ_0 and temperature T_0. Each source S_L and S_R emits one electron and one hole during a period.

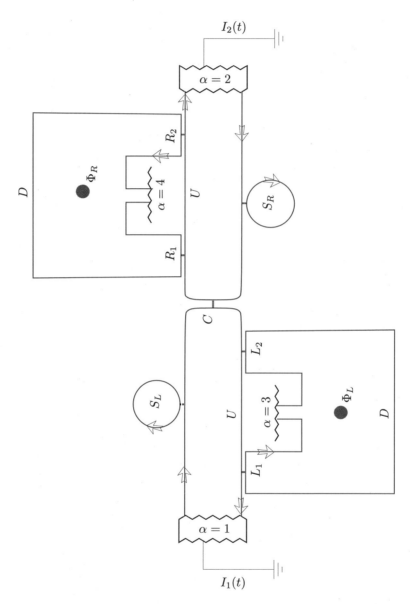

Fig. 9.8 A quantum electronic circuit comprising two single-particle sources S_L and S_R and two Mach–Zehnder interferometers with magnetic fluxes Φ_L and Φ_R.

We are interested in a zero-frequency cross-correlation function $\mathcal{P}_{12}$ for the currents I_1 and I_2. At zero temperature, $k_B T_0 = 0$, [see Eq. (6.16)]

$$\mathcal{P}_{12} = \frac{e^2}{2h} \sum_{q=-\infty}^{\infty} \text{sign}(q) \int_{\mu_0 - q\hbar\Omega_0}^{\mu_0} dE$$

(9.104)

$$\sum_{n,m=-\infty}^{\infty} \sum_{\gamma,\delta=1}^{4} S_{F,1\gamma}(E_n, E) S_{F,1\delta}^*(E_n, E_q) S_{F,2\delta}(E_m, E_q) S_{F,2\gamma}^*(E_m, E),$$

where $E_n = E + n\hbar\Omega_0$. Using $\hat{S}_{in}$ and summing over n and m we find

$$\mathcal{P}_{12} = \frac{e^2}{2h} \sum_{q=-\infty}^{\infty} \text{sign}(q) \int_{\mu_0 - q\hbar\Omega_0}^{\mu_0} dE$$

(9.105)

$$\sum_{\gamma,\delta=1}^{4} \left\{ S_{in,1\gamma}(E) S_{in,1\delta}^*(E_q) \right\}_q \left\{ S_{in,2\gamma}(E) S_{in,2\delta}^*(E_q) \right\}_q^*.$$

In the circuit under consideration there are no paths from contact 4 to contact 1 and from contact 3 to contact 2. Therefore, the relevant indices are $\gamma, \delta = 1, 2$. Thus the final expression for the current cross-correlator is

$$\mathcal{P}_{12} = \frac{e^2}{2h} \sum_{q=-\infty}^{\infty} \text{sign}(q) \int_{\mu_0 - q\hbar\Omega_0}^{\mu_0} dE \left\{ A_q + B_q + C_q + D_q \right\}, \quad (9.106a)$$

where

$$A_q = \left\{ S_{in,11}(E) S_{in,11}^*(E_q) \right\}_q \left\{ S_{in,21}(E) S_{in,21}^*(E_q) \right\}_q^*, \quad (9.106b)$$

$$B_q = \left\{ S_{in,11}(E) S_{in,12}^*(E_q) \right\}_q \left\{ S_{in,21}(E) S_{in,22}^*(E_q) \right\}_q^*, \quad (9.106c)$$

$$C_q = \left\{ S_{in,12}(E) S_{in,11}^*(E_q) \right\}_q \left\{ S_{in,22}(E) S_{in,21}^*(E_q) \right\}_q^*, \quad (9.106d)$$

$$D_q = \left\{ S_{in,12}(E) S_{in,12}^*(E_q) \right\}_q \left\{ S_{in,22}(E) S_{in,22}^*(E_q) \right\}_q^*. \quad (9.106e)$$

Note because of the integration over energy in Eq. (9.106a) only the quantities with $q \neq 0$ are relevant. Shifting $E_q \to E$ (under the integration over energy) and replacing $q \to -q$ (under the sum over q) one can show that the contributions due to A_q and D_q are real, while the contributions due to C_q and B_q are complex conjugates of each other.

When manipulating with Fourier coefficients we will use the following relations:

$$\{X(t)\}_q e^{iq\Omega_0\tau} = \{X(t-\tau)\}_q,$$

$$\{X(t)\}_q \{Y(t)\}_q^* = \{X(t-\tau)\}_q \{Y(t-\tau)\}_q^*. \qquad (9.107)$$

Before presenting expressions for the scattering matrix elements we introduce some definitions. We assume that the kinematic phase $\varphi_{\mathcal{L}}(E)$ acquired by an electron with energy E along the trajectory $\mathcal{L}$ with length $L_{\mathcal{L}}$ is linear in energy,

$$\varphi_{\mathcal{L}}(E) = \varphi_{\mathcal{L}} + (E - \mu_0)\tau_{\mathcal{L}}/\hbar, \qquad (9.108)$$

where $\varphi_{\mathcal{L}} = k_F L_{\mathcal{L}}$ and $\tau_{\mathcal{L}}$ is independent of energy and is the time spent by an electron within the trajectory $\mathcal{L}$. We will label each trajectory with a three-character lower index, where the first character is the number of the destination contact, the second indicates the branch of the corresponding MZI, and the third indicates the source of electrons. For instance, the label $\mathcal{L} = 2UL$ indicates a trajectory starting at the left single-particle source S_L, passing across the upper branch of the (right) MZI, and finishing at contact 2. We will describe an MZI's branch as upper (with lower index U) or down (with lower index D) if an electron going through this branch encircles the magnetic flux counterclockwise or clockwise, respectively, see Fig. 9.8.

It is convenient to introduce an interferometer imbalance time $\Delta\tau_j$, $j = L, R$ and a time delay $\Delta\tau_{LR}$,

$$\Delta\tau_L = \tau_{1Uj} - \tau_{1Dj}, \quad \Delta\tau_R = \tau_{2Uj} - \tau_{2Dj}, \quad \Delta\tau_{LR} = \tau_{\alpha YL} - \tau_{\alpha YR}, \quad (9.109)$$

where $\alpha = 1, 2$, $Y = U, D$, and $j = L, R$. The quantity $\Delta\tau_{LR}$ characterizes the asymmetry in the position of the sources S_L and S_R with respect to the central QPC. With these definitions some differences, which we need later, are

$$\tau_{1UL} - \tau_{1DR} = \Delta\tau_L + \Delta\tau_{LR}, \quad \tau_{1UR} - \tau_{1DL} = \Delta\tau_L - \Delta\tau_{LR},$$

$$\tau_{2UL} - \tau_{2DR} = \Delta\tau_R + \Delta\tau_{LR}, \quad \tau_{2UR} - \tau_{2DL} = \Delta\tau_R - \Delta\tau_{LR}. \qquad (9.110)$$

The magnetic flux, Φ_j, $j = L, R$, through the corresponding MZI is the sum of the fluxes associated with its upper and lower branches,

$$\Phi_j = \Phi_{jU} + \Phi_{jD}. \tag{9.111}$$

Each MZI has two quantum point contacts, which we label as $j1$ and $j2$, $j = L, R$. Without loss of generality we choose the scattering matrices for these QPCs as

$$\hat{S}_{j\alpha} = \begin{pmatrix} \sqrt{R_{j\alpha}} & i\sqrt{T_{j\alpha}} \\ i\sqrt{T_{j\alpha}} & \sqrt{R_{j\alpha}} \end{pmatrix}. \tag{9.112}$$

Here $\alpha = 1, 2$. For the central quantum point contact C connecting the two branches of the circuit we use a scattering matrix of the same form but with the index $j\alpha$ being replaced by the index C.

We assume also that the dwell time τ_j for electrons in each single-particle source, $j = L$ for S_L and $j = R$ for S_R, is short compared to the period of the drive [see Eq. (8.56)],

$$\Omega_0 \tau_j \ll 1. \tag{9.113}$$

Therefore, for the left and right single-particle sources we can use the frozen scattering amplitudes, which we denote as $S_L(t, E)$ and $S_R(t, E)$, respectively. In particular, within this approximation one can use $S_j(t, E) \approx S_j(t, E_n)$. In what follows we use $S_j(t, E) \approx S_j(t, \mu_0) \equiv S_j(t)$. The amplitudes $S_j(t)$ are given in Eq. (9.8) with θ_r, Γ_τ, and $t_0^{(\mp)}$ being replaced by θ_{rj}, $\Gamma_{\tau j}$, and $t_{0j}^{(\mp)}$, respectively.

We are interested in the regime when the interferometer imbalance times are large compared to the duration of wave packets but small compared to the period of the drive,

$$\mathcal{T} \gg \Delta\tau_L, \Delta\tau_R \gg \Gamma_{\tau L}, \Gamma_{\tau R}. \tag{9.114}$$

Then there is no single-particle interference effect that could result in the magnetic-flux dependence of the current cross-correlator. On the other hand if

$$\Delta\tau_L \pm \Delta\tau_R = 0, \tag{9.115}$$

then the interference of two-particle amplitudes makes $\mathcal{P}_{12}$ dependent on $\Phi_L \pm \Phi_R$.

9.6.2 Scattering matrix elements

Calculating the scattering matrix elements we take into account that an electron can arrive at a given contact by following different trajectories. For instance, we have

$$S_{F,11}(E_n, E) = S_{F,1U1}(E_n, E) + S_{F,1D1}(E_n, E),$$

$$S_{F,1U1}(E_n, E) = \sqrt{R_C R_{L1} R_{L2}}\, e^{i2\pi \frac{\Phi_{LU}}{\Phi_0}} e^{i\varphi_{1UL}(E_n)} S_{L,n},$$

$$= \sqrt{R_C R_{L1} R_{L2}}\, e^{i2\pi \frac{\Phi_{LU}}{\Phi_0}} e^{i\varphi_{1UL}(E)} e^{in\Omega_0 \tau_{1UL}} S_{L,n},$$

$$S_{F,1D1}(E_n, E) = -\sqrt{R_C T_{L1} T_{L2}}\, e^{-i2\pi \frac{\Phi_{LD}}{\Phi_0}} e^{i\varphi_{1DL}(E_n)} S_{L,n},$$

$$= -\sqrt{R_C T_{L1} T_{L2}}\, e^{-i2\pi \frac{\Phi_{LD}}{\Phi_0}} e^{i\varphi_{1DL}(E)} e^{in\Omega_0 \tau_{1DL}} S_{L,n},$$

where $\varphi_{\mathcal{L}}(E)$ is given in Eq. (9.108). After the inverse Fourier transformation, we arrive at the following:

$$S_{in,11}(t, E) = \sqrt{R_C}\bigg\{\sqrt{R_{L1} R_{L2}}\, e^{i2\pi \frac{\Phi_{LU}}{\Phi_0}} e^{i\varphi_{1UL}(E)} S_L(t - \tau_{1UL}, E)$$

$$-\sqrt{T_{L1} T_{L2}}\, e^{-i2\pi \frac{\Phi_{LD}}{\Phi_0}} e^{i\varphi_{1DL}(E)} S_L(t - \tau_{1DL}, E)\bigg\}. \tag{9.116a}$$

By analogy we can find the other scattering matrix elements we need to calculate Eqs. (9.106):

$$S^*_{in,11}(t, E_q) = \bigg\{\sqrt{R_{L1} R_{L2}}\, e^{-i2\pi \frac{\Phi_{LU}}{\Phi_0}} e^{-i\varphi_{1UL}(E)} e^{-iq\Omega_0 \tau_{1UL}} S^*_L(t - \tau_{1UL}, E)$$

$$-\sqrt{T_{L1} T_{L2}}\, e^{i2\pi \frac{\Phi_{LD}}{\Phi_0}} e^{-i\varphi_{1DL}(E)} e^{-iq\Omega_0 \tau_{1DL}} S^*_L(t - \tau_{1DL}, E)\bigg\}\sqrt{R_C}, \tag{9.116b}$$

$$S_{in,12}(t, E) = i\sqrt{T_C}\bigg\{\sqrt{R_{L1} R_{L2}}\, e^{i2\pi \frac{\Phi_{LU}}{\Phi_0}} e^{i\varphi_{1UR}(E)} S_R(t - \tau_{1UR}, E)$$

$$-\sqrt{T_{L1} T_{L2}}\, e^{-i2\pi \frac{\Phi_{LD}}{\Phi_0}} e^{i\varphi_{1DR}(E)} S_R(t - \tau_{1DR}, E)\bigg\}, \tag{9.116c}$$

$$S^*_{in,12}(t, E_q) = \left\{ \sqrt{R_{L1}R_{L2}}\, e^{-i2\pi\frac{\Phi_{LU}}{\Phi_0}} e^{-i\varphi_{1UR}(E)} e^{-iq\Omega_0 \tau_{1UR}} S^*_R(t - \tau_{1UR}, E) \right.$$

$$\left. - \sqrt{T_{L1}T_{L2}}\, e^{i2\pi\frac{\Phi_{LD}}{\Phi_0}} e^{-i\varphi_{1DR}(E)} e^{-iq\Omega_0 \tau_{1DR}} S^*_R(t - \tau_{1DR}, E) \right\} \left(-i\sqrt{T_C}\right), \tag{9.116d}$$

$$S_{in,21}(t, E) = i\sqrt{T_C} \left\{ \sqrt{R_{R1}R_{R2}}\, e^{i2\pi\frac{\Phi_{RU}}{\Phi_0}} e^{i\varphi_{2UL}(E)} S_L(t - \tau_{2UL}, E) \right.$$

$$\left. - \sqrt{T_{R1}T_R}\, e^{-i2\pi\frac{\Phi_{RD}}{\Phi_0}} e^{i\varphi_{2DL}(E)} S_L(t - \tau_{2DL}, E) \right\}, \tag{9.116e}$$

$$S^*_{in,21}(t, E_q) = \left\{ \sqrt{R_{R1}R_{R2}}\, e^{-i2\pi\frac{\Phi_{RU}}{\Phi_0}} e^{-i\varphi_{2UL}(E)} e^{-iq\Omega_0 \tau_{2UL}} S^*_L(t - \tau_{2UL}, E) \right.$$

$$\left. - \sqrt{T_{R1}T_{R2}}\, e^{i2\pi\frac{\Phi_{RD}}{\Phi_0}} e^{-i\varphi_{2DL}(E)} e^{-iq\Omega_0 \tau_{2DL}} S^*_L(t - \tau_{2DL}, E) \right\} \left(-i\sqrt{T_C}\right), \tag{9.116f}$$

$$S_{in,22}(t, E) = \sqrt{R_C} \left\{ \sqrt{R_{R1}R_{R2}}\, e^{i2\pi\frac{\Phi_{RU}}{\Phi_0}} e^{i\varphi_{2UR}(E)} S_R(t - \tau_{2UR}, E) \right.$$

$$\left. - \sqrt{T_{R1}T_{R2}}\, e^{-i2\pi\frac{\Phi_{RD}}{\Phi_0}} e^{i\varphi_{2DR}(E)} S_R(t - \tau_{2DR}, E) \right\}, \tag{9.116g}$$

$$S_{in,22}(t, E_q) = \left\{ \sqrt{R_{R1}R_{R2}}\, e^{i2\pi\frac{\Phi_{RU}}{\Phi_0}} e^{i\varphi_{2UR}(E)} e^{iq\Omega_0 \tau_{2UR}} S_R(t - \tau_{2UR}, E) \right.$$

$$\left. - \sqrt{T_{R1}T_{R2}}\, e^{-i2\pi\frac{\Phi_{RD}}{\Phi_0}} e^{i\varphi_{2DR}(E)} e^{iq\Omega_0 \tau_{2DR}} S_R(t - \tau_{2DR}, E) \right\} \sqrt{R_C}. \tag{9.116h}$$

Using the above equations we can calculate the cross-correlator $\mathcal{P}_{12}$ and analyze its dependence on the magnetic fluxes Φ_L and Φ_R.

9.6.3 Current cross-correlator

We consider separately the quantities A_q, B_q, C_q, and D_q in Eq. (9.106a).

9.6.3.1 Partial contributions

First we calculate the quantities A_q, D_q and the corresponding contributions $\mathcal{P}_{12}^{(A)}$, $\mathcal{P}_{12}^{(D)}$ to the cross-correlator. Substituting Eqs. (9.116) into Eq. (9.106b) we find for $q \neq 0$

$$A_q = R_C T_C \zeta_L \zeta_R \sum_{i=1}^{4} A_{i,q}, \quad \zeta_j = \sqrt{R_{j1} R_{j2} T_{j1} T_{j2}}, \quad j = L, R, \quad (9.117)$$

where

$$A_{1,q} = e^{i 2\pi \frac{\Phi_L + \Phi_R}{\Phi_0}} e^{i \frac{E}{\hbar}(\Delta \tau_L + \Delta \tau_R)} \{S_L(t - \Delta \tau_L) S_L^*(t)\}_q \{S_L(t + \Delta \tau_R) S_L^*(t)\}_q^*,$$

$$A_{2,q} = e^{-i 2\pi \frac{\Phi_L + \Phi_R}{\Phi_0}} e^{-i \frac{E}{\hbar}(\Delta \tau_L + \Delta \tau_R)} \{S_L(t + \Delta \tau_L) S_L^*(t)\}_q \{S_L(t - \Delta \tau_R) S_L^*(t)\}_q^*,$$

$$A_{3,q} = e^{i 2\pi \frac{\Phi_L - \Phi_R}{\Phi_0}} e^{i \frac{E}{\hbar}(\Delta \tau_L - \Delta \tau_R)} \{S_L(t - \Delta \tau_L) S_L^*(t)\}_q \{S_L(t - \Delta \tau_R) S_L^*(t)\}_q^*,$$

$$A_{4,q} = e^{-i 2\pi \frac{\Phi_L - \Phi_R}{\Phi_0}} e^{-i \frac{E}{\hbar}(\Delta \tau_L - \Delta \tau_R)} \{S_L(t + \Delta \tau_L) S_L^*(t)\}_q \{S_L(t + \Delta \tau_R) S_L^*(t)\}_q^*.$$

Notice the sums $A_{1,q} + A_{2,-q}$ and $A_{3,q} + A_{4,-q}$ become real (only) after integrating over energy in Eq. (9.106a). The Fourier coefficients are

$$\{S_L(t \mp \Delta \tau_L) S_{0L}^*(t)\}_q = -s_{L,q} \begin{cases} e^{i q \Omega_0 t_{0L}^{(-)}} e^{\pm i q \Omega_0 \Delta \tau_L} + e^{i q \Omega_0 t_{0L}^{(+)}}, & q > 0, \\ e^{i q \Omega_0 t_{0L}^{(+)}} e^{\pm i q \Omega_0 \Delta \tau_L} + e^{i q \Omega_0 t_{0L}^{(-)}}, & q < 0, \end{cases}$$

$$\{S_L(t \pm \Delta \tau_R) S_{0L}^*(t)\}_q^* = -s_{L,q} \begin{cases} e^{-i q \Omega_0 t_{0L}^{(-)}} e^{\pm i q \Omega_0 \Delta \tau_R} + e^{-i q \Omega_0 t_{0L}^{(+)}}, & q > 0, \\ e^{-i q \Omega_0 t_{0L}^{(+)}} e^{\pm i q \Omega_0 \Delta \tau_R} + e^{-i q \Omega_0 t_{0L}^{(-)}}, & q < 0, \end{cases}$$

where $s_{L,q} = 2\Omega_0 \Gamma_{\tau L} e^{-|q| \Omega_0 \Gamma_{\tau L}}$. And the corresponding products are

$$\{S_L(t - \Delta \tau_L) S_{0L}^*(t)\}_q \{S_L(t + \Delta \tau_R) S_{0L}^*(t)\}_q^* = s_{L,q}^2$$

$$\times \begin{cases} 1 + e^{i q \Omega_0 (\Delta \tau_L + \Delta \tau_R)} + e^{i q \Omega_0 \left(t_{0L}^{(-)} - t_{0L}^{(+)} + \Delta \tau_L\right)} + e^{-i q \Omega_0 \left(t_{0L}^{(-)} - t_{0L}^{(+)} - \Delta \tau_R\right)}, & q > 0, \\ 1 + e^{i q \Omega_0 (\Delta \tau_L + \Delta \tau_R)} + e^{i q \Omega_0 \left(t_{0L}^{(-)} - t_{0L}^{(+)} + \Delta \tau_R\right)} + e^{-i q \Omega_0 \left(t_{0L}^{(-)} - t_{0L}^{(+)} - \Delta \tau_L\right)}, & q < 0, \end{cases}$$

$$\{S_L(t+\Delta\tau_L)S_{0L}^*(t)\}_q \{S_L(t-\Delta\tau_R)S_{0L}^*(t)\}_q^* = s_{L,q}^2$$

$$\times \begin{cases} 1 + e^{-iq\Omega_0(\Delta\tau_L+\Delta\tau_R)} + e^{iq\Omega_0\left(t_{0L}^{(-)}-t_{0L}^{(+)}-\Delta\tau_L\right)} + e^{-iq\Omega_0\left(t_{0L}^{(-)}-t_{0L}^{(+)}+\Delta\tau_R\right)}, & q>0, \\ 1 + e^{-iq\Omega_0(\Delta\tau_L+\Delta\tau_R)} + e^{iq\Omega_0\left(t_{0L}^{(-)}-t_{0L}^{(+)}-\Delta\tau_R\right)} + e^{-iq\Omega_0\left(t_{0L}^{(-)}-t_{0L}^{(+)}+\Delta\tau_L\right)}, & q<0, \end{cases}$$

$$\{S_L(t-\Delta\tau_L)S_{0L}^*(t)\}_q \{S_L(t-\Delta\tau_R)S_{0L}^*(t)\}_q^* = s_{L,q}^2$$

$$\times \begin{cases} 1 + e^{iq\Omega_0(\Delta\tau_L-\Delta\tau_R)} + e^{iq\Omega_0\left(t_{0L}^{(-)}-t_{0L}^{(+)}+\Delta\tau_L\right)} + e^{-iq\Omega_0\left(t_{0L}^{(-)}-t_{0L}^{(+)}+\Delta\tau_R\right)}, & q>0, \\ 1 + e^{iq\Omega_0(\Delta\tau_L-\Delta\tau_R)} + e^{iq\Omega_0\left(t_{0L}^{(-)}-t_{0L}^{(+)}-\Delta\tau_R\right)} + e^{-iq\Omega_0\left(t_{0L}^{(-)}-t_{0L}^{(+)}-\Delta\tau_L\right)}, & q<0, \end{cases}$$

$$\{S_L(t+\Delta\tau_L)S_{0L}^*(t)\}_q \{S_L(t+\Delta\tau_R)S_{0L}^*(t)\}_q^* = s_{L,q}^2$$

$$\times \begin{cases} 1 + e^{-iq\Omega_0(\Delta\tau_L-\Delta\tau_R)} + e^{iq\Omega_0\left(t_{0L}^{(-)}-t_{0L}^{(+)}-\Delta\tau_L\right)} + e^{-iq\Omega_0\left(t_{0L}^{(-)}-t_{0L}^{(+)}-\Delta\tau_R\right)}, & q>0, \\ 1 + e^{-iq\Omega_0(\Delta\tau_L-\Delta\tau_R)} + e^{iq\Omega_0\left(t_{0L}^{(-)}-t_{0L}^{(+)}+\Delta\tau_R\right)} + e^{-iq\Omega_0\left(t_{0L}^{(-)}-t_{0L}^{(+)}+\Delta\tau_L\right)}, & q<0. \end{cases}$$

Taking into account Eqs. (9.54), (9.55) and the presence of integration over energy in Eq. (9.106a) we conclude that the quantity A_q (after summing over q) results in a noticeable contribution to the cross-correlator only if it does not oscillate in energy. This is the case under the conditions given in Eq. (9.115). Then we calculate

$$\mathcal{P}_{12}^{(A)} = 4\mathcal{P}_0 \zeta_L \zeta_R \cos\left(2\pi\frac{\Phi_L \pm \Phi_R}{\Phi_0}\right). \tag{9.118}$$

The quantity D_q leads to the same contribution, $\mathcal{P}_{12}^{(D)} = \mathcal{P}_{12}^{(A)}$. Then the corresponding part of the cross-correlator, $\mathcal{P}_{12}^{(A+D)} = \mathcal{P}_{12}^{(A)} + \mathcal{P}_{12}^{(D)}$, is

$$\mathcal{P}_{12}^{(A+D)} = 8\mathcal{P}_0 \zeta_L \zeta_R \cos\left(2\pi\frac{\Phi_L \pm \Phi_R}{\Phi_0}\right). \tag{9.119}$$

Next we calculate B_q. Using Eqs. (9.116) we calculate the corresponding products of the scattering amplitudes in Eq. (9.106c):

$$S_{in,11}(E)S^*_{in,12}(E_q) = -i\sqrt{T_C R_C}\, e^{i\frac{E\Delta\tau_{LR}}{\hbar}}\Bigg\{$$

$$R_{L1}R_{L2}\, e^{-iq\Omega_0 \tau_{1UR}} S_L(t-\tau_{1UL})S^*_R(t-\tau_{1UR})$$

$$+T_{L1}T_{L2}\, e^{-iq\Omega_0\tau_{1DR}} S_L(t-\tau_{1DL})S^*_R(t-\tau_{1DR})$$

$$-\zeta_L\, e^{i\frac{E\Delta\tau_L}{\hbar}} e^{i2\pi\frac{\Phi_L}{\Phi_0}} e^{-iq\Omega_0\tau_{1DR}} S_L(t-\tau_{1UL})S^*_R(t-\tau_{1DR})$$

$$-\zeta_L\, e^{-i\frac{E\Delta\tau_L}{\hbar}} e^{-i2\pi\frac{\Phi_L}{\Phi_0}} e^{-iq\Omega_0\tau_{1UR}} S_L(t-\tau_{1DL})S^*_R(t-\tau_{1UR})\Bigg\},$$

$$S_{in,21}(E)S^*_{in,22}(E_q) = i\sqrt{T_C R_C}\, e^{i\frac{E\Delta\tau_{LR}}{\hbar}}\Bigg\{$$

$$R_{R1}R_{R2}\, e^{-iq\Omega_0\tau_{2UR}} S_L(t-\tau_{2UL})S^*_R(t-\tau_{2UR})$$

$$+T_{R1}T_{R2}\, e^{-iq\Omega_0\tau_{2DR}} S_L(t-\tau_{2DL})S^*_R(t-\tau_{2DR})$$

$$-\zeta_R\, e^{i\frac{E\Delta\tau_R}{\hbar}} e^{i2\pi\frac{\Phi_R}{\Phi_0}} e^{-iq\Omega_0\tau_{2DR}} S_L(t-\tau_{2UL})S^*_R(t-\tau_{2DR})$$

$$-\zeta_R\, e^{-i\frac{E\Delta\tau_R}{\hbar}} e^{-i2\pi\frac{\Phi_R}{\Phi_0}} e^{-iq\Omega_0\tau_{2UR}} S_L(t-\tau_{1DL})S^*_R(t-\tau_{1UR})\Bigg\}.$$

The Fourier coefficients are

$$\left\{S_{in,11}(E)S^*_{in,12}(E_q)\right\}_q = -i\sqrt{T_C R_C}\, e^{i\frac{E\Delta\tau_{LR}}{\hbar}}\Bigg\{$$

$$\{R_{L1}R_{L2}+T_{L1}T_{L2}\}\left\{S_L(t-\Delta\tau_{LR})S^*_R(t)\right\}_q$$

$$-\zeta_L\, e^{i\frac{E\Delta\tau_L}{\hbar}} e^{i2\pi\frac{\Phi_L}{\Phi_0}} \left\{S_L(t-\Delta\tau_L-\Delta\tau_{LR})S^*_R(t)\right\}_q$$

$$-\zeta_L\, e^{-i\frac{E\Delta\tau_L}{\hbar}} e^{-i2\pi\frac{\Phi_L}{\Phi_0}} \left\{S_L(t+\Delta\tau_L-\Delta\tau_{LR})S^*_R(t)\right\}_q\Bigg\},$$

$$\left\{S_{in,21}(E)S_{in,22}^*(E_q)\right\}_q^* = -i\sqrt{T_C R_C}\, e^{-i\frac{E\Delta\tau_{LR}}{\hbar}} \Biggl\{$$

$$\{R_{R1}R_{R2} + T_{R1}T_{R2}\}\{S_L(t-\Delta\tau_{LR})S_R^*(t)\}_q^*$$

$$-\zeta_R e^{-i\frac{E\Delta\tau_R}{\hbar}} e^{-i2\pi\frac{\Phi_R}{\Phi_0}} \{S_L(t-\Delta\tau_R-\Delta\tau_{LR})S_R^*(t)\}_q^*$$

$$-\zeta_R e^{i\frac{E\Delta\tau_R}{\hbar}} e^{i2\pi\frac{\Phi_R}{\Phi_0}} \{S_L(t+\Delta\tau_R-\Delta\tau_{LR})S_R^*(t)\}_q^* \Biggr\}.$$

Then the quantity B_q, Eq. (9.106c), is

$$B_q = -R_C T_C \left\{ B_{0,q} + \zeta_L \zeta_R \sum_{i=1}^{4} B_{i,q} + \sum_{i=5}^{6} B_{i,q} \right\}, \qquad (9.120)$$

where

$$B_{0,q} = T_{MZI}^{(L,0)} T_{MZI}^{(R,0)} \left|\{S_L(t-\Delta\tau_{LR})S_R^*(t)\}_q\right|^2,$$

$$B_{1,q} = e^{i2\pi\frac{\Phi_L+\Phi_R}{\Phi_0}} e^{i\frac{E}{\hbar}(\Delta\tau_L+\Delta\tau_R)} \{S_L(t-\Delta\tau_L-\Delta\tau_{LR})S_R^*(t)\}_q$$

$$\times \{S_L(t+\Delta\tau_R-\Delta\tau_{LR})S_R^*(t)\}_q^*,$$

$$B_{2,q} = e^{-i2\pi\frac{\Phi_L+\Phi_R}{\Phi_0}} e^{-i\frac{E}{\hbar}(\Delta\tau_L+\Delta\tau_R)} \{S_L(t+\Delta\tau_L-\Delta\tau_{LR})S_R^*(t)\}_q$$

$$\times \{S_L(t-\Delta\tau_R-\Delta\tau_{LR})S_R^*(t)\}_q^*,$$

$$B_{3,q} = e^{i2\pi\frac{\Phi_L-\Phi_R}{\Phi_0}} e^{i\frac{E}{\hbar}(\Delta\tau_L-\Delta\tau_R)} \{S_L(t-\Delta\tau_L-\Delta\tau_{LR})S_R^*(t)\}_q$$

$$\times \{S_L(t-\Delta\tau_R-\Delta\tau_{LR})S_R^*(t)\}_q^*,$$

$$B_{4,q} = e^{-i2\pi\frac{\Phi_L-\Phi_R}{\Phi_0}} e^{-i\frac{E}{\hbar}(\Delta\tau_L-\Delta\tau_R)} \{S_L(t+\Delta\tau_L-\Delta\tau_{LR})S_R^*(t)\}_q$$

$$\times \{S_L(t+\Delta\tau_R-\Delta\tau_{LR})S_R^*(t)\}_q^*,$$

$$B_{5,q} = -T_{MZI}^{(L,0)} \zeta_R \left\{ S_L(t - \Delta\tau_{LR}) S_R^*(t) \right\}_q \Bigg\{$$

$$e^{-i\frac{E\Delta\tau_R}{\hbar}} e^{-i2\pi\frac{\Phi_R}{\Phi_0}} \left\{ S_L(t - \Delta\tau_R - \Delta\tau_{LR}) S_R^*(t) \right\}_q^*$$

$$+ e^{i\frac{E\Delta\tau_R}{\hbar}} e^{i2\pi\frac{\Phi_R}{\Phi_0}} \left\{ S_L(t + \Delta\tau_R - \Delta\tau_{LR}) S_R^*(t) \right\}_q^* \Bigg\},$$

$$B_{6,q} = -T_{MZI}^{(R,0)} \zeta_L \left\{ S_L(t - \Delta\tau_{LR}) S_R^*(t) \right\}_q^* \Bigg\{$$

$$e^{i\frac{E\Delta\tau_L}{\hbar}} e^{i2\pi\frac{\Phi_L}{\Phi_0}} \left\{ S_L(t - \Delta\tau_L - \Delta\tau_{LR}) S_R^*(t) \right\}_q$$

$$+ e^{-i\frac{E\Delta\tau_L}{\hbar}} e^{-i2\pi\frac{\Phi_L}{\Phi_0}} \left\{ S_L(t + \Delta\tau_L - \Delta\tau_{LR}) S_R^*(t) \right\}_q \Bigg\},$$

where $T_{MZI}^{(j,0)} = R_{j1}R_{j2} + T_{j1}T_{j2}$. The quantities $T_{MZI}^{(j,0)}$ and ζ_j [see Eq. (9.117)] define the transmission probability, $T_{MZI}^{(j)}(E) = T_{MZI}^{(j,0)} - 2\zeta_j \cos(2\pi\Phi_L/\Phi_0 + E\Delta\tau_j/\hbar)$, for an electron with energy E to propagate through the interferometer j from the central QPC to contact $1\,(2)$ for $j = L\,(R)$.

We see from Eq. (9.120) that the term $B_{0,q}$ always contributes. It results in (the part of) the cross-correlator similar to that in Section 9.4.1. The difference is only in an additional factor $T_{MZI}^{L,0} T_{MZI}^{R,0}$ due to the interferometers and in a time delay $\Delta t_{L,R}$ that appears in a non-adiabatic regime. Other terms in Eq. (9.120) contribute only in the case where they lose oscillating dependence on the energy E: terms $B_{1,q}$ to $B_{4,q}$ contribute if $\Delta\tau_L = \pm\Delta\tau_R$. While the terms $B_{5,q}$ and $B_{6,q}$ (both or either of them) contribute to the current cross-correlator in the case of symmetrical interferometers, $\Delta\tau_L = 0$ or $\Delta\tau_R = 0$. Alternatively, all the terms do contribute in the adiabatic regime. Therefore, with Eqs. (9.114), (9.115) the relevant coefficients are $B_{0,q}$ and $B_{1,q}$ to $B_{4,q}$, which we combine as

$$B_{\pm,q} = e^{i2\pi\frac{\Phi_L \pm \Phi_R}{\Phi_0}} \left| \left\{ S_L(t - \Delta\tau_L - \Delta\tau_{LR}) S_R^*(t) \right\}_q \right|^2$$

$$+ e^{-i2\pi\frac{\Phi_L \pm \Phi_R}{\Phi_0}} \left| \left\{ S_L(t + \Delta\tau_L - \Delta\tau_{LR}) S_R^*(t) \right\}_q \right|^2.$$

(9.121)

Note the sign "+" or "−" is chosen depending on the sign ("+" or "−")

used in Eq. (9.115). For the geometry given in Fig. 9.8 $\Delta\tau_L = \Delta\tau_R < 0$. Therefore, in Eq. (9.115) the sign "−" should be kept.

Taking into account that $C_q^* = B_q$, we can write the relevant coefficients as

$$B_{0,q} + C_{0,q} = 2T_{MZI}^{(L,0)} T_{MZI}^{(R,0)} \left| \{S_L(t - \Delta\tau_{LR}) S_R^*(t)\}_q \right|^2, \quad (9.122a)$$

$$B_{\pm,q} + C_{\pm,q} = 2\cos\left(2\pi \frac{\Phi_L \pm \Phi_R}{\Phi_0}\right) \Big\{ \quad (9.122b)$$

$$\left| \{S_L(t - \Delta\tau_L - \Delta\tau_{LR}) S_R^*(t)\}_q \right|^2 + \left| \{S_L(t + \Delta\tau_L - \Delta\tau_{LR}) S_R^*(t)\}_q \right|^2 \Big\}.$$

Let us consider that part $\mathcal{P}_{12}^{(B+C,0)}$ of the cross-correlator due to $B_{0,q}$ and $C_{0,q}$. With Eq. (9.122a) the integration in Eq. (9.106) is trivial. Then summing over q by analogy with Section 9.4.1 we calculate

$$\mathcal{P}_{12}^{(B+C,0)} = -2\mathcal{P}_0 T_{MZI}^{(L,0)} T_{MZI}^{(R,0)} \left\{ \gamma\left(\Delta t_{L,R}^{(-,-)} + \Delta\tau_{LR}\right) + \gamma\left(\Delta t_{L,R}^{(+,+)} + \Delta\tau_{LR}\right) \right\}, \quad (9.123)$$

where the damping factor $\gamma(\Delta t)$ is given in Eq. (9.58).

Similarly we calculate the part due to $B_{\pm,q}$ and $C_{\pm,q}$:

$$\mathcal{P}_{12}^{(B+C,\Phi)} = -2\mathcal{P}_0 \zeta_L \zeta_R \cos\left(2\pi \frac{\Phi_L \pm \Phi_R}{\Phi_0}\right) \Big\{ \quad (9.124)$$

$$\gamma\left(\Delta t_{L,R}^{(-,-)} - \Delta\tau_L + \Delta\tau_{LR}\right) + \gamma\left(\Delta t_{L,R}^{(+,+)} - \Delta\tau_L + \Delta\tau_{LR}\right)$$

$$+\gamma\left(\Delta t_{L,R}^{(-,-)} + \Delta\tau_L + \Delta\tau_{LR}\right) + \gamma\left(\Delta t_{L,R}^{(+,+)} + \Delta\tau_L + \Delta\tau_{LR}\right) \Big\}.$$

9.6.3.2 Total equation and its analysis

Using Eqs. (9.119), (9.123), and (9.124) we find the total current cross-correlation function, $\mathcal{P}_{12} = \mathcal{P}_{12}^{(A+D)} + \mathcal{P}_{12}^{(B+C,0)} + \mathcal{P}_{12}^{(B+C,\Phi)}$:

$$\mathcal{P}_{12} = -2\mathcal{P}_0 T_{MZI}^{(L,0)} T_{MZI}^{(R,0)} \left\{ \gamma \left(\Delta t_{L,R}^{(-,-)} + \Delta \tau_{LR} \right) + \gamma \left(\Delta t_{L,R}^{(+,+)} + \Delta \tau_{LR} \right) \right\}$$

$$+ 2\mathcal{P}_0 \zeta_L \zeta_R \cos \left(2\pi \frac{\Phi_L \pm \Phi_R}{\Phi_0} \right) \left\{ 4 \right.$$

(9.125)

$$-\gamma \left(\Delta t_{L,R}^{(-,-)} - \Delta \tau_L + \Delta \tau_{LR} \right) - \gamma \left(\Delta t_{L,R}^{(-,-)} + \Delta \tau_L + \Delta \tau_{LR} \right)$$

$$\left. -\gamma \left(\Delta t_{L,R}^{(+,+)} - \Delta \tau_L + \Delta \tau_{LR} \right) - \gamma \left(\Delta t_{L,R}^{(+,+)} + \Delta \tau_L + \Delta \tau_{LR} \right) \right\}.$$

Notice, the suppression of the magnetic-flux independent contribution and the appearance of a contribution dependent on the magnetic flux occur under different conditions.

If the particles emitted by the sources S_L and S_R propagate through the circuit without collisions between themselves, then we have

$$\mathcal{P}_{12} = -4\mathcal{P}_0 T_{MZI}^{(L,0)} T_{MZI}^{(R,0)}, \qquad (9.126)$$

(compare with Eq. (9.74) for an adiabatic regime without interferometers). Here the factor 4 reflects the presence of four particles (two electrons and two holes) emitted by the two sources during a pumping period. The factor $T_{MZI}^{(j,0)} = R_{j1}R_{j2} + T_{j1}T_{j2}$ is the probability for an electron (a hole) to pass through the interferometer j. In the non-adiabatic regime under consideration, Eq. (9.114), this probability is the sum of the probabilities for passing through each arm of an interferometer: The probability $R_{j1}R_{j2}$ is for the arm U, and the probability $T_{j1}T_{j2}$ is for the arm D; see Fig. 9.8.

To analyze the effect of particle collisions we consider the sources emitting wave packets with the same shape, $\Gamma_{\tau L} = \Gamma_{\tau R}$. If two electrons emitted during a period collide at the central QPC, $\Delta t_{L,R}^{(-,-)} + \Delta \tau_{LR} = 0$, then the correlator is suppressed: $\mathcal{P}_{12} = -2\mathcal{P}_0 T_{MZI}^{(L,0)} T_{MZI}^{(R,0)}$. If in addition two holes collide, $\Delta t_{L,R}^{(+,+)} + \Delta \tau_{LR} = 0$, then it is suppressed down to zero: $\mathcal{P}_{12} = 0$. We already discussed this effect in previous sections.

9.6.3.3 Magnetic-flux dependent correlator

An interesting effect arises if two electrons (or two holes) can collide at the interferometer's exit, i.e., at the quantum point contact $L1$ ($R2$) for the interferometer L (R); see Fig. 9.8. Because of Eq. (9.115) the collision conditions are satisfied for both interferometers simultaneously. For definiteness we consider an electron contribution and assume the following conditions hold

$$\Delta t_{L,R}^{(-;-)} - \Delta \tau_L + \Delta \tau_{LR} = 0, \qquad \Delta \tau_L = \Delta \tau_R. \tag{9.127}$$

Then we find that the current cross-correlator depends on the magnetic fluxes of distant interferometers

$$\mathcal{P}_{12}^{(e)} = -2\mathcal{P}_0 \left\{ T_{MZI}^{(L,0)} T_{MZI}^{(R,0)} - \zeta_L \zeta_R \cos\left(2\pi \frac{\Phi_L - \Phi_R}{\Phi_0}\right) \right\}. \tag{9.128}$$

This non-local effect is due to two-particle correlations being a consequence of erasing which-path information for electrons arriving simultaneously at contacts 1 and 2. As was shown in Ref. [218], these correlations are quantum, since they violate the Bell inequalities [254].

To clarify the origin of this effect and to relate the magnetic-flux dependent part of $\mathcal{P}_{12}^{(e)}$ to the two-electron probability $\mathcal{N}_{12}$ we analyze in detail the propagation of two electrons through the circuit.

Let us consider two electrons going to the same, say L, interferometer. From Eq. (9.127) we have $\tau_{1DC} + \tau_{CL} + t_{0L}^{(-)} = \tau_{1UC} + \tau_{CR} + t_{0R}^{(-)}$. This means that an electron emitted by the source S_L and going along the down arm of the left interferometer, the path $\mathcal{L}_{1DL}$, meets an electron emitted by the source S_R and going along the upper arm of the same interferometer, the path $\mathcal{L}_{1UR}$. Therefore, past the quantum point contact $L1$ we do not know where each electron originated. The same happens if two electrons go to the interferometer R: Again due to Eq. (9.127) $\tau_{2DC} + \tau_{CL} + t_{0L}^{(-)} = \tau_{2UC} + \tau_{CR} + t_{0R}^{(-)}$. Therefore, an electron emitted by the source S_L and going along the down arm of the right interferometer, the path $\mathcal{L}_{2DL}$, and an electron emitted by the source S_R and going along the upper arm of the same interferometer, the path $\mathcal{L}_{2UR}$, lose their which-path information after the quantum point contact $R2$. We stress that these events do not manifest themselves in the cross-correlator $\mathcal{P}_{12}$. We considered them only with the purpose of showing the existence of two pairs of single-particle trajectories, $\mathcal{L}_{1DL}$, $\mathcal{L}_{1UR}$ and $\mathcal{L}_{2DL}$, $\mathcal{L}_{2UR}$, responsible for losing which-path information.

From these single-particle trajectories one can compose two-particle trajectories corresponding to the particles going to different interferometers. These trajectories are the following: $\mathcal{L}_a^{(2)} = \mathcal{L}_{1DL} \mathcal{L}_{2UR}$ and $\mathcal{L}_b^{(2)} =$

$\mathcal{L}_{2DL}\mathcal{L}_{1UR}$. With Eq. (9.127) the trajectories $\mathcal{L}_a^{(2)}$ and $\mathcal{L}_b^{(2)}$ correspond to two-particle indistinguishable events: They have the same initial and final states. The final state is characterized by the places where the two electrons appeared and the times when they appeared at them. Electrons going along these trajectories are responsible for the magnetic-flux dependence of the cross-correlator $\mathcal{P}_{12}^{(e)}$. Since there are a number of different two-particle trajectories, the amplitude $\mathcal{A}^{(2)}$ for the mentioned trajectories defines only a part of the two-particle probability, which we denote as $\mathcal{N}_{12}^{(2)} = |\mathcal{A}^{(2)}|^2$. Since the amplitude $\mathcal{A}^{(2)}$ comprises contributions from two indistinguishable trajectories, it is the Slater determinant,

$$\mathcal{A}^{(2)} = \det \begin{vmatrix} \mathcal{A}_{1DL} & \mathcal{A}_{1UR} \\ \mathcal{A}_{2DL} & \mathcal{A}_{2UR} \end{vmatrix}, \qquad (9.129)$$

with the following single-particle amplitudes,

$$\mathcal{A}_{1DL} = -\sqrt{R_C T_{L2} T_{L1}}\, e^{-i2\pi \frac{\Phi_{LD}}{\Phi_0}}\, e^{ik_F L_{1DL}},$$

$$\mathcal{A}_{1UR} = i\sqrt{T_C R_{L2} R_{L1}}\, e^{i2\pi \frac{\Phi_{LU}}{\Phi_0}}\, e^{ik_F L_{1UR}},$$

$$\mathcal{A}_{2DL} = -i\sqrt{T_C T_{R1} T_{R2}}\, e^{-i2\pi \frac{\Phi_{RD}}{\Phi_0}}\, e^{ik_F L_{2DL}},$$

$$\mathcal{A}_{2UR} = \sqrt{R_C R_{R1} R_{R2}}\, e^{i2\pi \frac{\Phi_{RU}}{\Phi_0}}\, e^{ik_F L_{2UR}}.$$

After squaring we find

$$\mathcal{N}_{12}^{(2)} = R_C^2 T_{L1} T_{L2} R_{R1} R_{R2} + T_C^2 R_{L1} R_{L2} T_{R1} T_{R2} \\ + 2 R_C T_C \zeta_L \zeta_R \cos\left(2\pi \frac{\Phi_L - \Phi_R}{\Phi_0}\right). \qquad (9.130)$$

Note in the above the difference of the magnetic fluxes is because we chose $\Delta\tau_L = \Delta\tau_R$, Eq. (9.127), see the explanation after Eq. (9.121).

Using Eq. (9.32) one can check that the magnetic-flux dependence of the two-electron probability $\mathcal{N}_{12}^{(2)}$, Eq. (9.130), explains completely the magnetic-flux dependence of the current cross-correlator $\mathcal{P}_{12}^{(e)}$, Eq. (9.128).

Note the two-particle interference in mesoscopic conductors has been discussed before [4, 5, 255, 256] and has already been observed experimentally [257, 258]. The Aharonov–Bohm effect due to two-particle interference has also been discussed [259, 260] and observed [261]. For a short review see Ref. [262]. What is essentially new and important in the example given above is the tunability allowing us to switch on and off the two-particle correlations in a controllable manner [218].

Bibliography

[1] Landauer, R. (1957). Spatial Variation of Currents and Fields Due to Localized Scatterers in Metallic Conduction, *IBM J. Res. Develop.* **1**, pp. 223–231.

[2] Landauer, R. (1970). Electrical resistance of disordered one-dimensional lattices, *Phil. Mag.* **21**, pp. 863–867.

[3] Landauer, R. (1975). Residual Resistivity Dipoles, *Z. Phys. B.* **21**, pp. 247–254.

[4] Büttiker, M. (1990). Scattering theory of thermal and excess noise in open conductors, *Phys. Rev. Lett.* **65**, pp. 2901–2904.

[5] Büttiker, M. (1992). Scattering theory of current and intensity noise correlations in conductors and wave guides, *Phys. Rev. B* **46**, pp. 12485–12507.

[6] Büttiker, M. (1993). Capacitance, admittance, and rectification properties of small conductors, *J. Phys. Condensed Matter* **5**, pp. 9361–9378.

[7] Imry, Y. (1986). Physics of mesoscopic systems, in: Grinstein, G., Mazenco, G. (eds.), *Directions in Condensed Matter Physics* (World Scientific, Singapore), pp. 101–163.

[8] Imry, Y. (1997). *Introduction to Mesoscopic Physics* (Oxford University Press, New York, Oxford).

[9] Büttiker, M. and Moskalets, M. (2010). From Anderson Localization to Mesoscopic Physics, in: Abrahams, E. (ed.), *50 Years of Anderson localization* (World Scientific, Singapore), pp. 169–190.

[10] Beenakker, C. W. J. (1997). Random-matrix theory of quantum transport, *Rev. Mod. Phys.* **69**, N 3, pp. 731–808.

[11] Fisher, D. S. and Lee, P. A. (1981). Relation between conductivity and transmission matrix, *Phys. Rev. B* **23**, pp. 6851–6854.

[12] Datta, S. (1995). *Electronic Transport in Mesoscopic Systems* (Cambridge University Press, Cambridge).

[13] Arrachea, L. and Moskalets, M. (2006). Relation between scattering-matrix and Keldysh formalisms for quantum transport driven by time-periodic fields, *Phys. Rev. B* **74**, 24, p. 245322 (13).

[14] Landau, L. D. and Lifshits, E. M. (1981). *Quantum Mechanics: Non-Relativistic Theory* (Butterworth-Heinemann, Oxford).

[15] Landau, L. D. and Lifshits, E. M. (1980). *Statistical Physics, Pt. 1* (Butterworth-Heinemann, Oxford).
[16] Friedel, J. (1952). The distribution of electrons round impurities in monovalent metals, *Phil. Mag.* **43**, 337, pp. 153–189.
[17] Taniguchi, T. and Büttiker, M. (1999). Friedel phases and phases of transmission amplitudes in quantum scattering systems, *Phys. Rev. B* **60**, 19, pp. 13814–13823.
[18] Büttiker, M. (1986). Four-terminal phase-coherent conductance, *Phys. Rev. Lett.* **57**, N 14, pp. 1761–1764.
[19] Sánchez, D. and Büttiker, M. (2004). Magnetic-field symmetry on nonlinear mesoscopic transport, *Phys. Rev. Lett* **93**, N 10, p. 106802 (4).
[20] Spivak, B. and Zyuzin, A. (2004). Signature of the Electron-Electron Interaction in the Magnetic-Field Dependence of Nonlinear I-V Characteristics in Mesoscopic Systems, *Phys. Rev. Lett* **93**, N 22, p. 226801 (3).
[21] Büttiker, M., Imry, Y., and Azbel, M. Y. (1984). Quantum oscillations in one-dimensional normal-metal rings, *Phys. Rev. A* **30**, pp. 1982–1989.
[22] Büttiker, M. (1985). Small normal-metal loop coupled to an electron reservoir, *Phys. Rev. B* **32**, 3, pp. 1846–1849.
[23] Griffith, J. S. (1953). A free-electron theory of conjugated molecules. Part 1. Polycyclic hydrocarbons, *Trans. Faraday. Soc.* **49**, pp. 345–351.
[24] Büttiker, M., Imry, Y., Landauer, R., and Pinhas, S. (1985). Generalized many channel conductance formula with application to small rings, *Phys. Rev. B* **31**, pp. 6207–6215.
[25] Büttiker, M. (1985). Role of quantum coherence in series resistors, *Phys. Rev. B* **33**, pp. 3020–3026.
[26] Büttiker. M. (1988). Coherent and sequential tunneling in series barriers, *IBM J. Res. Develop.* **32**, 1, pp. 63–75.
[27] Beenakker, C. W. J. and Michaelis, B. (2005). Stub model for dephasing in a quantum dot, *J. Phys. A: Math. Gen.* **38**, pp. 10639–10646.
[28] Anderson, P. W. and Lee, P. A. (1980). The Thouless conjecture for a one-dimensional chain, *Suppl. Prog. Theor. Phys.* **69**, pp. 212–219.
[29] Cheung, H.-F., Gefen, Y., Riedel, E. K., and Shih, W.-H. (1988). Persistent currents in small one-dimensional metal rings, *Phys. Rev. B* **37**, 11, pp. 6050–6062.
[30] Landau, L. D. and Lifshits, E. M. (1980). *Statistical Physics, Pt. 2* (Butterworth-Heinemann, Oxford).
[31] Blanter, Ya. M. and Büttiker, M. (2000). Shot noise in mesoscopic conductors, *Physics Reports* **336**, pp. 1–166.
[32] Schottky, W. (1918). Über spontane Stromschwankungen in verschiedenen Elektrizitätsleitern, *Ann. Phys. (Leipzig)* **57**, pp. 541–567.
[33] Kulik, I. O. and Omelyanchouk, A. N. (1984). *Fiz. Nizk. Temp.* **10**, p. 158 [*Sov. J. Low Temp. Phys.* **10**, p. 305].
[34] Lesovik, G. B. (1989). Excess quantum noise in 2D ballistic point contacts, *Pis'ma v ZhETF* **49**, 9, pp. 513–515 [*JETP Lett.* **49**, 9, pp. 592–594].

[35] Khlus, V. A. (1987). Current and voltage fluctuations in microjunctions between normal metals and superconductors, *Zh. Eksp. Teor. Fiz.* **93**, p. 2179 [*Sov. Phys. JETP* **66**, p. 1243].
[36] Gardiner, C. W. and Zoller, P. (2000). *Quantum Noise* (Springer, New York).
[37] Dirac, P. A. M. (1926). On the theory of quantum mechanics, *Proceedings of the Royal Society, Series A* **112**, pp. 661–677.
[38] Shirley, J. H. (1965). Solution of the Schrödinger equation with a Hamiltonian periodic in time, *Phys. Rev.* **138**, 4B, pp. 979–987.
[39] Platero, G. and Aguado, R. (2004). Photon-assisted transport in semiconductor nanostructures, *Physics Reports* **395**, pp. 1–157.
[40] Moskalets, M. and Büttiker, M. (2002). Floquet scattering theory of quantum pumps, *Phys. Rev. B* **66**, 20, p. 205320 (10).
[41] Moskalets, M. and Büttiker, M. (2004). Adiabatic quantum pump in the presence of external ac voltages, *Phys. Rev. B* **69**, 20, p. 205316 (12).
[42] Moskalets, M. and Büttiker, M. (2005). Magnetic-field symmetry of pump currents of adiabatically driven mesoscopic structures, *Phys. Rev. B* **72**, 3, p. 035324 (11).
[43] Büttiker, M., Thomas, H., and Prêtre, A. (1994). Current partition in multiprobe conductors in the presence of slowly oscillating external potentials, *Z. Phys. B* **94**, pp. 133–137.
[44] Moskalets, M. and Büttiker, M. (2007). Time-resolved noise of adiabatic quantum pumps, *Phys. Rev. B* **75**, 3, p. 035315 (11).
[45] Moskalets, M. and Büttiker, M. (2008). Dynamic scattering channels of a double barrier structure, *Phys. Rev. B* **78**, 3, p. 035301 (12).
[46] Splettstoesser, J., Ol'khovskaya, S., Moskalets, M., and Büttiker, M. (2008). Electron counting with a two-particle emitter, *Phys. Rev. B* **78**, 20, p. 205110 (5).
[47] Martinez, D. F. and Reichl, L. E. (2001). Transmission properties of the oscillating δ-function potential, *Phys. Rev. B* **64**, 24, p. 245315 (9).
[48] Sadreev, A. F. and Davlet-Kildeev, K. (2007). Electron transmission through an ac biased quantum point contact, *Phys. Rev. B* **75**, 23, p. 235309 (6).
[49] Wigner, E. P. (1955). Lower limit for the energy derivative of the scattering phase shift, *Phys. Rev.* **98**, 1, pp. 145–147.
[50] Smith, F. T. (1960). Lifetime matrix in collision theory, *Phys. Rev.* **118**, 1, pp. 349–356.
[51] Büttiker, M. and Landauer, R. (1982). Traversal time for tunneling, *Phys. Rev. Lett.* **49**, 23, pp. 1739–1742.
[52] Büttiker, M. and Landauer, R. (1986). Traversal time for tunneling, *IBM J. Res. Develop.* **30**, 5, pp. 451–454.
[53] Yafaev, D. R. (1992). *Mathematical Scattering Theory* (AMS).
[54] de Carvalho, C. A. A. and Nussenzveig, H. M. (2002). Time delay, *Physics Reports* **364**, 2, pp. 83–174.

[55] Wang, B., Wang, J., and Guo, H. (2003). Current plateaus of nonadiabatic charge pump: Multiphoton assisted processes, *Phys. Rev. B* **68**, 15, p. 155326 (7).

[56] Wagner, M. (1994). Quenching of resonant transmission through an oscillating quantum well, *Phys. Rev. B* **49**, 23, pp. 16544–16547.

[57] Wagner, M. (1995). Photon-assisted transmission through an oscillating quantum well: A transfer-matrix approach to coherent destruction of tunneling, *Phys. Rev. A* **51**, 1, pp. 798–808.

[58] Gutzwiller, M. C. (1971). Periodic orbits and classical quantization conditions, *J. Math. Phys.* **12**, 3, pp. 343–358.

[59] Jalabert, R. A., Baranger, H. U., and Stone, A. D. (1990). Conductance fluctuations in the ballistic regime: A probe of quantum chaos? *Phys. Rev. Lett.* **65**, 19, pp. 2442–2445.

[60] Martínez-Mares, M., Lewenkopf, C. H., and Mucciolo, E. R. (2004). Statistical fluctuations of pumping and rectification currents in quantum dots, *Phys. Rev. B* **69**, 8, p. 085301 (12).

[61] Rahav, S. and Brouwer, P. (2006). Semiclassical theory of a quantum pump, *Phys. Rev. B* **74**, 20, p. 205327 (13).

[62] Chung, S.-W. V, Moskalets, M., and Samuelsson, P. (2007). Quantum pump driven fermionic Mach–Zehnder interferometer, *Phys. Rev. B* **75**, 11, p. 115332 (10).

[63] Yang, M. and Li, S.-S. (2004). Device for charge- and spin-pumped current generation with temperature-induced enhancement, *Phys. Rev. B* **70**, 19, p. 195341 (5).

[64] Yang, M. and Li, S.-S. (2005). Level-oscillation-induced pump effect in a quantum dot with asymmetric constrictions, *Phys. Rev. B* **71**, 12, p. 125307 (4).

[65] Moskalets, M. V. (1999). Persistent current in a one-dimensional ring with a weak link, *Physica E* **5**, pp. 124–135.

[66] Moskalets, M. V. (1997). Interference phenomena and ballistic transport in one-dimensional ring, *Fiz. Nizk. Temp.* **23**, 10, pp. 1098–1105 [*Sov. Low Temp. Phys.* **23**, 10, pp. 824–829].

[67] Moskalets, M. V. (1998). Temperature dependence of the kinetic coefficients of interference ballistic structures, *Zh. Eksp. Teor. Fiz.* **114**, 5, pp. 1827–1835 [*Sov. Phys. JETP* **87**, 5, pp. 991–995].

[68] Moskalets, M. V. (1998). Temperature-induced current in a one-dimensional ballistic ring with contacts, *Europhys. Lett.* **41**, 2, pp. 189–194.

[69] Brouwer, P. W. (1998). Scattering approach to parametric pumping, *Phys. Rev. B* **58**, 16, pp. R10135–R10138.

[70] Thouless, D. J. (1983). Quantization of particle transport, *Phys. Rev. B* **27**, 10, pp. 6083–6087.

[71] Hekking, F. and Nazarov, Y. V. (1991). Pauli pump for electrons, *Phys. Rev. B* **44**, 16, pp. 9110–9113.

[72] Spivak, B., Zhou, F., and Beal-Monod, M. T. (1995). Mesoscopic mechanisms of the photovoltaic effect and microwave absorption in granular metals, *Phys. Rev. B* **51**, 19, pp. 13226–13230.

[73] Stafford, C. A. and Wingreen, N. S. (1996). Resonant photon-assisted tunneling through a double quantum dot: An electron pump from spatial Rabi oscillations, *Phys. Rev. Lett.* **76**, 11, pp. 1916–1919.
[74] Aleiner, I. L. and Andreev, A. V. (1998). Adiabatic charge pumping in almost open dots, *Phys. Rev. Lett.* **81**, 6, pp. 1286–1289.
[75] Zhou, F., Spivak, B., and Altshuler, B. (1999). Mesoscopic mechanism of adiabatic charge transport, *Phys. Rev. Lett.* **82**, 3, pp. 608–611.
[76] Wagner, M. and Sols, F. (1999). Subsea electron transport: pumping deep within the Fermi sea,*Phys. Rev. Lett.* **83**, 21, pp. 4377–4380.
[77] Simon, S. H. (2000). Proposal for a quantum Hall pump, *Phys. Rev. B* **61**, 24, pp. R16327–R16330.
[78] Wei, Y., Wang, J., and Guo, H. (2000). Resonance-assisted parametric electron pump, *Phys. Rev. B* **62**, 15, pp. 9947–9950.
[79] Avron, J. E., Elgart, A., Graf, G. M., and Sadun, L. (2000). Geometry, statistics, and asymptotics of quantum pumps, *Phys. Rev. B* **62**, 16, pp. R10618–R10621.
[80] Sharma, P. and Chamon, C. (2001). Quantum pump for spin and charge transport in a Luttinger liquid, *Phys. Rev. Lett.* **87**, 9, p. 096401 (17).
[81] Avron, J. E., Elgart, A., Graf, G. M., and Sadun, L. (2001). Optimal quantum pumps, *Phys. Rev. Lett.* **87**, 23, p. 236601 (4).
[82] Vavilov, M. G., Ambegaokar, V., and Aleiner, I. L. (2001). Charge pumping and photovoltaic effect in open quantum dots, *Phys. Rev. B* **63**, 19, p. 195313 (12).
[83] Polianski, M. L. and Brouwer, P. W. (2001). Pumped current and voltage for an adiabatic quantum pump, *Phys. Rev. B* **64**, 7, p. 075304 (6).
[84] Blaauboer, M. and Heller, E. J. (2001). Statistical distribution of Coulomb blockade peak heights in adiabatically pumped quantum dots, *Phys. Rev. B* **64**, 24, p. 241301(R) (4).
[85] Tang, C. S. and Chu, C. S. (2001). Nonadiabatic quantum pumping in mesoscopic nanostructures, *Solid State Communications* **120**, pp. 353–357.
[86] Wang, B., Wang, J., and Guo, H. (2002). Parametric pumping at finite frequency, *Phys. Rev. B* **65**, 7, p. 073306 (4).
[87] Zhu, S.-L. and Wang, Z. D. (2002). Charge pumping in a quantum wire driven by a series of local time-periodic potentials, *Phys. Rev. B* **65**, 15, p. 155313 (5).
[88] Moskalets, M. and Büttiker, M. (2002). Dissipation and noise in adiabatic quantum pumps, *Phys. Rev. B* **66**, 3, p. 035306 (9).
[89] Kim, S. W. (2002). Floquet scattering in parametric electron pumps,*Phys. Rev. B* **66**, 23, p. 235304 (6).
[90] Moskalets, M. and Büttiker, M. (2003). Hidden quantum pump effects in quantum coherent rings, *Phys. Rev. B* **68**, 7, p. 075303 (8).
[91] Cohen, D. (2003). Quantum pumping in closed systems, adiabatic transport, and the Kubo formula, *Phys. Rev. B* **68**, 15, p. 155303 (15).
[92] Moskalets, M. and Büttiker, M. (2003). Quantum pumping: Coherent rings versus open conductors, *Phys. Rev. B* **68**, 16, p. 161311(R) (4).

[93] Zhou, H.-Q., Cho, S. Y., and McKenzie, R. H. (2003). Gauge fields, geometric phases, and quantum adiabatic pumps, *Phys. Rev. Lett.* **91**, 18, p. 186803 (4).

[94] Cohen, D. (2003). Quantum pumping and dissipation: From closed to open systems, *Phys. Rev. B* **68**, 20, p. 201303(R) (4).

[95] Avron, J. E., Elgart, A., Graf, G. M., and Sadun, L. (2004). Transport and dissipation in quantum pumps, *J. of Stat. Phys.* **116**, pp. 425–473.

[96] Faizabadi, E. and Ebrahimi, F. (2004). Charge pumping in quantum wires, *J. Phys.: Condens. Matter* **16**, pp. 1789–1802.

[97] Zhou, H.-Q., Lundin, U., Cho, S. Y., and McKenzie, R. H. (2004). Measuring geometric phases of scattering states in nanoscale electronic devices, *Phys. Rev. B* **69**, 11, p. 113308 (4).

[98] Shin, D. and Hong, J. (2004). Electron transport in the Aharonov–Bohm pump, *Phys. Rev. B* **70**, 7, p. 073301 (4).

[99] Blaauboer, M. (2005). Quantum pumping and nuclear polarization in the integer quantum Hall regime, *Europhys. Lett.* **69**, 1, pp. 109–115.

[100] Zhou, H.-Q., Lundin, U., and Cho, S. Y. (2005). Geometric phases of scattering states in a ring geometry: adiabatic pumping in mesoscopic devices, *J. Phys.: Condens. Matter* **17**, pp. 1059–1066.

[101] Splettstoesser, J., Governale, M., König, J., and Fazio, R. (2005). Adiabatic pumping through interacting quantum dots, *Phys. Rev. Lett.* **95**, 24, p. 246803 (4).

[102] Governale, M., Taddei, F., Fazio, R., and Hekking, F. W. J. (2005). Adiabatic pumping in a superconductor-normal-superconductor weak link, *Phys. Rev. Lett.* **95**, 25, p. 256801 (4).

[103] Sela, I. and Cohen, D. (2006). Operating a quantum pump in a closed circuit, *J. Phys. A: Math. Gen.* **39**, pp. 3575–3592.

[104] Mahmoodian, M. M., Braginsky, L. S., and Entin, M. V. (2006). One-dimensional two-barrier quantum pump with harmonically oscillating barriers: Perturbative, strong-signal, and nonadiabatic regimes, *Phys. Rev. B* **74**, 12, p. 125317 (6).

[105] Banerjee, S., Mukherjee, A., Rao, S., and Saha, A. (2007). Adiabatic charge pumping through a dot at the junction of N quantum wires, *Phys. Rev. B* **75**, 15, p. 153407 (4).

[106] Hwang, N. Y., Kim, S. C., Park, P. S., and Eric Yang, S.-R. (2008). Pumping in quantum dots and non-Abelian matrix Berry phases, *Solid State Commun.* **145**, pp. 515–519.

[107] Qi, X.-L., Hughes T. L., and Zhang, S.-C. (2008). Fractional charge and quantized current in the quantum spin Hall state, *Nature Physics* **4**, pp. 273–276.

[108] Winkler, N., Governale, M., and König, J. (2009). Diagrammatic real-time approach to adiabatic pumping through metallic single-electron devices, *Phys. Rev. B* **79**, 23, p. 235309 (11).

[109] Zhu, R. and Chen, H. (2009). Quantum pumping with adiabatically modulated barriers in graphene, *Appl. Phys. Lett.* **95**, 12, p. 122111 (3).

[110] Prada, E., San-Jose, P., and Schomerus, H. (2009). Quantum pumping in graphene, *Phys. Rev. B* **80**, 24, p. 245414 (5).

[111] Luo, S.-L. and Wei, Y.-D. (2009). Properties of graphene based parametric pump, *Chin. Phys. Lett.* **26**, 11, p. 117202 (4).

[112] Zhu, R. and Berakdar, J. (2010). Spin-dependent pump current and noise in an adiabatic quantum pump based on domain walls in a magnetic nanowire, *Phys. Rev. B* **81**, 1, p. 014403 (7).

[113] Riwar, R.-P. and Splettstoesser, J. (2010). Charge and spin pumping through a double quantum dot, *Phys. Rev. B* **82**, 20, p. 205308 (14).

[114] Zhang, Q., Chan, K. S., and Lin, Z. (2011). Spin current generation by adiabatic pumping in monolayer graphene, *Appl. Phys. Lett.* **98**, 3, p. 032106 (3).

[115] Sinitsyn, N. A. (2009). The stochastic pump effect and geometric phases in dissipative and stochastic systems, *J. Phys. A: Math. Theor.* **42**, 19, p. 193001 (33).

[116] Switkes, M., Marcus, C. M., Campman, K., and Gossard, A. C. (1999). An adiabatic quantum electron pump, *Science* **283**, pp. 1905–1908.

[117] Höhberger, E. M., Lorke, A., Wegscheider, W., and Bichler, M. (2001). Adiabatic pumping of two-dimensional electrons in a ratchet-type lateral superlattice, *Appl. Phys. Lett.* **78**, 19, pp. 2905–2907.

[118] Watson, S. K., Potok, R. M., Marcus, C. M., and Umansky, V. (2003). Experimental realization of a quantum spin pump, *Phys. Rev. Lett.* **91**, N 25, p. 258301 (4).

[119] Liu, K.-M., Umansky, V., and Hsu, S.-Y. (2010). Time dependent electric fields generated DC currents in a large gate-defined open dot, *Jap. J. Appl. Phys.* **49**, p. 114001 (4).

[120] Niu, Q. (1990). Towards a quantum pump of electric charges, *Phys. Rev. Lett.* **64**, 15, pp. 1812–1815.

[121] Liu, C. and Niu, Q. (1993). Nonadiabatic effect in a quantum charge pump, *Phys. Rev. B* **47**, 19, pp. 13031–13034.

[122] Makhlin, Y. and Mirlin, A. D. (2001). Counting statistics for arbitrary cycles in quantum pumps, *Phys. Rev. Lett.* **87**, 27, p. 276803 (4).

[123] Levinson, Y., Entin-Wohlman, O., and Wölfle, P. (2001). Pumping at resonant transmission and transferred charge quantization, *Physica A* **302**, pp. 335–344.

[124] Kashcheyevs, V., Aharony, A., and Entin-Wohlman, O. (2004). Resonance approximation and charge loading and unloading in adiabatic quantum pumping, *Phys. Rev. B* **69**, 19, p. 195301 (9).

[125] Blumenthal, M. D., Kaestner, B., Li, L., Giblin, S., Janssen, T. J. B. M., Pepper, M., Anderson, D., Jones, G., and Ritchie, D. A. (2007). Gigahertz quantized charge pumping, *Nature Physics* **3**, pp. 343–347.

[126] Fujiwara, A., Nishiguchi, K., and Ono, Y. (2008). Nanoampere charge pump by single-electron ratchet using silicon nanowire metal-oxide-semiconductor field-effect transistor, *Appl. Phys. Lett.* **92**, 4, p. 042102 (3).

[127] Kaestner, B., Kashcheyevs, V., Amakawa, S., Blumenthal, M. D., Li, L., Janssen, T. J. B. M., Hein, G., Pierz, K., Weimann, T., Siegner, U., and Schumacher, H. W. (2008). Single-parameter nonadiabatic quantized charge pumping, *Phys. Rev. B* **77**, 15, p. 153301 (4).

[128] Miyamoto, S., Nishiguchi, K., Ono, Y., Itoh, K. M., and Fujiwara, A. (2010). Resonant escape over an oscillating barrier in a single-electron ratchet transfer, *Phys. Rev. B* **82**, 3, p. 033303 (4).

[129] Fève, G., Mahé, A., Berroir, J.-M., Kontos, T., Plaçais, B., Glattli, D. C., Cavanna, A., Etienne, B., and Jin, Y. (2007). An on-demand coherent single-electron source, *Science* **316**, pp. 1169–1172.

[130] Kouwenhoven, L. P., Johnson, A. T., van der Vaart, N. C., Harmans, C. J. P. M., and Foxon, C. T. (1991). Quantized current in a quantum-dot turnstile using oscillating tunnel barriers, *Phys. Rev. Lett.* **67**, 12, pp. 1626–1629.

[131] Keller, M. W., Martinis, J. M., Zimmerman, N. M., and Steinbach, A. H. (1996). Accuracy of electron counting using a 7-junction electron pump, *Appl. Phys. Lett.* **69**, 12, pp. 1804–1806.

[132] Berg, E., Levin, M., and Altman, E. (2011). Quantized pumping and topology of the phase diagram for a system of interacting bosons, *Phys. Rev. Lett.* **106**, 11, p. 110405 (4).

[133] Niu, Q. (1986). Quantum adiabatic particle transport, *Phys. Rev. B* **34**, 8, pp. 5093–5100.

[134] Onoda, S., Murakami, S., and Nagaosa, N. (2004). Topological nature of polarization and charge pumping in ferroelectrics, *Phys. Rev. Lett.* **93**, 16, p. 167602 (4).

[135] Fu, L. and Kane, C. L. (2006). Time reversal polarization and a Z2 adiabatic spin pump, *Phys. Rev. B.* **64**, 19, p. 195312 (13).

[136] Chern, C. H., Onoda, S., Murakami, S., and Nagaosa, N. (2007). Quantum charge pumping and electric polarization in Anderson insulators, *Phys. Rev. B.* **76**, 3, p. 035334 (15).

[137] Leone, R., Lévy, L. P., and Lafarge, P. (2008). Cooper-pair pump as a quantized current source, *Phys. Rev. Lett.* **100**, 11, p. 117001 (4).

[138] Teo, J. C. Y. and Kane, C. L. (2010). Topological defects and gapless modes in insulators and superconductors, *Phys. Rev. B.* **82**, 11, p. 115120 (26).

[139] Meidan, D., Micklitz, T., and Brouwer, P. W. (2010). Optimal topological spin pump, *Phys. Rev. B.* **82**, 16, p. 161303(R) (4).

[140] Maruyama, I. and Hatsugai, Y. (2009). Non-adiabatic effect on Laughlins argument of the quantum Hall effect, *J. Phys.: Conf. Ser.* **150**, 2, p. 022055 (4).

[141] Graf, G. M. and Ortelli, G. (2008). Comparison of quantization of charge transport in periodic and open pumps, *Phys. Rev B* **77**, 3, p. 033304 (3).

[142] Bräunlich, G., Graf, G. M., and Ortelli, G. (2010). Equivalence of topological and scattering approaches to quantum pumping, *Commun. Math. Phys.* **295**, pp. 243–259.

[143] Reydellet, L.-H., Roche, P., Glattli, D. C., Etienne, B., and Jin, Y. (2003). Quantum partition noise of photon-created electron-hole pairs, *Phys. Rev Lett.* **90**, 17, p. 176803 (4).
[144] Rychkov, V. S., Polianski, M. L., and Büttiker, M. (2005). Photon-assisted electron-hole shot noise in multiterminal conductors, *Phys. Rev. B* **72**, 15, p. 155326 (9).
[145] Polianski, M. L., Samuelsson, P., and Büttiker, M. (2005). Shot noise of photon-excited electron-hole pairs in open quantum dots, *Phys. Rev. B* **72**, 16, p. 161302(R) (4).
[146] Moskalets, M. and Büttiker, M. (2001). Effect of inelastic scattering on parametric pumping, *Phys. Rev. B* **64**, 20, p. 201305(R) (4).
[147] Büttiker, M. and Moskalets, M. (2006). Scattering theory of dynamic electrical transport, *Lecture Notes in Physics* **690**, p. 33–44.
[148] Cremers, J. N. H. J. and Brouwer, P. W. (2002). Dephasing in a quantum pump, *Phys. Rev. B* **65**, 11, p. 115333 (7).
[149] Wagner, M. (2000). Probing Pauli blocking factors in quantum pumps with broken time-reversal symmetry, *Phys. Rev. Lett.* **85**, 1, pp. 174–177.
[150] Kohler, S., Lehmann, J., and Hänggi, P. (2005). Driven quantum transport on the nanoscale, *Physics Reports* **406**, pp. 379–443.
[151] DiCarlo, L., Marcus, C. M., and Harris, J. S., Jr. (2003). Photocurrent, rectification, and magnetic field symmetry of induced current through quantum dots, *Phys. Rev. Lett.* **91**, 24, p. 246804 (4).
[152] Vavilov, M. G., DiCarlo, L., and Marcus, C. M. (2005). Photovoltaic and rectification currents in quantum dots, *Phys. Rev. B* **71**, 24, p. 241309(R) (4).
[153] Arrachea, L. (2005). Green-function approach to transport phenomena in quantum pumps, *Phys. Rev. B* **72**, 12, p. 125349 (11).
[154] Foa Torres, L. E. F. (2005). Mono-parametric quantum charge pumping: Interplay between spatial interference and photon-assisted tunneling, *Phys. Rev. B* **72**, 24, p. 245339 (7).
[155] Agarwal, A. and Sen, D. (2007). Non-adiabatic charge pumping by an oscillating potential, *Phys. Rev. B* **76**, 23, p. 235316 (8).
[156] Cavaliere, F., Governale, M., and König, J. (2009). Nonadiabatic pumping through interacting quantum dots, *Phys. Rev. Lett.* **103**, 13, p. 136801 (4).
[157] Gu, Y., Yang, Y. H., Wang, J., and Chan, K. S. (2009). Single-parameter charge pump in a zigzag graphene nanoribbon, *J. Phys. Condens. Matter* **21**, p. 405301 (6).
[158] Soori, A. and Sen, D. (2010). Nonadiabatic charge pumping by oscillating potentials in one dimension: Results for infinite system and finite ring, *Phys. Rev. B* **82**, 11, p. 115432 (15).
[159] Kaestner, B., Kashcheyevs, V., Hein, G., Pierz, K., Siegner, U., and Schumacher, H. W. (2008). Robust single-parameter quantized charge pumping, *Appl. Phys. Lett.* **92**, 19, p. 192106 (3).

[160] Wright, S. J., Blumenthal, M. D., Gumbs, G., Thorn, A. L., Pepper, M., Janssen, T. J. B. M., Holmes, S. N., Anderson, D., Jones, G. A. C., Nicoll, C. A., and Ritchie, D. A. (2008). Enhanced current quantization in high-frequency electron pumps in a perpendicular magnetic field, *Phys. Rev. B* **78**, 23, p. 233311 (4).

[161] Kaestner, B., Leich, C., Kashcheyevs, V., Pierz, K., Siegner, U., and Schumacher, H. W. (2009). Single-parameter quantized charge pumping in high magnetic fields, *Appl. Phys. Lett.* **94**, 1, p. 012106 (3).

[162] Wright, S. J., Blumenthal, M. D., Pepper, M., Anderson, D., Jones, G. A. C., Nicoll, C. A., and Ritchie, D. A. (2009). Parallel quantized charge pumping, *Phys. Rev. B* **80**, 11, p. 113303 (3).

[163] Giblin, S. P., Wright, S. J., Fletcher, J. D., Kataoka, M., Pepper, M., Janssen, T. J. B. M., Ritchie, D. A., Nicoll, C. A., Anderson, D., and Jones, G. A. C. (2010). An accurate high-speed single-electron quantum dot pump, *New J. Phys.* **12**, p. 073013 (8).

[164] Mirovsky, P., Kaestner, B., Leicht, C., Welker, A. C., Weimann, T., Pierz, K., and Schumacher, H. W. (2010). Synchronized single electron emission from dynamical quantum dots, *Appl. Phys. Lett.* **97**, 25, p. 252104 (3).

[165] Qi, X.-L. and Zhang, S.-C. (2009). Field-induced gap and quantized charge pumping in a nanoscale helical wire, *Phys. Rev. B* **79**, 23, p. 235442 (6).

[166] Zólyomi, V., Oroszlány, L., and Lambert, C. J. (2009). Quantum pumps formed of double walled carbon nanotubes, *Phys. Status Solidi B* **246**, 11-12, pp. 2650–2653.

[167] Oroszlány, L., Zólyomi, V., and Lambert, C. J. (2010). Carbon nanotube Archimedes screws, *ACS Nano* **4**, 12, pp. 7363–7366.

[168] Jääskeläinen, M., Corvino, F., Search, C. P., and Fessatidis, V. (2008). Quantum pumping of electrons by a moving modulated potential, *Phys. Rev. B* **77**, 15, p. 155319 (8).

[169] Das, K. K. and Opatrný, T. (2010). What is quantum in quantum pumping: The role of phase and asymmetries, *Phys. Lett. A* **374**, pp. 485–490.

[170] Prêtre, A., Thomas, H., and Büttiker, M. (1996). Dynamic admittance of mesoscopic conductors: Discrete-potential model, *Phys. Rev B* **54**, 11, pp. 8130–8143.

[171] Brouwer, P. W. and Büttiker, M. (1997). Charge-relaxation and dwell time in the fluctuating admittance of a chaotic cavity, *Europhys. Lett.* **37**, 7, pp. 441–446.

[172] Büttiker, M. (2002). Charge densities and charge noise in mesoscopic conductors, *Pramana-J. Phys.* **58**, 2, pp. 241–257.

[173] Jauho, A.-P., Wingreen, N. S., and Meir, Y. (1994). Time-dependent transport in interacting and noninteracting resonant-tunneling systems, *Phys. Rev. B* **50**, 8, pp. 5528–5544.

[174] Pedersen, M. H. and Büttiker, M. (1998). Scattering theory of photon-assisted electron transport, *Phys. Rev. B* **58**, 19, pp. 12993–13006.

[175] Brouwer, P. W. (2001). Rectification of displacement currents in an adiabatic electron pump, *Phys. Rev. B* **63**, 12, p. 121303(R) (2).

[176] Arrachea, L. (2005). Symmetry and environment effects on rectification mechanisms in quantum pumps, *Phys. Rev. B* **72**, 12, p. 121306(R) (4).

[177] Benjamin, C. (2006). Detecting a true quantum pump effect, *Eur. Phys. J. B* **52**, pp. 403–410.

[178] Moskalets, M. and Büttiker, M. (2004). Floquet scattering theory for current and heat noise in large amplitude adiabatic pumps, *Phys. Rev. B* **70**, 24, p. 245305 (15).

[179] Polianski, M. L., Vavilov, M. G., and Brouwer, P. W. (2002). Noise through quantum pumps, *Phys. Rev. B* **65**, 24, p. 245314 (9).

[180] Polianski, M. L. and Brouwer, P. W. (2003). Scattering matrix ensemble for time-dependent transport through a chaotic quantum dot, *J. Phys. A: Math. Gen.* **36**, pp. 3215–3236.

[181] Vavilov, M. G. (2005). Quantum chaotic scattering in time-dependent external fields: random matrix approach, *J. Phys. A: Math. Gen.* **38**, pp. 10587–10611.

[182] Levitov, L. S. and Lesovik, G. B. (1993). Charge distribution in quantum shot noise, *Pis'ma v ZhETF* **58**, 3, pp. 225–230 [*JETP Lett.* **58**, 3, pp. 230–235].

[183] Ivanov, D. A. and Levitov, L. S. (1993). Statistics of charge fluctuations in quantum transport in an alternating field, *Pis'ma v ZhETF* **58**, 6, pp. 450–456 [*JETP Lett.* **58**, 6, pp. 461–468].

[184] Esposito, M., Harbola, U., and Mukamel, S. (2009). Nonequilibrium fluctuations, fluctuation theorems, and counting statistics in quantum systems, *Rev. Mod. Phys.* **81**, 4, pp. 1665–1702.

[185] Andreev, A. and Kamenev, A. (2000). Counting statistics of an adiabatic pump, *Phys. Rev. Lett.* **85**, 6, pp. 1294–1297.

[186] Levitov, L. S. Counting statistics of charge pumping in an open system, cond-mat/0103617 (unpublished).

[187] Andreev, A. V. and Mishchenko, E. G. (2001). Full counting statistics of a charge pump in the Coulomb blockade regime, *Phys. Rev. B* **64**, 23, p. 233316 (4).

[188] Muzykantskii, B. A. and Adamov, Y. (2003). Scattering approach to counting statistics in quantum pumps, *Phys. Rev. B* **68**, 15, p. 155304 (9).

[189] Camalet, S., Lehmann, J., Kohler, J., and Hänggi, P. (2003). Current noise in ac-driven nanoscale conductors, *Phys. Rev. Lett.* **90**, 21, p. 210602.

[190] Abanov, A. G. and Ivanov, D. A. (2009). Factorization of quantum charge transport for noninteracting fermions, *Phys. Rev. B* **79**, 20, p. 205315 (9).

[191] Ivanov, D. A. and Abanov, A. G. (2010). Phase transitions in full counting statistics for periodic pumping, *Europhys. Lett.* **92**, 3, p. 37008 (5).

[192] Wang, B., Wang, J., and Guo, H. (2004). Shot noise of spin current, *Phys. Rev. B* **69**, 15, p. 153301 (4).

[193] Camalet, S., Kohler, J., and Hänggi, P. (2004). Shot noise control in ac-driven nanoscale conductors, *Phys. Rev. B* **70**, 15, p. 155326.

[194] Strass, M., Hänggi, P., and Kohler, S. (2005). Nonadiabatic electron pumping: Maximal current with minimal noise, *Phys. Rev. Lett.* **95**, 13, p. 130601 (4).

[195] Li, C., Yu, Y., Wei, Y., and Wang, J. (2007). Statistical analysis for current fluctuations in a disordered quantum pump, *Phys. Rev. B* **76**, 23, p. 235305 (5).
[196] Maire, N., Hohls, F., Kaestner, B., Pierz, K., Schumacher, H. W., and Haug, R. J. (2008). Noise measurement of a quantized charge pump, *Appl. Phys. Lett.* **92**, 8, p. 082112 (3).
[197] Samuelsson, P. and Büttiker, M. (2005). Dynamic generation of orbital quasiparticle entanglement in mesoscopic conductors, *Phys. Rev. B* **71**, 24, p. 245317 (5).
[198] Beenakker, C. W. J., Titov, M., and Trauzettel, B. (2005). Optimal spin-entangled electron-hole pair pump, *Phys. Rev. Lett.* **94**, 18, p. 186804 (4).
[199] Moskalets, M. and Büttiker, M. (2006). Multiparticle correlations of an oscillating scatterer, *Phys. Rev. B* **73**, 12, p. 125315 (6).
[200] Das, K. K., Kim, S., and Mizel, A. (2006). Controlled flow of spin-entangled electrons via adiabatic quantum pumping, *Phys. Rev. Lett.* **97**, 9, p. 096602 (4).
[201] Beenakker, C. W. J. (2005). Electron-hole entanglement in the Fermi sea, in *Quantum Computers, Algorithms and Chaos, Proceedings of the International School of Physics Enrico Fermi, Course CLXII, Varenna, 2005* (IOS, Amsterdam, 2006), pp. 307–347; see also e-print arXiv:cond-mat/0508488.
[202] Abrikosov, A. A. (1988). *Fundamentals of the Theory of Metals* (Elsevier Science, Amsterdam).
[203] Arrachea, L. and Moskalets, M. (2010). Energy transport and heat production in quantum engines, in: Sattler, K. D. (ed.), *Handbook of Nanophysics: Nanomedicine and Nanorobotics* (CRC Press, Taylor & Francis Group).
[204] Wang, B. and Wang, J. (2002). Heat current in a parametric quantum pump, *Phys. Rev. B* **66**, 12, p. 125310 (4).
[205] Wei, Y., Wan, L., Wang, B., and Wang, J. (2004). Heat current and spin current through a carbon-nanotube-based molecular quantum pump, *Phys. Rev. B* **70**, 4, p. 045418 (9).
[206] Fransson, J. and Galperin, M. (2010). Inelastic scattering and heating in a molecular spin pump, *Phys. Rev. B* **81**, 7, p. 075311 (8).
[207] Humphrey, T. E., Linke, H., and Newbury, R. (2001). Pumping heat with quantum ratchets, *Physica E* **11**, pp. 281–286.
[208] Segal, D. and Nitzan, A. (2006). Molecular heat pump, *Phys. Rev. E* **73**, 02, p. 026109 (9).
[209] Arrachea, L., Moskalets, M., and Martin-Moreno, L. (2007). Heat production and energy balance in nanoscale engines driven by time-dependent fields, *Phys. Rev. B* **75**, 24, p. 245420 (5).
[210] Rey, M., Strass, M., Kohler, S., Hänggi, P., and Sols, F. (2007). Nonadiabatic electron heat pump, *Phys. Rev. B* **76**, 8, p. 085337 (4).
[211] Cuansing, E. C. and Wang, J.-S. (2010). Tunable heat pump by modulating the coupling to the leads, *Phys. Rev. B* **82**, 2, p. 021116 (9).
[212] Chamon, C., Mucciolo, E. R., Arrachea, L., and Capaz, R. B. (2011). Heat pumping in nanomechanical systems, *Phys. Rev. Lett.* **106**, 13, p. 135504 (4).

[213] Bauer, G. E. W., Bretzel, S., Brataas, A., and Tserkovnyak, Y. (2010). Nanoscale magnetic heat pumps and engines, *Phys. Rev. B* **81**, 2, p. 024427 (11).
[214] Büttiker, M., Thomas, H., and Prêtre, A. (1993). Mesoscopic capacitors, *Phys. Lett. A* **180**, 4,5, pp. 364–369.
[215] Moskalets, M. and Büttiker, M. (2009). Heat production and current noise for single- and double-cavity quantum capacitors, *Phys. Rev. B* **80**, 8, p. 081302 (4).
[216] Gabelli, J., Fève, G., Berroir, J.-M., Plaçais, B., Cavanna, A., Etienne, B., Jin, Y., and Glattli, D. C. (2006). Violation of Kirchhoff's Laws for a coherent RC circuit, *Science* **313**, pp. 499–502.
[217] Ol'khovskaya, S., Splettstoesser, J., Moskalets, M., and Büttiker, M. (2008). Shot noise of a mesoscopic two-particle collider, *Phys. Rev. Lett.* **101**, 16, p. 166802 (4).
[218] Splettstoesser, J., Moskalets, M., and Büttiker, M. (2009). Two-particle non-local Aharonov–Bohm effect from two single-particle emitters, *Phys. Rev. Lett.* **103**, 7, p. 076804 (4).
[219] Mahé, A., Parmentier, F. D., Bocquillon, E., Berroir, J.-M., Glattli, D. C., Kontos, T., Plaçais, B., Fève, G., Cavanna, A., and Jin, Y. (2010). Current correlations of an on-demand electron source as an evidence of single particle emission, *Phys. Rev. B* **82**, 20, p. 201309 (4).
[220] Albert, M., Flindt, C., and Büttiker, M. (2010). Accuracy of the quantum capacitor as a single-electron source, *Phys. Rev. B* **82**, 4, p. 041407(R) (4).
[221] Moskalets, M., Samuelsson, P., and Büttiker, M. (2008). Quantized dynamics of a coherent capacitor, *Phys. Rev. Lett.* **100**, 8, p. 086601 (4).
[222] Fertig, H. A. (1988). Semiclassical description of a two-dimensional electron in a strong magnetic field and an external potential, *Phys. Rev. B* **38**, 2, pp. 996–1015.
[223] Chung, V. S.-W., Samuelsson, P. and Büttiker, M. (2005). Visibility of current and shot noise in electrical Mach–Zehnder and Hanbury Brown Twiss interferometers, *Phys. Rev. B* **72**, 12, p. 125320 (13).
[224] van der Vaart, N. C., van de Ruyter, Stevenick, M. P., Kouwenhoven, L. P., Johnson, A. T., Nazarov, Y. V., Harmans, C. J. P. M., and Foxon, C. T. (1994). Time-resolved tunneling of single electrons between Landau levels in a quantum dot, *Phys. Rev. Lett.* **73**, 2, pp. 320–323.
[225] McClure, D. T., Zhang, Y., Rosenow, B., Levenson-Falk, E. M., Marcus, C. M., Pfeiffer, L. N., and West, K. W. (2009). Edge-state velocity and coherence in a quantum Hall Fabry–Pérot interferometer. *Phys. Rev. Lett.* **103**, 20, p. 206806 (4).
[226] Stern, F. (1972). Low temperature capacitance of inverse silicon MOS devices in high magnetic fields, *IBM Research Report*, RC 3758.
[227] Smith, T. P., Goldberg, B. B., Stiles, P. J., and Heiblum, M. (1985). Direct measurement of the density of states of a two-dimensional electron gas, *Phys. Rev. B* **32**, 4, pp. 2696–2699.

[228] Smith, T. P., Wang, W. I., and Stiles, P. J. (1986). Two-dimensional density of states in the extreme quantum limit, *Phys. Rev. B* **34**, 4, pp. 2995–2998.

[229] Luryi, S. (1987). Quantum capacitance devices, *Appl. Phys. Lett.* **52**, 6, pp. 501–503.

[230] Lafarge, P., Pothier, H., Williams, E. R., Esteve, D., Urbina, C., and Devoret, M. H. (1991). Direct observation of macroscopic charge quantization, *Z. Phys. B – Condensed Matter* **85**, pp. 327–332.

[231] Ashoori, R. C., Stormer, H. L., Weiner, J. S., Pfeiffer, L. N., Pearton, S. J., Baldwin, K. W., and West, K. W. (1992). Single-electron capacitance spectroscopy of discrete quantum levels, *Phys. Rev. Lett.* **68**, 20, pp. 3088–3091.

[232] Field, M., Smith, C. G., Pepper, M., Brown, K. M., Linfield, E. H., Grimshaw, M. P., Ritchie, D. A., and Jones, G. A. C. (1996). Coulomb blockade as a noninvasive probe of local density of states, *Phys. Rev. Lett.* **77**, 2, pp. 350–353.

[233] Büttiker, M. (2000). Time-dependent transport in mesoscopic structures, *J. Low Temp. Phys.* **118**, 5/6, pp. 519–542.

[234] Nigg, S. E., Lopez, R. and Büttiker, M. (2006). Mesoscopic charge relaxation, *Phys. Rev. Lett.* **97**, 20, p. 206804 (4).

[235] Ringel, Z., Imry, Y., and Entin-Wohlman, O. (2008). Delayed currents and interaction effects in mesoscopic capacitors, *Phys. Rev. B* **78**, 16, p. 165304 (8).

[236] Mora, C. and Le Hur, K. (2010). Universal resistances of the quantum RC circuit, *Nature Physics* **6**, pp. 697–701.

[237] Hamamoto, Y., Jonckheere, T., Kato, T., and Martin, T. (2010). Dynamic response of a mesoscopic capacitor in the presence of strong electron interactions, *Phys. Rev. B* **81**, 15, p. 153305 (4).

[238] Nigg, S. E. and Büttiker, M. (2008). Quantum to classical transition of the charge relaxation resistance of a mesoscopic capacitor, *Phys. Rev. B* **77**, 8, p. 085312 (10).

[239] Fu, Y. and Dudley, S. C. (1993). Quantum inductance within linear response theory, *Phys. Rev. Lett.* **70**, 1, pp. 65–68.

[240] Gabelli, J., Fève, G., Kontos, T., Berroir, J.-M., Plaçais, B., Glattli, D.C., Etienne, B., Jin, Y., and Büttiker, M. (1997). Relaxation time of a chiral quantum R-L circuit, *Phys. Rev. Lett.* **98**, 16, p. 166806 (4).

[241] Wang, J., Wang, B., and Guo, H. (2007). Quantum inductance and negative electrochemical capacitance at finite frequency in a two-plate quantum capacitor, *Phys. Rev. B* **75**, 15, p. 155336 (5).

[242] Begliarbekov, M., Strauf, S., and Search, C. P. (2011). Quantum inductance and high frequency oscillators in graphene nanoribbons, *Nanotechnology* **22**, 16, p. 165203 (8).

[243] Christen, T. and Büttiker, M. (1996). Low-frequency admittance of quantized Hall conductors, *Phys. Rev. B* **53**, 4, pp. 2064–2072.

[244] Keeling, J., Shytov, A. V., and Levitov, L. S. (2008). Coherent particle transfer in an on-demand single-electron source, *Phys. Rev. Lett.* **101**, 19, p. 196404 (4).
[245] Sasaoka, K., Yamamoto, T., and Watanabe, S. (2010). Single-electron pumping from a quantum dot into an electrode, *Appl. Phys. Lett.* **96**, 10, p. 102105 (3).
[246] Splettstoesser, J., Governale, M., König, J., and Büttiker, M. (2010). Charge and spin dynamics in interacting quantum dots, *Phys. Rev. B* **81**, 16, p. 165318 (5).
[247] Breit, G. and Wigner, E. (1934). Capture of slow neutrons, *Phys. Rev.* **49**, 7, pp. 519–531.
[248] Battista, F. and Samuelsson, P. (2011). Proposal for non-local electron-hole turnstile in the Quantum Hall regime, *Phys. Rev. B* **83**, 12, p. 125324 (5).
[249] Keeling, J., Klich, I., and Levitov, L. S. (2006). Minimal excitation states of electrons in one-dimensional wires, *Phys. Rev. Lett.* **97**, 11, p. 116403 (4).
[250] Sherkunov, Y., Zhang, J., d'Ambrumenil, N., and Muzykantskii, B. (2009). Optimal electron entangler and single-electron source at low temperatures, *Phys. Rev. B* **80**, 4, p. 041313(R) (4).
[251] Hong, C. K., Ou, Z. Y., and Mandel, L. (1987). Measurements of subpicosecond time intervals between two photons by interference, *Phys. Rev. Lett.* **59**, 18, pp. 2044–2046.
[252] Glauber, R. J. (1963). The quantum theory of optical coherence, *Phys. Rev.* **130**, 6, pp. 2529–2539.
[253] Moskalets, M. and Büttiker, M. (2011). Spectroscopy of electron flows with single- and two-particle emitters, *Phys. Rev. B* **83**, 3, p. 035316 (11).
[254] Bell, J. S. (1966). On the problem of hidden variables in quantum mechanics, *Rev. Mod. Phys.* **38**, 3, pp. 447–452.
[255] Martin, T. and Landauer, R. (1992). Wave-packet approach to noise in multichannel mesoscopic systems, *Phys. Rev. B* **45**, 4, pp. 1742–1755.
[256] Büttiker, M. (1992). Role of scattering amplitudes in frequency-dependent current fluctuations in small conductors, *Phys. Rev. B* **45**, 7, pp. 3807–3810.
[257] Henny, M., Oberholzer, S., Strunk, C., Heinzel, T., Ensslin, K., Holland, M., and Schönenberger, C. (1999). The Fermionic Hanbury Brown and Twiss Experiment, *Science* **284**, pp. 296–298.
[258] Oliver, W. D., Kim, J., Liu, R. C., and Yamamoto, Y. (1999). Hanbury Brown and Twiss-Type Experiment with electrons, *Science* **284**, pp. 299–301.
[259] Samuelsson, P., Sukhorukov, E. V., and Büttiker, M. (2004). Two-particle Aharonov–Bohm effect and entanglement in the electronic Hanbury Brown Twiss setup, *Phys. Rev. Lett* **92**, 2, pp. 026805 (4).
[260] Samuelsson, P., Neder, I., and Büttiker, M. (2009). Reduced and projected two-particle entanglement at finite temperatures, *Phys. Rev. Lett* **102**, 10, pp. 106804 (4).

[261] Neder, I., Ofek, N., Chung, Y., Heiblum, M., Mahalu, D., and Umansky, V. (2007). Interference between two indistinguishable electrons from independent sources, *Nature* **448**, pp. 333–337.

[262] Splettstoesser, J., Samuelsson, P., Moskalets, M., and Büttiker, M. (2010). Two-particle Aharonov–Bohm effect in electronic interferometers, *J. Phys. A: Math. Theor.* **43**, 35, p. 354027 (9).

Index

adiabaticity, 80
   parameter, 80
      linear response regime, 210
      non-linear regime, 210
approximation
   single-electron, 9

capacitance
   differential, 198
conductance
   frequency dependent, 191
   frozen, 158
   matrix elements, 18
   quantum, 18
   thermoelectric, 20
current decay time, 200
current spectral density, 120

density of states, 11
   dynamic, 197
   frozen
      chiral ring, 197
   partial
      frozen, 131
   stationary, 194
distribution function
   equilibrium, 13
   non-equilibrium, 15, 113

equivalent electric circuit, 192
   parameters
      high temperatures, 192
      zero temperature, 193
equivalent electrical circuit
   non-linear regime, 197
   parameters
      non-linear regime, 198

heat flow
   generated, 169
   quantized emission regime, 211
   transferred, 170

mesoscopic capacitor, 175

noise
   shot, 36, 46
      photon-assisted, 159
      quantized, 212
      suppression effect, 232
   thermal, 35, 46
      non-equilibrium, 158
      quasi-equilibrium, 158

particle
   incident, 9
   scattered, 9
probability
   joint detection, 234
   single-particle, 215
   two-particle, 215

quantized
   alternating current, 111, 207

emission regime, 210
shot noise, 212
quantum
    Archimedes screw, 127
    capacitance, 194
    inductance, 194
    pump, 111
    pump effect, 111
        existence condition, 115, 119
        quantized emission regime, 111

reflection probability, 22
resistance
    differential, 198
    relaxation
        quantum, 181

scattering channel, 21
    dynamic, 109
scattering matrix, 3
    anomalous, 83, 133
        double-barrier potential, 104
        point-like potential, 91

dimension
    1×1, 21
    2×2, 22
    3×3, 23
    double-barrier potential, 102
    dual, 84
    Floquet, 71
        chiral ring, 182
    frozen, 78
    point-like potential, 90
spectral current density, 132
    partial, 132

transmission probability, 23
two-particle
    correlation function
        correlated particles, 236
        uncorrelated particles, 235
    correlations, 233